AF759346

Materials and Manufacture in Ancient Mesopotamia: The evidence of Archaeology and Art

Metals and metalwork, glazed materials and glass.

by P. R. S. Moorey

BAR International Series 237
1985

UNIV DUNELM
18. JUL. 1985

B.A.R.

5, Centremead, Osney Mead, Oxford OX2 0ES, England.

930
BRI
(237)

~~059.13~~
BR1
(237)

GENERAL EDITORS

A.R. Hands, B.Sc., M.A., D.Phil.
D.R. Walker, M.A.

~~059.13~~
MOOREY
AML 849/37
BRITISH
Archaeological reports
International ser.,
supplementary
(237)

BAR- S237, 1985: 'Materials and Manufacture in Ancient Mesopotamia: The evidence of Archaeology and Art'

Price £ 18.00 post free throughout the world. Payments made in dollars must be calculated at the current rate of exchange and $3.00 added to cover exchange charges. Cheques should be made payable to B.A.R. and sent to the above address.

© P.R.S.Moorey, 1985

ISBN 0 86054 306 4

For details of all B.A.R. publications in print please write to the above address. Information on new titles is sent regularly on request, with no obligation to purchase.

Volumes are distributed from the publisher. All B.A.R. prices are inclusive of postage by surface mail anywhere in the world.

Printed in Great Britain

"Pour bien savoir les choses, il en faut savoir le détail; et, comme il est presque infini, nos connaissances sont toujours superficielles et imparfaites".

La Rochefoucauld, Maximes, 106.

<u>Contents</u>

ACKNOWLEDGEMENTS

As the introduction makes clear, anyone who seeks to write a book of this kind is constantly trespassing into territory well outside his field of competence. I owe a special debt of gratitude to colleagues in the British Museum who read the rough drafts of the text and commented on them: Dr. John Curtis and Dr. Julian Reade of the Department of Western Asiatic Antiquities; Miss Mavis Bimson, Dr. Paul Craddock and Dr. Michael Tite of the Research Laboratory. They should not be held responsible for what survives, as I did not always take their advice; but without their guidance there would be even more scientific solecisms and more errors of commission and omission than there are. One of the many privileges of working in Oxford is the opportunity it offers for constant contact with the Research Laboratory for Archaeology and the History of Art. Over a period of twenty years I have benefitted considerably from the advice and expertise of Dr. Martin Aitken, Dr. Robert Hedges, Dr. E.T. Hall, and Dr. Mark Pollard, and from co-operation with two visiting members of the Laboratory, Dr. Francis Schweizer of Geneva and Professor Alexander Kaczmarczyk of Tufts University, Boston, U.S.A., who also kindly commented on drafts of the glazed materials section. Mrs. Pam Vandiver of M.I.T., through her work on the Ashmolean's Egyptian faience, did much to stimulate my interest in faience at a crucial moment.

A handbook like this depends ultimately on first class library resources and the support of library staff. This I have richly enjoyed in the Ashmolean Library. My colleagues in the Ashmolean Museum, my research students, and a long succession of visiting scholars, have all contributed more than they know by the stimulus of question and argument over many years. The existence of this book is in no small part the result of my inability to find even basic answers to specific inquiries without research in obscure literature. Often, when this had been done, it only allowed me to offer a highly conditional or negative answer. Although I am very conscious of the absence of the philological aspects of my subject, their virtual exclusion is deliberate and fully explained in the introduction. There are places, however, where it has been impossible to omit this aspect of the problem and then I have been most grateful to Dr. Jeremy Black, Dr. Stephanie Dalley and to Professor O.R. Gurney for guidance; for what survived their scrutiny I alone am responsible.

I am grateful to Dr. Tamara Stech, and her colleagues, for information on the current "Sumerian Metals Project" in the University of Pennsylvania at Philadelphia; to Dr. Paul Craddock of the British Museum Laboratory for information on unpublished analyses; and to Dr. Dan Barag and Mr. Terence Mitchell for permission to read the manuscript of Dr. Barag's forthcoming catalogue of glass in the Western Asiatic Department of the British Museum.

Mrs. Ruth Flanagan skilfully typed, and retyped, a series of draft manuscripts and Mrs. Pat Jacobs did the artwork.

LIST OF MAPS AND TABLES

INTRODUCTION

> "A systematic and critical inventory of the technological achievements of Mesopotamian civilization is needed to establish the degree to which Mesopotamian man succeeded in mastering nature, an essential aspect of the man-nature relationship in any civilization. This inventory will remain woefully inadequate, however, for two reasons: the paucity of the evidence available and the lack of scholarly interest in material culture". Oppenheim, 1978, 646.

The following studies are the archaeological basis for a history of craft and technology in ancient Mesopotamia not in any way an attempt to write that history. I have endeavoured to provide Oppenheim's "systematic and critical inventory of the technological achievements of Mesopotamian civilization" in metal, glazed materials and glass, as a preliminary to broader surveys. Their eventual integration with the relevant documentary evidence for the wider socio-economic environment in which they were developed and flourished (the only proper framework for a history of technology) is a project of quite another order. Save in very rare and specific cases, the data on material culture and the documentary evidence from Mesopotamia remain partially published and scattered, and are often available only in a form intelligible to specialists. Even in Classical Studies, where the whole range of documentation is so much richer and has been much more closely studied in modern times, the history of technology has had until quite recently a relatively low academic status, perhaps because there has long been a tendency to treat the subject in isolation and without due regard to social context and economic processes (cf. White, 1984, 6ff).

The problem is greater and even more complex for the pre-classical civilizations of the Old World, where at present the writing of any kind of economic history is, with good reason, rarely attempted and when it is attempted turns as often on questions of philology as on those of economy. It is not so much a matter of absent evidence, though that is a problem never to be underestimated, as of scattered and incoherent source material which has never been examined even at the most primitive level of analysis and synthesis. Some of the groundwork for a history of technology has been done for ancient Egypt by scholars such as Lucas and Harris (1962); but even there broader surveys of craft and technology in a socio-economic context are still awaited. In Western Asia, as Oppenheim (1978) made clear, there is nothing comparable to Lucas and Harris (1962), though pioneer works of synthesis like Meissner (1920/1925) and Partington (1935) still have much to offer the discriminating reader able to check their sources in modern editions. Singer's monumental History of Technology I (Oxford, 1954) barely mentioned glass and glazed materials, and was restricted in its treatment of metals and metal-working, whilst in every case little attention was paid to problems of technological innovation and development under particular socio-economic conditions. Forbes, Studies in Ancient Technology failed to make use of the steadily growing body of information provided by archaeology in Western Asia and have other serious drawbacks (see also p.xxiii here).

The surviving material culture of ancient Mesopotamia, in contrast to Egypt, is conspicuously poorer, though the textual evidence is potentially greater. In the soil of Mesopotamia, and adjacent regions, organic materials for the most part perished rapidly; ecological factors encouraged stability of location with the regular reconstruction of buildings destroying rather than preserving what had gone before. As essential raw materials were scarce, they were regularly recycled if at all feasible. Richly furnished tombs are as exceptional in the archaeological record as are unplundered settlements preserved by fire or natural disaster for the archaeologist's spade. Even what little has survived has rarely been the subject of proper scientific investigation. No one has ever been in a position comparable to that of Lucas in Egypt (as chemist to the Antiquities Service, 1923-32; honorary consultant 1932-45) and it is unlikely that one man with his range of expertise ever will be.

The following essays, for at this stage they could not hope to be anything more, seek only to provide the foundations for a study of the technological achievements of ancient Mesopotamia. As I am not a philologist they are based primarily on the evidence of archaeology and of art. The advantages of combining philological skills with a well-informed knowledge of ancient art and archaeology are evident from the revision of Lucas by Harris (1962). Nor am I a scientist. This, more serious drawback for my purposes here, has to some extent been remedied by the counsel of many scientific friends, now involved with archaeometry, and by diligent and (I trust) profitable reading in the ever increasing literature of this subject. In this generation of proliferating information no single scholar can hope to combine the formidable range of learning ideally required for a task of this kind. Still, as Oppenheim appreciated, it is much needed and, at a preliminary stage, there are distinct advantages in one scholar, rather than a team, undertaking it. I have proceeded on the clear assumption that what follows will do no more than offer a provisional basis for others, with specific specialist qualifications, to modify and extend in the light of fresh, more accurate data.

It has, of course, been impossible to avoid reference, from time to time, to ancient textual sources. Whenever this was necessary I have cited the standard dictionaries and the most respected, if not always necessarily the most recent, treatment of a particular problem. I am aware of the hazards of even so cautious an approach, since technical vocabulary in all ages and places, above all in dead languages, is notoriously difficult. In ancient Mesopotamia the textual evidence for materials and industries is potentially very rich, but, with rare exceptions, it has not been a popular branch of inquiry in recent years. There are signs that this is changing. I hope that by assembling here a representative sample of the archaeological and modern scientific evidence, often scarce even for the most durable of materials, I will have contributed in two ways. First, by weighing up the material evidence, I may have assisted in clarifying the issues involved in settling the meaning of a particular word or phrase. Annaku (p.124) is, perhaps, the best known instance of the problem. Second, in assembling what is known in a critical spirit, I may have cleared a certain amount of dead wood and indicated more clearly, by highlighting what is not known, where the full weight of research might now best be directed.

In referring to ancient terms I have always been aware that relatively few ancient names of materials, or processes, are known for certain and that readers must be directed to those publications, when they exist, where they may follow up the premises upon which a current translation is based. When using ancient written sources as evidence for the origin of raw materials, potentially one of their most obvious contributions to a study of this kind, special problems arise. The stated source is by no means necessarily the place of origin. Ancient terminology, as with the English word "turkey" for a domestic fowl of American origin, may well be misleading. Although some stated sources are clearly collection points to which traders from a variety of more distant places brought goods for onward transmission (as with ancient Dilmun), this is not always so obvious. Transit zones gave their names to commodities or raw materials brought from much further away. Above all ancient terminology, often reflecting as it does a scribe's very inadequate knowledge of distant places and strange goods, may not be relied upon for the degree of precision assumed in more modern records of commerce and industry. Texts relating to gift exchanges between rulers also have their own pitfalls, since they sometimes involved prestige exchanges of identical products, confusing the issue if this is not fully appreciated. In the "Amarna Letters", for instance, ivory from Cyprus, where it was not produced, travels to Egypt, which had easy access to the vast ivory resources of Africa (Liverani, 1972). Lists of booty, cleverly exploited by Jankowska (1969) in her study of the economy of the Neo-Assyrian Empire, have evidence to offer in the quest for sources; but again it is often oblique and has to be assessed in the light of all the data available.

My approach throughout has been pragmatic. In general, the information available to me is not nearly so full as that Lucas commanded for Egypt, even for those materials and industries archaeologically evident in Mesopotamia. There are, of course, whole aspects of material culture, primarily involving organic materials, which in Egypt are archaeologically documented, but in Mesopotamia are exclusively the territory of the philologist. The material evidence for them perished long ago. I have, of necessity, largely omitted these from my proposed survey, of which this is the first part. In future studies I plan to treat materials like clay, stones and pigments, bone, ivory and shell, not included here. The decision to treat metals, glazed materials and glass together first was deliberate. These crafts are among the best documented and studied, their technologies are interactive, and they raise at the outset many of the general problems confronting students of materials and manufacture in Mesopotamia.

Expectations of such a work as this have changed considerably since Lucas produced his book. New theoretical approaches to archaeological research, a growing sophistication in the application of the natural sciences to archaeology, and a greater interest in the replication of ancient craft methods, demand a broader perspective. In an attempt to compensate for a less varied range of evidence, or for less detailed scientific study of certain aspects of it, I have paid more attention to some secondary problems with which Lucas chose not to concern himself.

This has meant that there is no standard form of entry, though there is a general consistency. The entries tend to vary as the available information, or the nature of the material, dictates. Questions appropriate to gold, for instance, may not be relevant to iron, and *vice versa*. It may be possible to say much more of glazed quartz bodies (faience) than of glazed baked clay, not for any academic preference, but simply because one is generally better known and more widely recognized than the other. In all cases these variations have conditioned the form of the relevant essay.

If the presentation varies, the three fundamental problems investigated generally do not. First, there is the question of sources and support technology: whence came the raw materials, if imported, and how were they processed? Second, what were the forms, and if relevant, the decoration, imposed on them by ancient craftsmen? Third, what were the primary processes of manufacture? Since there has never been a systematic historical survey of the crafts of ancient Mesopotamia, considerable attention has to be paid to the development of crafts and in many cases the information is presented within an historical framework.

Each of these primary questions presents its own difficulties. Source studies are at present very undeveloped. This is not surprising in view of the great complexity of devising sufficiently reliable methods for the scientific "finger-printing" of metals and minerals. Even when they exist, there is usually an absence of geological samples from all the potential sources or doubts about the secure separation of materials from distinct source zones. Such information is vital in the long run for a full understanding of life in ancient Mesopotamia, dependent as it was on imported raw materials. In the absence of reliable information from scientific investigations, typological and stylistic analyses are regularly invoked to reconstruct cultural or commercial contacts between Mesopotamia and those of her neighbours whence other indicators suggest the raw materials might have come. I have tried whenever possible to take account of this research.

There has been a tendency among archaeologists in recent years to pass beyond typological and source analysis in quest of hard information on socio-economic questions in periods, or regions, where no textual evidence of an appropriate kind is available. This is a singularly interesting, but highly problematic, aspect of material culture studies which cannot be ignored in any work of the kind proposed here. In a number of pioneering papers on specific raw materials relevant to Mesopotamia, notably obsidian (Renfrew *et al.*,

1976), steatite/chlorite (Kohl, 1979) and Iranian copper (Heskel and Lamberg-Karlovsky, 1980), explanatory models have been proposed to elucidate trading mechanisms, cultural transmission and consequent socio-economic transformations. In every case alternative explanations, some more, some less defensible, have been argued. As one who takes all models, hypotheses and interpretations to be little more than structured questions, not potential answers, there will be those who find my references to this aspect of the matter inadequate. But, as models are by their very nature over-simplifications, they tend to sacrifice details in the interests of structure and the very business of a handbook, such as this attempts to be, is with details, though not necessarily unstructured. I have striven, as far as possible, for the essentials by digesting rather than rejecting the details; the uneven nature of the text will make clear how difficult this often proved to be.

In this kind of research it is most important to remember at every stage that in studying ancient production processes much of what is said here, as in all modern textbooks, is framed with the benefit of scientific hindsight. The knowledge acquired by ancient craftsmen was purely empirical, with endless possibilities for variation, not only in the composition of fired objects, but also in the manipulation of materials: there was not, to take a well-known example, a single "lost" art of granulation (pp.89-90). Successful solutions to chemical and physical problems were eventually arrived at by trial and error over long periods of time, by careful selection from amongst the raw materials available in one time or place, and by the native ingenuity of generations of craftsmen passed on from master craftsman to apprentice. Even when technical treatises first appear, and they are a medieval phenomenon (much more recent if those written for the instruction of literate craftsmen are included), it is not easy for modern experts to follow the technical processes involved. The same word may be used for different processes and _vice versa_.

It is also as well to remember that current evaluations of ancient craft practices are as much at the mercy of prevailing research fashions and academic pre-occupations as they are of the nature of the surviving material evidence. It is not only increase in the information available that may change ideas about past craft practices. The spotlight of scientific research passes to and fro. Long standing interests in composition may suddenly give way to an exclusive concern for aspects of production, so that much more may become known of the one than the other in a particular case. This variation has been particularly evident in research on faience (pp.142ff.). Comparative ethnographical study (Watson, 1979; Kramer, 1982), never so much neglected in the history of technology as it was for a long period in the study of archaeology, presents its own problems. The best information is provided by studies from the same general area as is under study in antiquity. For the Near East two works are particularly relevant and valuable, Dalman's monumental _Arbeit und Sitte in Palästina_ (1928-42), and Wulff's _The Traditional Crafts of Persia_ (1966). They offer stimulus to research by providing data for hypotheses about possible ancient craft techniques to be carefully tested by modern experiment, as notably in the case of faience production (pp.158ff.).

Certain practical considerations have particularly conditioned my approach in writing. I have striven to be concise and clear, whilst attempting to make the frame of reference as wide as possible in the interests of a variety of readers. I hope that the text will be useful both to specialists in other aspects of ancient Near Eastern culture and to the increasing number of people concerned with studies of ancient materials and industries in other parts of the world who will not be so familiar with the history of this particular region in antiquity. The geographical limits of this investigation have been loosely defined. Although the focus of these studies is broadly the area now embraced by Iraq, northeast Syria, and southwest Iran (Khuzistan), I have not been restricted to this region in case of individual pieces of information useful to the development of these essays. The chronological range is from the Neolithic period, the ninth millennium B.C., to the conquests of Alexander the Great in the third quarter of the fourth century B.C. A quarter of a century ago Harris found in revising Lucas that expense prohibited the kind of illustration most appropriate to studies of this kind; it is even more so today. Their absence has not seriously impaired the utility of Lucas (1962) and I trust the same may be true here. The text is written so as to direct readers as easily as possible to the most appropriate source of illustrations.

The last word is best left with Samuel Johnson who, in the preface to his famous English dictionary (1755), inimitably expressed the most recurrent fears of any writer engaged upon a book of this kind:

> "...that what is obvious is not always known, and what is known is not always present; that sudden fits of inadvertency will surprize vigilance, slight avocations will seduce attention, and casual eclipses of the mind will darken learning; and that the writer shall often in vain trace his memory at the moment of need, for that which yesterday he knew with intuitive readiness, and which will come uncalled into his thoughts tomorrow".

Ashmolean Museum, Oxford — October, 1984

NOTES FOR THE READER

(i) Chronological Terminology:

Absolute dates before about 3000 B.C. are very insecurely based; other periods, until the fall of Nineveh, c.612 B.C., are approximations.

Period		Date
Umm Dabaghiyah/Ceramic Jarmo	)Early	c. 6000 B.C.
Hassuna/Samarra/Halaf periods	)Prehistoric	c.5500-4500 B.C.
´Ubaid period	)	c.4500-3800 B.C.
Uruk IV (Proto-literate a-b	)Late	c.3800-3300 B.C.
Uruk III (Proto-literate c-d/ Jamdat Nasr)	)Prehistoric	c.3300-3000 B.C.
Early Dynastic I		c.3000-2750 B.C.
Early Dynastic II		c.2750-2600 B.C.
Early Dynastic III		c.2600-2350 B.C.
Akkadian (or Sargonic)		c.2350-2100 B.C.
Ur III		c.2100-2000 B.C.
Isin-Larsa/Old Babylonian/Old Assyrian		c.2000-1600 B.C.
Kassite/Mitannian/Middle Babylonian/Middle Assyrian		c.1600-1000 B.C.
Neo-Assyrian		c.1000-612 B.C.
Neo-Babylonian		c. 612-539 B.C.
Achaemenid Persian		c. 539-330 B.C.
Seleucid		c. 330-150 B.C.

The chronology for individual kings cited in this book is that given by J.A. Brinkman in A. Leo Oppenheim, Ancient Mesopotamia (revised edition, Chicago, 1977), 335ff. This is based, for convenience sake, on the "Middle Chronology" for the First Dynasty of Babylon (c.1894-1595 B.C.). The "High Chronology" (c.1950-1651 B.C.) remains probable; support for the "Low Chronology" (c.1830-1531 B.C.) has rather diminished in recent years.

The archaeological terminology for Mesopotamia, as that for dynastic Egypt, is strongly marked by the local historical tradition; very broadly speaking the terminology used in Syro-Palestine and Turkey may be compared through the following time horizons: Neolithic (c.7000-4500 B.C.); Chalcolithic (c.4500-3300 B.C.); Early Bronze Age (c.3300-2000 B.C.); Middle Bronze Age (c.2000-1550 B.C.); Late Bronze Age (c.1550-1150 B.C.); Iron Age (c.1150-550 B.C.).

(ii) Weight systems:

Reference is occasionally made in the text to weights by their ancient terminology. The Babylonian system was sexagesimal: 60 minas equalled 1 talent and 1 mina was divided into 60 shekels. 1 mina weighed approximately 500 grams (Powell, 1973). There was a significant difference in the Bahrein (Dilmun) - Indus system, in which minas were combined into hundreds and thousands not multiples of sixty. The Dilmun mina weighed approximately 1,350 grams (Roaf, 1982).

(iii) Problems of historical geography:

This is clearly not an appropriate place for lengthy discussions of historical geography, but, to avoid cluttering the source studies with recurrent annotations, a few regions constantly cited by their ancient names are briefly considered here:

A: Within Iran (primarily third millennium B.C.):

(a) Aratta, whose exact location is still controversial, was renowned in the Sumerian literary tradition as a fabulously wealthy place to the east of Sumer, which had an intimate relationship with Uruk, related in the "Enmerkar cycle" of epics (cf. Wilcke, 1969; Cohen, 1973). These connections were traditionally dated to the reign of Enmerkar, second ruler of the first dynasty of Uruk; archaeologically speaking in Diyala Early Dynastic II. But the Sumerian literary tradition associated the contest between Enmerkar and the ruler of Aratta with the invention of writing (Cohen, 1973, 136-7, lines 502-6), suggesting a tradition older by centuries. At least one archaic text of the later fourth millennium B.C. from Uruk may contain a reference to Aratta (Green, 1980, 16-17). In the epic tradition Aratta was renowned for its precious materials and its skills in manufacturing them (cf. Wilcke, 1969, 127-8; Pettinato, 1972).

1. "Gold, silver, copper, tin, lumps of lapis lazuli, [and mountain stones] - all these were not brought down from their mountain (sources)" (Cohen, 1973, 112, lines 18-19).

2. "O my sister, for Uruk may Aratta, fashion artfully gold (and) silver on my behalf, [Let them cut] the pure lapis lazuli from the lumps" (Ibid., 113, lines 38-40).

According to this tradition Aratta was near a navigable river (Cohen, 1973, 56) and was in a highland zone, east of Anshan, known to have been at modern Tepe Malyan in Iranian Fars (Reiner, 1973). The references to crossing "five, six, seven mountains" in moving east from Anshan to Aratta may be merely formulaic and of no relevance to the quest for Aratta's location; nor is it clear whether crossing ma (n) dum territory refers to "flat land" or to a specific geographical name (Cohen, 1973, 53-4, 57; cf. Wilcke, 1969, 29ff.). Majidzadeh (1976) has argued for a location of Aratta in the vicinity of modern Kerman, where the archaeological site of Shahdad has produced richly equipped third millennium graves; Hansman (1978) proposed a more easterly focal point, round the archaeological site of Shahr-i Sokhta in Seistan. For both scholars the reference to lapis lazuli and its manufacture was important; but it is not clear from the epics whether Aratta was a source zone or a transit zone for this mineral. Even if it were a source zone, it does not have to be near Badakhshan in Afghanistan. Medieval texts refer to a lapis lazuli source in the Kerman mountains (Majidzadeh, 1976, 109) and a vein of low quality lapis lazuli is known near Quetta in Pakistan (Jarrige, pers. comm.). Aratta as a place is not yet known in third millennium non-literary texts; rare late references in economic texts are probably to a city of this name in Babylonia (Cohen, 1973, 61, n.90). The name became an epithet for "abundance" and "glory" (Ibid., 55, n.67).

(b) Marhaši, sometimes BARḪAŠI (Sumerian) (Paraḫšum in Akkadian), has for long been conventionally placed in the mountains northwest or north of Elam and east of the Diyala river. Steinkeller (1982) has recently advanced a case for regarding it as an important part of central southern Iran, to the east of Anshan (Tepe Malyan) in Fars.

(c) Šimaški: an inscription of Shu-Sin (c.2037-2029 B.C.) indicates that Šimaški was a region, comprised of six named lands, Zabšali conspicuous amongst them (Edzard, 1959-60, 9). It has commonly been located to the north of Khuzistan and/or Fars (Stolper, 1982); Vallat´s (1980) suggestion that it might have been in Kerman or eastern Fars is not sustained by the available evidence.

(d) Tukrish: there is no direct evidence in the available textual sources for the location of Tukrish, beyond indications that it lay east of the Tigris in the Zagros region. An identification with Luristan (Groneberg, 1980, 239), following a suggestion of Bottero´s (1957, 312), does not carry conviction in view of the recorded commodities coming from Tukrish in the second millennium B.C. and current archaeological knowledge of Luristan at that time. A region somewhere north of Luristan, in modern Kurdistan or western Azerbaijan, seems more probable.

B: The Gulf and the route to the Indus Valley:

(a) Dilmun (Tilmun) was first identified with the island of Bahrein by Sir Henry Rawlinson (1880). This conclusion has never been seriously challenged, though scholars continuously attempt to refine the equation: "...we must agree that Rawlinson´s identification of Dilmun with Bahrain is the guess that fits the facts, though we must keep in mind that Dilmun was more than just Bahrein; it was also a still unidentified area of the Arabian coast" (Bibby, 1972, 62).

Howard-Carter (1981, 210) was quite explicit: "It must be emphasized strongly that the legendary Dilmun of the Sumerians, although certainly a physical reality, was not located in the same place as the Dilmun which played such an important role in the economy of later periods". Using the latest geological information on the early history of the Gulf, she argues for an initial location in the region of Qurneh in Iraq. Since no objects certainly of Mesopotamian origin have been identified either on Bahrein or Failaka in contexts dated before 2200 B.C., it is only after that time, Howard-Carter argues, that they became identified as Dilmun.

Potts (1983), who is not so concerned with the legendary tradition, argued for a rather different series of changes within the historic period. Dilmun is attested in the texts from Jamdat Nasr, possibly in the even earlier Uruk IV texts, and then with some

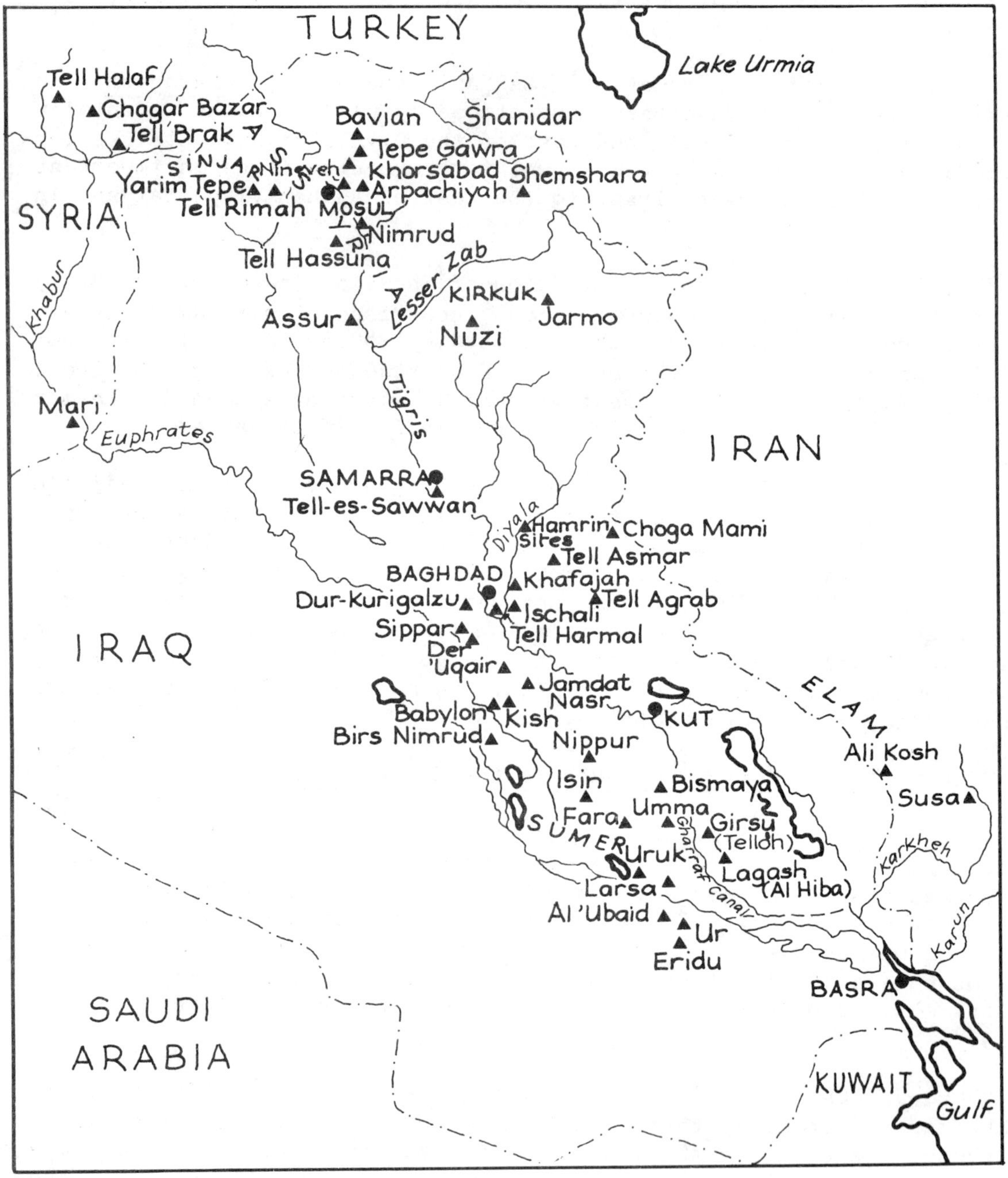

Map 1 Mesopotamia : main sites mentioned in the text

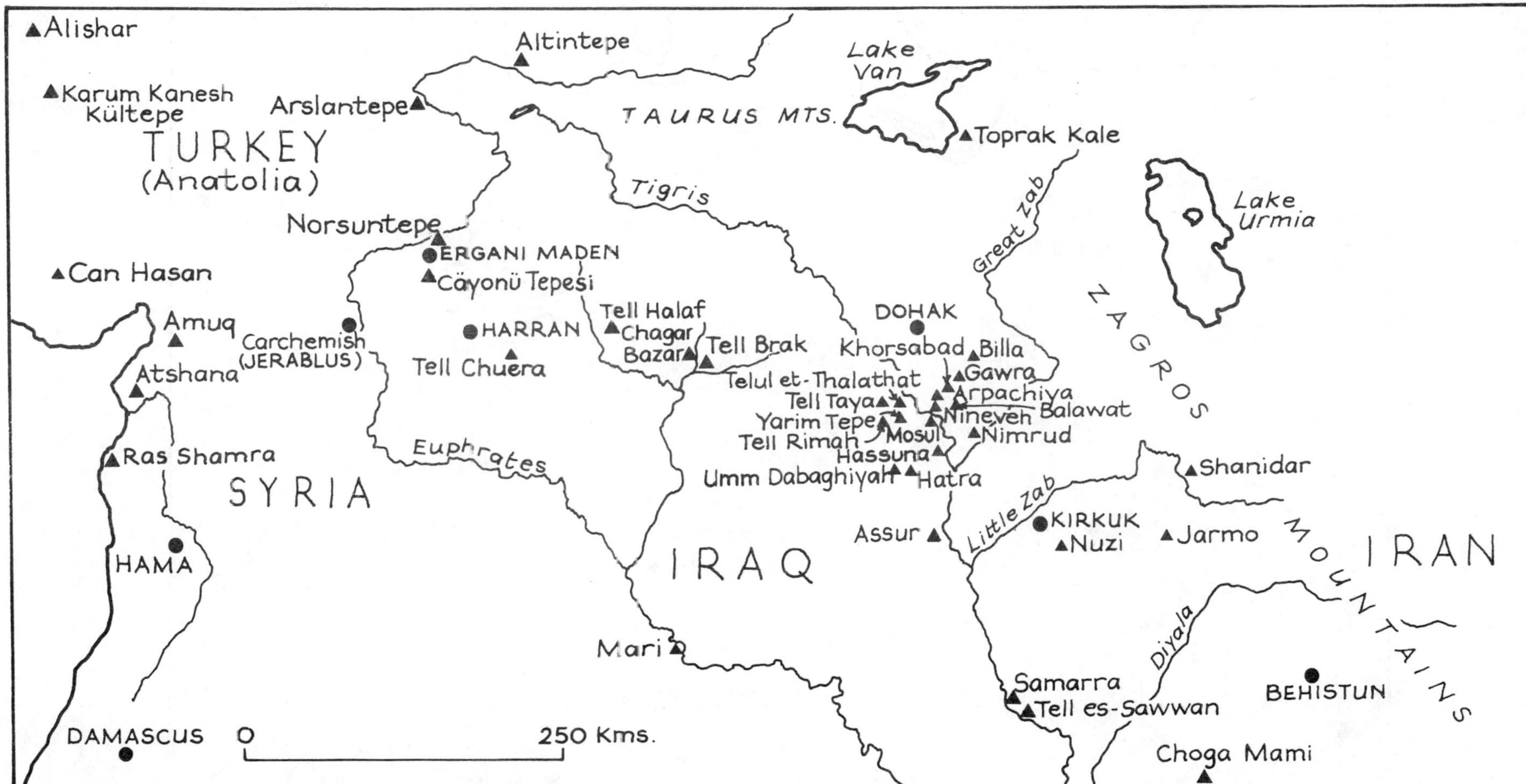

Map 2 Mesopotamia and its northern periphery

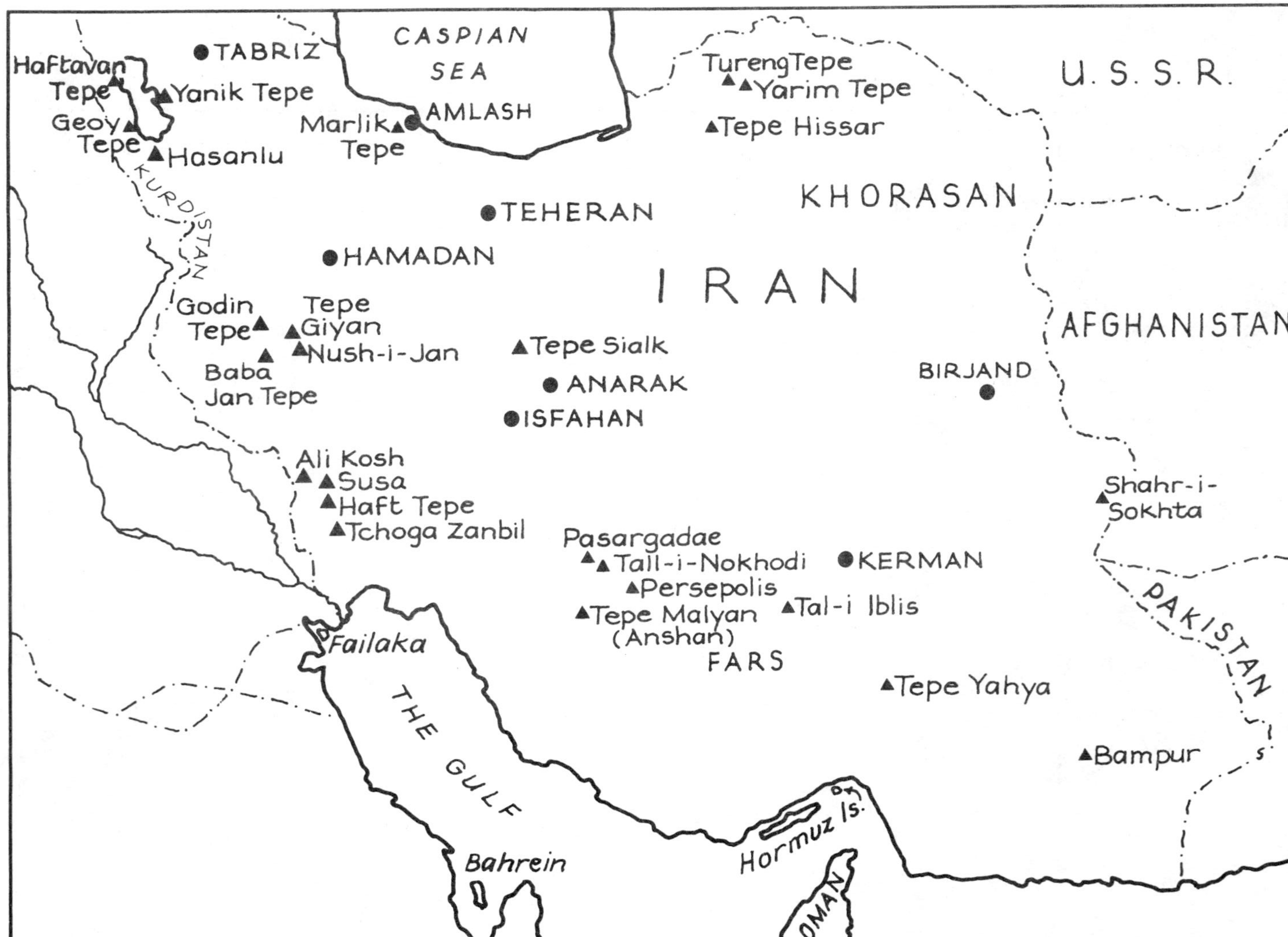

Map 3 Iran : main sites mentioned in the text

regularity in Early Dynastic III (Potts, 1983, 15). Potts believes, on archaeological grounds, that in the later fourth and earlier third millennium B.C. the Mesopotamian references to *Dilmun* apply to eastern Arabia rather than to Bahrein, which only became the centre of *Dilmun* in the later third and earlier second millennium B.C. In the last centuries of the third millennium Failaka was colonized from *Dilmun*, probably from Bahrein itself, and then became part of the community known in Mesopotamia as *Dilmun*. In the Neo-Assyrian period the focus of *Dilmun* was again in Bahrein.

Texts of the later third and early second millennium B.C. list the following items as reaching Ur from *Dilmun*: lapis lazuli, carnelian, and various other unidentified semi-precious stones; ivory and ivory objects; copper; silver; "fish-eyes" (pearls or perhaps agate); red gold; white corals; various woods; dates. Save for the dates and the "fish-eyes", if they are indeed pearls, all the other commodities came to *Dilmun* from elsewhere for onward shipment (cf. Edzard et al., 1977, 157-8; Edzard and Farber, 1974, 193; Groneberg, 1980, 237 as *Tilmun* in all cases).

Any summary of the evidence for the locations of *Magan* and *Meluḫḫa* must begin with a statement applying to them both. Most students of the subject are agreed that in texts of the first millennium B.C. *Magan* and *Meluḫḫa* correspond approximately to Egypt and Nubia/Ethiopia respectively (Potts, 1982). In texts of the third and earlier second millennium B.C. these identifications are generally thought to be unlikely. It has been widely argued that there was a shift in toponymy between this early period and Neo-Assyrian times and that originally the two names corresponded to regions more directly accessible to Sumer down the Gulf and to the east, not to the west. Two distinguished Sumerologists (Jacobsen, 1960, 184, n.18; Kramer, 1963, 276), do not accept this shift, believing that the two names always referred to regions in Africa.

(b) *Magan*, when referred to in early texts, is now usually believed to have been a region at the mouth of the Gulf, embracing much of modern Oman and the opposite coastal regions of Iran, possibly including Makran, though some would include this area in *Meluḫḫa* (cf. Leemans, 1960; Gelb, 1970; Ratnagar, 1981). So long as there is no archaeological evidence that the Umm an-Nar culture extended to the Iranian side of the Gulf, the Makran = *Magan* equation remains unresolved. *Magan*'s boats reached Sumer and the Akkadian kings claimed to have campaigned there and taken booty, which included stone vessels bearing inscriptions to this effect (Hirsch, 1963, 18). Timber and wooden objects, a type of onion (?), copper (Weisgerber, 1984), ivory, gold dust, carnelian and other semi-precious stones, diorite, red ochre, as well as goats, were all said to come from *Magan*, but never lapis lazuli. Some of these goods, notably carnelian and ivory, were being shipped from further east, some, like copper and diorite, were local. Occasional references in texts of the Ur III period mention men from *Magan* in Sumer.

(c) *Meluḫḫa* is now usually located in the Indus valley region when mentioned in early texts, though it might be of wider reference covering the shores of what in modern times has been called the Indian Ocean. It was a land of sea-farers, whose boats are mentioned in Mesopotamian sources. Its exports included timber and wooden furniture, copper, gold dust, lapis lazuli, carnelian, birds, including the peacock, and such manufactured objects as multicoloured ivory birds, a carnelian monkey and a "red dog". People from Meluhha were also settled in Mesopotamia (cf. During-Caspars, 1978; Gelb, 1970; Sollberger, 1970; Hansman, 1973; Parpola, 1977; Ratnagar, 1981).

C: Within Syria, eastern Turkey and Iran (primarily first millennium B.C.):

There are a series of regional names, largely in Neo-Assyrian texts, which require elucidation to understand references to the sources assembled by Jankowska (1969) cited in the following pages under her name (cf. especially for the historical geography, Hawkins, 1982).

Bit-Ḫalupe: the region between the Euphrates and the lower Habur, north of Tell ʿAshara, in Syria.

Bit-Zamani: the region north and northeast of Diyabakir in Turkey.

Hamath: the region of modern Hama in Syria.

Ḫattina (correctly Pattin): the Amuq region of Syria, otherwise known in antiquity as Unqi.

K(Q)ummuḫ: the region north of Samsat, to the west of the Euphrates, in Turkey.

Laqe: region to the southwest of Bit-Ḫalupe on the other side of the Euphrates in Syria.

Muṣaṣir: the religious centre of ancient Urartu, southwest of Lake Urmia near Rowanduz in Iraq (cf. Boehmer, 1973).

Nairi: a region, between the Upper Euphrates and Lake Van in eastern Turkey, so-named in second millennium and later sources, when this area is better known as the kingdom of Urartu.

Que (Ḫume): in modern Cilicia, Turkey; the region of modern Adana.

Shubria: region south of modern Mus, in Turkey, northwards to Lake Van.

Tabal: south-east corner of the Turkish plateau.

Zamua: near the headquarters of the Diyala river in the Zagros mountains.

(iv) Previous literature on the subject:

Some explanation is required at the outset about certain books which the reader might expect to find cited here more often than they are.

R. Campbell Thompson's A Dictionary of Assyrian Chemistry and Geology (1936) was a great pioneering work, but it is now largely superseded by the relevant sections of the standard English and German language dictionaries of Akkadian and, in a specific case, by Oppenheim's (1970) study of the so-called glass texts (see pp.224ff here). To the same period belongs J.R. Partington's Origins and Development of Applied Chemistry (1935), in itself much indebted to the work of Berthelot. This is a remarkable compendium of information in which Egypt and the Near East are well represented. It has been for me an invaluable source of reference for studies made before the early 1930s. Inevitably many of its conclusions and detailed discussions have been replaced by fresh philological research or by new discoveries in archaeology and science. Partington was as much indebted as I am to an excellent work which in many ways is still not superseded, though the absence of an English or French translation has always restricted its currency, Bruno Meissner's Babylonien und Assyrien, I-II (1920, 1925).

R.J. Forbes, Bibliographia antiqua: Philosophia naturalis (1940-50; supplement 1952) is a valuable and enduring compilation of references; but his Studies in Ancient Technology (1955-1972), are quite another matter. These works are widely available handbooks, designed for scholar and general reader alike, and are commonly cited. Though full of useful information difficult to find elsewhere, they suffer from a number of serious defficiencies. The use of philological evidence is notoriously unreliable; archaeological information is uncritically handled, if cited at all; there is often an imbalance in the historical evidence chosen for inclusion; and many references are either misleading or so imprecise as to be merely irritating. The "revised" editions are unfortunately no better, for as Landsberger (1965, 291, n.27) remarked of one of them "(it) is in a sense, up to date, but his results are so confused that they defy criticism".

The remarkable series of books on aspects of Mesopotamian daily life and material culture published by Armas Salonen in German between 1939 and 1976 are difficult for non-philologists to use. At the same time they adopt an approach to the philological evidence that may be seriously misleading, if the reader is not aware of it. In Oppenheim's (1978, 660, n.151) cautionary words: "Salonen's books in many respects constitute pioneering work that of necessity is word-orientated. They do not take into account the specific evidentiary value of the text types in which these words occur and the complexities of the socio-economic structures that created the documentation".

In 1978 Dayton devoted substantial parts of a large and well illustrated book to the nature and history of Egyptian Blue, and its closest relatives, in the ancient Near East and Egypt, under the title Minerals, Metals, Glazing and Man. Dayton, by arbitrarily lowering the dates of much Near Eastern and Egyptian archaeological material from the third to the second millennium B.C., and by virtually ignoring the prehistory of applied chemistry in the region, attempted to sustain the hypothesis that in the Near East metallurgy and the craft of glazing depended on Europe both for raw materials and for technology. The author's indiscriminate mixture of good and bad information, to sustain an indefensible and self-evidently confused case, vitiated a book whose bibliography is often useful and whose illustrations are for the moment a unique point-of-reference.

(v) Select Technical Glossary for Metalworking:

The following sections inevitably introduce the archaeologist and art-historian to a new vocabulary. The following concise glossary includes only the most recurrent terms in the general literature of the subject.

A: with particular reference to precious metals, copper and copper alloys (cf. particularly Coghlan, 1975; Maryon, 1949):

Annealing: the process of heating to red heat to remedy brittleness caused by hammering (see pp.39ff. here).

Brazing (Hard-Soldering) and Soft-Soldering: are joining techniques involving an agent (filler metal). Conventionally soldering refers to processes below 450°C; brazing to processes above 450°C. (see pp.87ff. here).

Casting: see pp.40ff. in the main text.

Casting-on: a method employed not only in repairing broken or damaged objects, but also in manufacture, to form joins between elements in copper/copper alloys, or between elements in copper/copper alloys and iron (see pp.40ff. in main text).

Chasing: hammering the metal down from the front to produce a low-relief design with linear margins.

Engraving: linear designs made on the surface with a sharp tool which removes the metal from the groove it cuts. It is not certain how much this was done before steel came widely into use for working tools (cf. Lowery et al., 1971).

Forging: the process by which a piece of metal is shaped through hammering either at its normal temperature (cold working) or when heated.

Punching: shape or decoration achieved with a patterned tool (or die) hammered into the surface of a piece of metal (cf. also Stamping).

Raising: production of a sheetmetal vessel by hammering from the outside. A disk of metal is worked over a special anvil (stake) with unusually long arms which can enter the vessel being made.

Repoussé: hammering the design up from the back of a piece of sheetmetal so that it appears in relief in the front; normally done with a blunt tool against a yielding surface.

Sinking: making a sheetmetal vessel by hammering the metal down into a suitably shaped depression cut into wood or a similar material.

Soldering: see Brazing above.

Stamping: producing a design by using a patterned tool (or die) from the back of a piece of sheetmetal (cf. also Punching).

Tracing: linear designs on the surface made with a slightly blunt tool which displaces and compresses, but does not remove the metal from the groove it makes (contrast Engraving).

B: with particular reference to iron technology (cf. particularly Stech-Wheeler et al., 1976, 108ff.; Waldbaum, 1983, 177ff.).

Austenite: a solid phase of carbon in iron, at temperatures over 710°C.

Bloomery iron: in antiquity the first product of smelting iron ore in charcoal. It is a relatively pure iron with small amounts of slag.

Carburization: the process of heating bloomery iron in direct contact with charcoal. Carbon is absorbed into the iron converting the areas in which they combine into steel.

Decarburization: the process for depleting surface layers of carbon. Carburized iron is heated in an oxidizing atmosphere so that the carbon in the surface layers combines with oxygen and is given off as gas.

Eutectoid steel: an alloy of iron with 0.8% carbon.

Martensite: the product arising when austenite is sharply cooled, as by plunging an object into cold water. It is hard and brittle.

Pearlite: the product arising when austenite is cooled more slowly. The name derives from the decomposed appearance of the austenite: alternate lamellae of ferrite (pure iron) and cementite (a compound of iron and carbon) with a "pearly" appearance under the optical microscope.

Quenching: plunging a hot metal into water or some other cooling medium; most commonly used for hot steel (austenite).

Tempering: the process for relieving the brittleness of martensite, accomplished by heating to a temperature in excess of 150°C.

Welding: heating two pieces of metal to be joined, almost to melting-point, and then hammering them together. In antiquity it could not have been used to join copper or copper alloys, only iron.

Widmanstätten structure: a microstructure characterized by a tendency of ferrite (pure iron) to form plates, during cooling from above 710°C, along certain crystallographic planes of the parent austenite.

(vi) Select Chemical Abbreviations:

Ag	Silver	Na	Sodium
Al	Aluminium	Ni	Nickel
As	Arsenic		
Au	Gold	P	Phosphorus
		Pb	Lead
B	Boron	Rb	Rubidium
Ba	Barium		
Bi	Bismuth	S	Sulphur
		Sb	Antimony
Ca	Calcium	Si	Silicon
Cl	Chlorine	Sn	Tin
Co	Cobalt	Sr	Strontium
Cr	Chromium		
Cu	Copper	Ti	Titanium
Fe	Iron	V	Vanadium
K	Potassium		
		Zn	Zinc
L	Lithium	Zr	Zirconium
Mg	Magnesium		
Mn	Manganese		
Mo	Molybdenum		

Chapter 1. METALS AND METALWORK

The main metals used in Mesopotamia in antiquity were copper, iron, lead, gold, and silver; tin was alloyed, more rarely used alone. Occasional instances of metal antimony have been reported, but their identity is equivocal. Bronze (copper and tin) was an important alloy; before the mid or later second millennium B.C. it may not have been as common as has been assumed. Arsenic as a metal was apparently not known. Arsenical coppers, however, were a significant feature of the copper industry for centuries; but whether strictly alloys or not has yet to be determined. Brass, an alloy of copper and zinc, appeared in Anatolia in the eight century B.C., but no certain example of it, manufactured in Mesopotamia, has yet been identified. Electrum, both natural and artificial, was employed and may well have been far more common than present identifications of "gold" and "silver" might suggest. Nickel and zinc as metals were not known.

I: ANTIMONY:

A number of Akkadian words, notably guḫlu (CAD.s.v.; AHw.s.v.), have been taken to mean antimony. Shamshi-Adad V (c.823-811 B.C.) mentions that on campaign to the east he crossed "the mountain of guḫlu-mineral" (Meissner, 1914). Guḫlu is also recorded as tribute, in Neo-Assyrian record, from the King of Judah, from Arabia and from Dilmun. Various other Akkadian words, used to denote cosmetics or ingredients in glass-making recipes, may also mean antimony (cf. Oppenheim, 1970, 19ff.). Limet (1960, 55ff.; Oppenheim, 1970, 21) argued that in metal texts of the Ur III period Sumerian SU.GAN (or SÚ.GAN might have been antimony (or arsenic). This word describes small quantities of some material ("dross") left unaccounted for in bronze making and, as Hallo (1963, 140) first argued, it is most unlikely to mean antimony (or arsenic) (cf. Röllig and Muhly, 1983, 350). Few chemical analyses of Mesopotamian eye cosmetics have yet been published. When they are available, they bear out the view expressed by Lucas (1962, 80-3, 196) for ancient Egypt, that antimony compound was very rarely used as a eye-paint (Oppenheim, 1970, 117, n.13; Bimson, 1980), so the ancient vocabulary for "antimony" may have been more restricted than is at present believed.

In view of what are taken to be textually attested uses of antimony compounds in Mesopotamia, it is particularly important to be clear about this metal's character. It occurs in nature chiefly as the sulphide (Stibnite) and, to a much lesser extent, in combination with other metallic sulphides. It is unlikely that the preparation of metallic antimony from stibnite was a process practised in antiquity. It was known in Europe only from the later Middle Ages. It is more likely that the metallic antimony of antiquity was the rare native metal, which is occasionally found as granular masses with a

UNIV DUNELM

TABLE I

The Ancient Terminology for Metals in Mesopotamia

	Sumerian	Akkadian
COPPER	URUDU	(w)erû
GOLD	KÙ.GI/GUŠKIN	ḫurāsu
IRON	KÙ.AN/AN.NA (3rd mill.) AN.BAR	parzillu
LEAD	A-BÁR	abāru
SILVER	KÙ-BABBAR	kaspu
TIN	AN.NA	annaku
BRONZE	ZABAR	siparru

(see particularly Limet, 1960, passim; for iron, Gurney in Maxwell-Hyslop, 1972).

tin-white colour and a metallic lustre in limestone or marble veins. Antimony occurs in small quantities in three Iranian mines, two near Anarak, the other about 100 miles north-west of Birjand (Harrison, 1968, 512). It has also been reported from the mountains of Kurdistan (Partington, 1935, 256) and Transcaucasia (Selimkhanov, 1975).

Antimony, as a trace element, has been recognized in copper and bronze objects from Mesopotamia and adjacent regions (Moorey, 1971, 298; Berthoud and Francaix, 1980, fig.4; Berthoud et al., 1981). Where antimony was available in association with copper ores it was exploited in prehistoric copper production in Europe (Davies, 1935) and in Caucasia (Kuftin, 1941). Antimony, like arsenic, has a hardening effect on copper, but in percentages up to about seven it is quite suitable for casting and a good temper may be given to cutting-tools by hot-forging. However, high percentages of antimony produce a brittle alloy and it is no competitor for tin-bronze (Charles, 1980, 171-2).

Instances of the use of metallic antimony for objects in Mesopotamia are very rare. Two examples have long been established in Assyriological literature, originally through a lecture by Herbert Gladstone (1892, 227):

> "M. Oppert indeed found at Khorsabad a tablet of metallic antimony, and M. Sarzec found at Tello part of a vase of pure antimony".

A remark made on the same occasion, in discussion, by Austen remains highly relevant:

"...we only know antimony at the present day as a highly brittle and crystalline metal, which could hardly be fashioned into a useful vase, and therefore this remarkable 'find' must represent the lost art of rendering antimony malleable..." Partington (1935, 256, citing Berthelot) said the metal consisted of "antimony free from copper, lead, bismuth and zinc, but with a trace of copper". Selimkhanov (1975), with the assistance of the Louvre laboratory, reinvestigated the Telloh fragment and published the following analysis:

Sb	Cu	Sn	Pb	Zn	As	Ag	Bi
Major	0.072%	0.002%	0.05%	0.003%	0.02%	0.017%	0.003%
Ni	Co	Fe	Mo	Li	Pb	Sr	Mg
0.002%	0.002%	0.01%	0.001%	0.0005%	0.001%	0.004%	0.009%
Na	K	Ca	Al	Mn	Ti	V	B
0.045%	0.019%	0.2%	0.01%	0.002%	0.0005%	0.01%	0.0005%

He attempted to relate the metal to Transcaucasian natural antimony and to local manufactured antimony; but the results were indecisive. It is not certain that the object in question was a vase. The antimony objects from Transaucasia are all small personal ornaments.

The supposed foundation tablet from Khorsabad involves a simple matter of mistranslation; there never was a metallurgical identification. No such object was found at Khorsabad; the tablet in question was of stone, probably magnesite (Ellis, 1968, 194, no. 78).

Beads of antimony have been reported from an Iron Age context at Hasanlu in north-west Iran (Dyson, 1964). Antimony combines readily with many other metals to form alloys; an alloy of tin and antimony makes a white metal of bluish tint that is malleable and ductile. In an archaeological context the eye might well confuse it with silver. Dayton (1978, 450) found that four beads from an Iron Age grave at Tell el-Farah (South) in Israel were made of 66% tin and 33% antimony, so this alloy may have been more prevalent in the Near East than is yet apparent. For the present there is no clear evidence for the intentional alloying of antimony with base metals in Mesopotamia.

II: ARSENIC

There is no analytical evidence for the ancient use of arsenic as a metal in Mesopotamia, nor has any word for it been reliably identified in Sumerian or Akkadian. So far as present evidence goes it seems neither to have been isolated as an independent element nor recognized as a native metal in ancient Mesopotamia. As an alloying element in copper, for which there is increasing evidence throughout the ancient Near East (see pp.15ff.), it may have occurred with copper ores. With smelting and reducing conditions, it formed natural alloys whose value was early recognized by observation and experience in copper casting. At a subsequent stage arsenical-copper alloys may have been artificially achieved by reducing suitable arsenic ores in contact with molten copper in a crucible.

The use of the brightly coloured arsenic sulphides, realgar (red) and orpiment (yellow) is a distinct matter relevant to the history of pigments and also of wax for writing boards, since the two were mixed for this purpose in Neo-Assyrian times (cf. Wiseman, 1955).

III: COPPER AND ITS ALLOYS

Copper was the most widely used of metals in Mesopotamia, where it was the cheapest of all until the Neo-Babylonian period, when iron became cheaper. Although, for the most part, only the manufacture of copper is represented in evidence from Mesopotamia, where it invariably arrived as a processed or semi-processed import, some general review of the preliminaries is necessary for a proper understanding of this metal´s crucial role in lowland metallurgy (cf. Röllig and Muhly, 1983).

Copper occurs naturally in small quantities in various parts of the world, the Near East among them (Coghlan, 1975, 18ff). Of copper ores the oxides, appearing in the weathered surface (gossan), are the most accessible in most regions and the simplest to work. Their rich colours would have made them particularly eye-catching to pioneer prospectors. This group includes cuprite and melaconite, the carbonates malachite (green basic carbonate; the richest in copper of all the ores) and azurite (blue basic carbonate) and the silicate chrysocolla (blue or bluish-green). More difficult to work are the secondary sulphides, from the parent ore, which is unweathered, and from a zone of secondary enrichment between the primary sulphides and the gossan. This group embraces chalcocite (copper glance) and covellite from the zone of secondary enrichment, and the iron sulphide minerals chalcopyrite (copper pyrites) and bornite (peacock ore). It has become increasingly clear to students of near Eastern metallurgy that certain copper-bearing ores of more restricted occurrence also played an important role, as it is probably through their use that the metals arsenic and antimony entered early copper objects. These ores include a sulphide tetrahedrite (fahlerz), containing iron and antimony; bournonite, containing lead and antimony; tennantite, containing iron and arsenic; and enargite, containing arsenic and copper. Recent research in Iran has also placed emphasis on certain native arsenides of copper, algodonite and domeykite, found close to the surface and included in copper-bearing gossans (Heskel and Lamberg-Karlovsky, 1980).

No less important for the history of metallurgy is the role of copper in alloys. An alloy is the result of combining intentionally or inadvertently two or more metals in the presence of heat. A deliberate alloy is usually made either through co-smelting various ores; through the combination of ores and metals; or through combining metals already processed separately (by smelting to the metallic state) as is most usual in an industry. As tin is of limited occurrence in nature, particularly in the Near East, it is conventionally assumed that tin-copper alloys (bronzes) are deliberate man-made combinations. Such also is the case with combinations of zinc and copper to make brass, which does not appear in the Near East until the mature Iron Age. Contrivances may not be so confidently identified in the case of arsenical coppers, which occur in nature more commonly. On the basis of geological evidence alone it is not possible to say whether an arsenical copper is a natural or an artificial alloy, so the term arsenical bronze, which implies the latter, is avoided here in order to retain the necessary element of ambiguity. However, even at the lowest level of selection of ores, there was a deliberate exercise of choice so the term alloy is permissible. Both with tin and arsenic the lower limits for an intentional alloy are arbitrarily set, usually at about 2 or 3% for tin (though much lower figures may reasonably be argued, see p.18) and at about 0.9% for arsenic, since most values in this case are either above this figure or merely parts per thousand (see p.17).

A. Support Technology

Although it is increasingly evident that the pattern of developments varies from region to region, it remains broadly true that the earliest coppers were the purest coppers, as they were either native metal (Patterson, 1971) or copper smelted from its oxide or carbonate ores. By hand picking specially pure pieces of the ore reasonably pure samples of copper (though less pure than the best of native coppers) can be produced. But, as the major copper deposits in Anatolia, Iran and Cyprus are of sulphide ore, their use may have been precipitated relatively early. By the earlier third millennium B.C., at the latest, sulphide ores were being smelted successfully in Anatolia (Zwicker, 1980) and Oman (Weisgerber, 1983).

All the preliminary stages in metal preparation (Coghlan, 1975, 26ff.), after mining, are likely to have taken place in the source zones where the considerable quantities of fuel needed would have been most readily available. Horne (1982) has examined the role of charcoal, and its production, in ancient Near Eastern metalworking. Taking 5 kg. of copper, enough for about 20 shaft-hole axeheads, as the potential product of a single smelt, she calculates a consumption of at least 100 kg. of charcoal. This would have taken 3.3 man days of labour to produce (at 30 kg. per man daily) and would have used 700 kg. of wood.

The processes of mining and basic smelting would generally have been remote from the knowledge and experience of Mesopotamian smiths. Gudea's reference to the mining of copper is exceptional (Falkenstein, 1966; Statue B:VI:21-3; Cylinder A:XVI.15ff; cf. C.A.D.s.v. ḫurru); if ore travelled at all, it is likely to have been as carefully selected pieces. In her study of the materials listed in Neo-Assyrian annals Jankowska (1969, 265) commented:

> "Highly significant is the predominance of ingots in the reserves of all the countries, whether it be precious metals, bronze, copper, tin or iron. Apparently it was usual only to smelt the ore in the mining regions...all further processing being done by the consumers, who had available craftsmen".

Although this may have been the general rule, there are isolated cases where it is possible to study in situ not only the composition of the unrefined metal straight from the smelting furnace, but also artefacts made from such metal in production centres adjacent to the mines. Timna in Israel, a mining region for copper where Late Bronze Age votives (of raw and manufactured metal) were placed in a temple, has proved ideal for such studies (Rothenberg, 1972; Conrad and Rothenberg, 1980; Craddock, 1980a). The information from Timna allows the composition of artefacts to be related directly to the parent ores and the smelting processes as revealed by excavations. The Timna analyses also show the purity and degree of consistency in artefact composition to be expected for copper from one ancient production centre. This information has been made more useful by modern smelting experiments using reconstructions of the ancient Timna furnaces (Tylecote, 1977). Such work is crucial as a control on research confined to artefacts alone, for after smelted metal has been removed to a manufacturing centre, it may be purified or perhaps mixed with metal from other sources and with scrap. Thus the true relationship to the parent ore and the smelting process is completely masked. Of the regions likely to have been supplying Mesopotamia with copper, Oman is the only one where studies of comparable intensity are at present proceeding (Weisgerber, 1981). Studies in other regions indicate that scrap may have been widely used from much earlier in the establishment of copper industries than has been commonly assumed.

Furnace development is a subject for which there is very little hard evidence from the ancient Near East. Despite many references to the contrary it is unlikely that copper was smelted in crucibles in antiquity.

Tylecote (1974) has argued that pure oxidized copper can be so converted but low grade oxide or sulphide minerals cannot and they were then by far the commonest available. The original Sumerian pictograph for the sign for a smith (SIMUG), in the middle of the fourth millennium B.C., appears to show the plan of a smelting furnace with attached blowpipes or tuyères (Falkenstein, 1936, no. 325). A number of words in the relevant ancient languages at present variously translated as "furnace", "kiln" or "brazier" have yet to be elucidated technically (Salonen, 1964). It is assumed at the present that kiškattû (C.A.D. s.v.) is the Akkadian word most likely to have denoted a metallurgical furnace.

It needs to be stressed that metallurgical furnaces differ from those of other fired materials in requiring that the metallic ores be placed in direct contact with the fuel (charcoal or coal) if they are to be reduced from their oxide, carbonate or sulphide states and to be separated from siliceous gangues (Tylecote, 1980). With rare exceptions, and even then there may be acute problems with close datings for industrial installations (Rothenberg et al., 1978; Tylecote, 1980), the early technology of smelting has been reconstructed by analogy and inference. Modern primitive methods (though they display a bewildering variety of possibilities) and common sense application of metallurgical chemistry (despite often poor understanding of many early processes) are the key resources in such discussions.

When their date may be reliably established, and that is all too rarely, the potential of slags for reconstructing smelting methods has been increasingly exploited (cf. Tylecote, 1980); but progress is inevitably slow. Iron ores were one of the prime fluxing agents for removing silica gangue in copper and lead smelting and left a fayalite or iron-silica slag (Cooke and Aschenbrenner, 1975). Norsun Tepe in south-east Anatolia, probably using ore from the Ergani Maden mine, is one of the few sites from which slag analyses are available (Zwicker, 1980). In all the examples examined, of the mid-fourth millennium B.C., chloride fluxes had been used to smelt copper ores. There was also evidence of trials with a more complex ore containing arsenic and antimony. By about 2800 B.C., as already noted, sulphide ores were being smelted on this site (see p.5).

It might be anticipated that one of the most obvious indicators of a commerce in metals into Mesopotamia would be ingots; but very few are yet known from controlled excavations. The earliest smelting furnaces are likely to have produced little more than scraps of metal in the slage. This would have been broken up to release the metal for melting in crucibles and casting into ingots (Tylecote, 1980, 193-6). Evidence outside Mesopotamia suggests that plano-convex or bun-shaped ingots were commonly used in the Bronze Age. Two examples were found in a context dated to the middle of the third millennium B.C. at Tell Chuera (Moortgat/Moortgat-Correns, 1978, 66ff., figs. 29a-b) and examples are reported from Oman (Weisgerber, 1981, fig.40). Bar ingots seem to have been used in the second millennium B.C. for general workshop purposes; they were being cast in the Tell edh-Dhibaʿi workshop (p.37) and two were found in the Giparu at Ur (Woolley, 1976, 56: U.6622, 6685; see also Woolley, 1962, 105: U.109, 115: U.9048). They may represent a secondary stage, perhaps made from larger ingots, or from scrap, to aid in the production of particular objects. But texts make clear that this was also regarded as a convenient shape for packaging in overland trade.

Ring-ingots (Limet, 1960, s.v. URUDU ḪAR) would have been easy to thread and transport, though they may more commonly have been produced for purposes of exchange than those of manufacture. It is probable that a number of the objects from Mesopotamian excavations catalogued as "heavy bracelets/anklets" were, in fact, ring-ingots.

In a widely ranging study Dayton (1974) has called attention to the great variety of copper and bronze rings encountered on excavations in Mesopotamia, and elsewhere in the ancient Near East. He attempted to relate their weights to a system based on equivalents in barley and wheat. The use of "ring money" in the region is well documented; but its role as a medium of exchange, passing by weight, has yet to be properly demonstrated in terms of the material evidence. The clear identity of ring money, as distinct from personal ornaments or the like, is not always readily apparent. Even when identity is securely established, weighing individual objects in search of an established standard may be to misconceive an essentially random system in which the total weight of pieces used as currency on a specific occasion was what mattered, rather than the weight of individual "rings".

Oxhide-shaped ingots were characteristic of the East Mediterranean Late Bronze Age trade in copper. There is an isolated example in Mesopotamia, from level I, of the twelfth century B.C., at Aqar Quf (Baghdad: IM 51170: DK 4-124, R.76; 45x33x6cm.). Such ingots were probably cast in this distinctive shape to facilitate handling and packing in transit (cf. Tylecote, 1980, 194-5); a mould for such an ingot has been excavated at Ras Ibn Hani in Syria (Lagarce, 1983, figs. 13, 15).

Both Sumerian and Akkadian have a varied terminology for distinguishing qualities of copper (Limet, 1960, 33ff.; C.A.D. s.v. erû), none of which may yet be identified with any precision among archaeological material (Röllig and Muhly, 1983, 345-6).

Even if furnacess for the extraction of metals from suitable minerals are unlikely to be found in excavations in Iraq, those suitable for the remelting or reforging of metal for making objects are certainly to be expected. They may not be easy to trace: "For the purpose of melting scraps of metal in crucibles, a ring of stone, a pile of hot charcoal, and a clay tuyère connected to bellows are all that would have been required. Nothing would have remained of this arrangement except, perhaps, the clay tuyère. It is possible to obtain sufficiently high temperatures for melting bronze (950°C) with a cylindrical furnace with a grate like a brazier. The air would have entered underneath the furnace assisted by the chimney effect of the cylindrical portion. Usually there would have been a side hole through which the crucible could have been placed in the hottest part of the fire a short distance above the grate. The grate would of course be of clay, pierced with holes like a pottery kiln" (Tylecote, 1980, 197). In the actual manufacture of tools, etc., in Sumerian towns minute quantities of wood and bundles of reeds were used. A text from Ur of the IIIrd Dynasty (Legrain, 1947, no.152) cites, over and over again, 2 minas of wood and 3 bundles of reeds (for workshops, see pp.36ff. here).

B. Sources

(i) *Textual evidence*:

Muhly (1973, 220ff., 1976, 104ff.) has thoroughly reviewed the ancient documentary sources for the use of copper and its trade in Mesopotamia, with extensive commentary on their relation to known deposits in the area. In the third millennium Sumerian texts list copper among the raw materials reaching Uruk from *Aratta* (Pettinato, 1977, 82-3, 128) and all three of the regions *Magan*, *Meluḫḫa* and *Dilmun* are associated with copper, but the latter only as an emporium (Limet, 1960, 85ff.). *Magan* was certainly a land producing the metal, since it is occasionally referred to as the "mountain of copper". It may also have been the source of finished bronze objects (Limet, 1972, 14-17).

In the early second millennium B.C. Mesopotamia appears to have lost direct contact with *Magan*, and with *Meluḫḫa*, also earlier mentioned in relation to copper. Copper now came through *Dilmun* and its traders. Gudea refers to mining copper in the mountain of *Kimaš* (Falkenstein, 1966, I, pp.50ff.: Statue B.VI: 21-3). This region is assumed to have been somewhere between the Jebel Hamrin and the Lesser Zab (Edzard and Farber, 1974, 100-1); an old identification with Ergani Maden in Anatolia is no longer regarded as tenable. If this location is correct, it may have been just an *entrepôt* for copper from mines deep in Iran, or it might be a direct reference to the copper mines visited by Layard in the Tiyari mountains, north of Amadiyeh (Layard, 1849, I, 223).

Since all these documentary sources are from southern Sumerian sites, the absence of references to Anatolian mines is not surprising, but it is a significant gap in the evidence. Muhly (1973, 208) has taken the available evidence to suggest Anatolia was not an exporter of copper in the third and early second millennium B.C.; but probability, if nothing else, requires that the case be left open in view of the meagre evidence. What information there is in the Old Assyrian texts indicates a copper trade within Anatolia, almost certainly facilitated by the Assyrian mercantile organization there, but gives no grounds for postulating shipments of copper back to Assur itself (Muhly, 1973, 206-8). Leemans (1960, 121ff.) showed that in the earlier second millennium B.C. the price of copper in the north was much higher than in the south and that copper from the north would not have been able to compete, in Babylonia, with copper from the Gulf.

As already noticed, after the Ur III period direct trade from Mesopotamia down the Gulf to *Magan* appears to stop and *Dilmun* becomes the primary *entrepôt* for all Gulf Trade, including metals, in the earlier second millennium B.C. Copper came this way until at least the eighteenth century B.C., when there is a break in the records almost exactly at the same time as the earliest surviving textual indications of copper from *Alashiya* (Cyprus) reaching Mari and Babylonia (C.A.D. s.v. *alašû*; Schaeffer, 1971, 547ff.; Millard, 1973). The Mari texts also refer to a type of copper qualified by the term te-ma-yu, for which Dossin (1970a, 39, n.1) suggested an association with Teima in Saudi Arabia, serving as an entrepôt for copper from the Wadi Arabah mines. Muhly (1976, 109) was sceptical; but it is not an impossible routing.

It has been argued that it was the collapse of the Indus valley civilization in the second quarter of the second millennium B.C. that brought to an end the flourishing Indus-Mesopotamian trade up the Gulf; but this has yet to be satisfactorily confirmed. Stray indicators suggest continuing, if intermittent, activity. In the middle of the fourteenth century B.C. a Babylonian official was stationed on Dilmun, whence he reported back on local threats to the date crop. Then Tukulti-Ninurta I of Assyria, c.1244-1208 B.C., after his sack of Babylon, assumed the title "King of Dilmun and Meluhha", emphasizing contemporary Babylonian interest in these regions, even if the full implications of the ancient names no longer applied (Brinkman, 1972, 275-6; 1976, 314). Danish excavators revealed considerable evidence for the occupation of Bahrein and Failaka in the Kassite period (Bibby, 1972, 358ff.). Otherwise documents contain no explicit references in the later second millennium to the sources of copper.

One of the most important sources in Neo-Assyrian times for the north was eastern Anatolia (copper probably from Ergani Maden), when the Nairi-lands constantly yielded massive amounts of metal, including copper, to the Neo-Assyrian kings (Muhly, 1973, 290ff.). Sargon II makes direct reference to the mining of copper in the mountain of Ba'al Sapuna, Gebel el-aqra^c, north of Ras Shamra in modern Syria (C.A.D. s.v. bušlu, 351b). In the Neo-Assyrian royal annals copper utensils were mentioned as booty more frequently and in more regions than any other goods; the Mediterranean coastal region and the Upper Euphrates were particularly rich in them. Jankowska (1969, 262-3) specifically commented:

> "The local manufacture of bronze objects of art in Muṣaṣir is recorded by the annals specifically (Luckenbill, 1927, 173); here copper was mined. But other centres too can indirectly be identified. Particularly multifarious was copperware in the tribute of Zamua, of Bīt-Zamāni, and of Carchemish (Luckenbill, 1926, 454, 466, 476, 501). While the handicrafts in Carchemish, a major transit centre, could have flourished on the base of imported raw materials, the treasuries in this city having constantly considerable stocks at their disposal, Bit-Zamani and Zamua were more likely to depend on local copper mining".

The role of Gulf sources in relation to Babylonia in the first millennium B.C. remains to be elucidated. In the time of Sennacherib (c.704-681 B.C.) Dilmunite craftsmen, together with their bronze products, went to Assyria; but whether as migrant experts or as tributeries is not clear (Luckenbill, 1927, section 438).

Two mid sixth century texts from Uruk, published by Oppenheim (1967, 236ff.), mention copper from Yamana (as also iron), which refers to Cyprus or some locality in south-eastern Anatolia colonized by Greeks.

Although attention is commonly concentrated on the metals' trade, the value of booty to Mesopotamia's metal industries at all times, and the craftsmen often taken with it, should never be underestimated, even though, by the nature of the surviving evidence, its contribution is now incalculable. Isolated indicators, one from the beginning, one from the end, of the period treated here, will suffice to establish the point. After his triumphs over Elam and Marhaši, Rimush (c.2278-70 B.C.) gave the temple of Enlil at Nippur 3,6000 minas of copper (Hirsch, 1963, Rimush b7 xxiv 49-62). Of the Neo-Assyrian kings only Sargon II (c.721-705 B.C.) explicitly refers to his mining activities in the Taurus mountains and the imports of metals (Lie, 1929, 233-4). He claimed that he had imported so much metal from Anatolia and northern Syria as tribute and booty that "they fixed the rate of silver in Assyria equal to that of copper". Even if this is a gross exaggeration, it points to the magnitude of such confiscations.

(ii) Modern mining areas:

Copper occurs widely in Anatolia, though only a few deposits are rated as worth large-scale exploitation (de Jesus, 1980, 21ff.; map 7). Today the most prevalent copper mineral is chalcopyrite; nodules of native copper may still be picked up in a number of copper producing areas. The only confirmed prehistoric mine is that at Kozlu, about 15 kms. south of the well known archaeological site of Horoztepe, in central, northern Turkey (Giles and Kuijpers, 1974; de Jesus, 1981). This was open at least as early as the first centuries of the third millennium B.C., when the miners were exploiting a copper sulphide deposit. The Ergani mine is the most famous in Anatolia and has certainly been open for a long time, though it is not yet known how long (de Jesus, 1980, 22). Here the ore is also found as a sulphide (Birgi, 1951); in antiquity, as recently, copper may have been exported from the mine area as "black copper" (copper with some iron, sulphur and arsenic) for refining and manufacture in nearby towns. Hittite inventory texts of the thirteenth century B.C. list copper from Ankuwa, in the vicinity of Alishar, and Kizzuwatna (Cilicia) (Košak, 1983, 195).

All modern surveys of Iran have also shown extensive copper resources, widely distributed across the country (Ladame 1945; Bazin and Hubner, 1969). The most satisfactory description of an ancient mine has been provided by research in the Veshnoveh area, between Qom and Kashan, in west-central Iran (Holzer and Momenzadeh, 1971). Here a series of low, narrow tunnels were traced following the natural configurations of the ore veins, with indications of mining by fire-setting and subsequent hammering with stone mauls. The copper occurs in the form of malachite, with some azurite and chalcocite, and specks of native copper. The earliest pottery found there was an isolated vessel of Sialk IV type (c.4000-3500 B.C.). Copper also occurs extensively in Afghanistan (Berthoud et al., 1977) and the Indian sub-continent (Muhly, 1973, 234ff.; Ratnagar, 1981, 96-7).

Since the pioneering studies of the metalwork discovered in excavations at Kish and Ur in the 1920s it has been argued that

Sumerian copper came from Oman, consequently identified with ancient Magan (Peake, 1928). Initially this case rested on the high nickel percentages both in Sumerian artefacts and in the copper of Oman. Although it is now recognized that nickel as a trace element may alone not be sufficient to sustain such a case, Berthoud (1979, 114) regards the percentages of nickel as one of the relevant trace elements. In recent years archaeological fieldwork in Oman has greatly intensified, with significant results (Berthoud and Cleuziou, 1983). Research has shown (Weisgerber, 1981, 1983) that sulphide ores such as chalcopyrite, bornite, together with pyrite and perhaps arsenic minerals, were mined and smelted in Oman. The sulphide ore deposits are comparable to those of Cyprus. After smelting in at least four stages to matte, the copper was finally refined in crucibles and traded as plano-convex ingots (Weisgerber, 1983, fig.3). Evidence of production is extensive enough to suggest an export trade, probably from the fourth millennium B.C. Both ores and local artefacts correlate in terms of nickel content, in artefacts from Ur, to support the old view that Oman was indeed a vital source of Sumerian copper. At Ur (Sumerian Metals Project, Philadelphia) 41% of the objects analysed contained over 0.9% nickel, with an average of 1.9%, and amounts up to 3.5%; by contrast only 0.03% of contemporary artefacts at Gawra contained over 0.9% nickel. Gawra seems thus to have drawn on another ore source, perhaps in Anatolia or Iran.

To the west of Mesopotamia, though Cyprus is renowned for its copper mines, problems still surround the chronology of their early exploitation and the methods employed. It is now increasingly argued that chalcopyrite was and is the only copper ore available in more than negligible quantities in Cyprus (Koucky and Steinberg, 1982; Maddin et al., 1983, 132). Despite assertions to the contrary, it seems modern Syria has no geologically attested copper sources (Muhly, 1973, 214). Palestine has important copper mines in the Wadi Arabah; those of Timna, in Israel, have received much the most intensive study (Rothenberg, 1972, 1978; Conrad and Rothenberg, 1980). Copper ores occur in Egypt, in Sinai and in the eastern desert; but it is unlikely that they ever directly served Mesopotamia (Lucas, 1962, 201ff.).

Taken together the ancient documentary sources and the distribution of potential copper sources indicate a coherent pattern. The major source areas of the region were those in Anatolia, Cyprus, Iran and the Gulf. The mines of Cyprus sporadically exported copper to Mesopotamia, probably through the major Syrian cities; Iran and the Gulf largely supplied the copper used in Sumer and Babylonia. The copper mines of Palestine and Egypt, Afghanistan and India, were primarily of local significance. In countries like Iran and Turkey isolating sources will always be difficult, since there may have been many small copper mines exploited so long as fuel was locally available, then abandoned until the wood cover was rejuvenated (cf. Allan, 1979, 35).

(iii) Scientific investigations:

Scientific study of metal objects found in Mesopotamia was encouraged from the outset of modern research in the region, notably by Layard (Percy in Layard, 1853, 670ff.). The earliest students of ancient metalwork were interested both in the methods of working and in the composition of alloys. Layard (1849, I, 220ff.) himself explored some of the mining regions immediately within the range of the major Assyrian cities; but it was not until the researches of Desch (1928-38; 1934) and his colleagues, in the 1920s and 1930s, on Sumerian copper and bronze objects, that the identification of the sources of Sumerian copper became a subject of primary interest. This research rested on the assumption that, since analyses of Sumerian artefacts had shown them to contain high percentages of nickel, then the source of the parent ore would be identified in the region yielding ore with a high nickel impurity. This in itself rests on the general proposition, crudely put, that the source of the copper used in any particular group of artefacts can be identified by comparison of their trace elements with those of the parent ore as it came from the mine. Fifty years later not only this special case, but the entire basis of such procedures, has met with increasingly rigorous criticism (cf. Thompson, 1958; Muhly, 1973, 339ff.; 1976, 115; Coghlan, 1975, 87ff.), despite ever more sophisticated methods of physical, chemical and statistical analysis, both of artefacts and ores, and a steadily accumulating body of highly diverse data. The most that would be agreed at present is that it may be possible to say that a certain object cannot have been made with metal from such-and-such an ore deposit. Coghlan (1975, 88) has judiciously summed up the question:

> "One must remember that the impurities pattern is based upon a combination of elements, but owing to certain oscillations or other variations which can arise in the ore-body, the presence or absence of a single element should not be regarded as decisive unless it is repeated over a number of specimens. Again, owing to variations in 'run-of-mine' ore, and in the smelting process, the impurities pattern of a metal must not be too closely related to quantative figures. In the case of trace-element impurities only marked oscillations in quantity should be considered as of true significance. Correlation by impurities patterns only truly applies to early copper and bronze artefacts before there may have been melting down of scrap metal, as may have been the case during and after the Middle Bronze Age. It is obvious that when scrap metals from various sources are melted together the impurities pattern obtained from the artefact would be highly misleading".

Recent developments in archaeometallurgy in general have marked a reaction to this traditional preoccupation with establishing a direct relationship between the metal used for an artefact and the ore source from which it derived. This change owes much to a fresh range of examination methods, beyond chemically-based analysis, for copper and copper alloyed objects. Potential information is maximised by combining analysis, measurement of segregation, phase identification, grain-size measurement and metallography. They allow for fuller reconstruction of the resources and techniques of manufacture in any time or place. Common impurity and alloying patterns may be established, indicating common technical traditions. Analysis of elements, like oxygen and sulphur, may further identify fossils of the manufacturing process. Lead isotope analysis may help with the identification of ores sources (see pp.112ff.). In this way changes in metal refining techniques, in alloying, in the number and range of metal sources, and in metalworking practices through time, may be more reliably elucidated (given a sufficiently broadly based investigation). Such programmes are only just emerging in the study of ancient Near Eastern metallurgy.

Recent research in key mining regions like Timna in Israel and Oman has emphasized how conclusions about ancient base metal industries will never be firmly based unless they take as much account of the primary, as of the secondary, production processes. If not controlled by as full a knowledge as possible of the original ore source (and the initial extractive processes employed there), conclusions about primary processes derived from artefactual studies alone run the danger of being seriously misleading.

A programme of research undertaken by Berthoud and others (1979; with Francaix 1980), fully aware of the basic problems of interpretation, set out to investigate the source zones for the copper-based industries of Lower Mesopotamia and Iran in the fourth and third millennia B.C., when native copper and copper smelted from simple oxide and carbonate ores are assumed to have been primarily in use (but cf. Oman, p.5). Berthoud concluded that additions of such elements as arsenic were made at a late stage in the production processes and therefore have no provenancing value, whereas manganese, chromium and nickel, together with the well-known indicators selenium, tellurium and silver have. Iranian and Omani ores were distinguished by the higher quantities of such elements as nickel, cobalt, vanadium and chromium in the latter. This research programme had two main aspects. First, careful investigation of potential ancient mines and systematic physio-chemical characterization of their ores; secondly, analysis, by the same methods as for the minerals, of artefacts from controlled excavations on key sites, using more than thirty trace-elements. In view of the crucial chemical changes implicit in the processes of metal and artefact production, this information may only be married up through specific thermodynamic and statistical models, upon which critical assessment of the work will inevitably turn. As a result of his research Berthoud summarized his conclusions as follows (in Berthoud and Cleuziou, 1983, 243):

Site	Mining Region
Susa A (Early Uruk Period), Iran.	Anarak (Central Iran).
Susa B (Late Uruk; Jamdat Nasr Period), Iran.	(Bardsir (Iran) No clear (Kashan (Iran) correspondence
Susa D (Early Dynastic), Iran) Umm an-Nar, Abu Dhabi.) Ur, Iraq.)	Oman
Tepe Yahya (IV B.C.), Iran.	Sheikh Ali (mine close to site).

C. Mechanical Properties and varieties:

Copper, as cast, is soft and unless hardened is not suitable for the cutting edges of weapons and tools, though for the production of vessels from sheetmetal this soft condition is crucial. Pure copper is hardened either by the introduction of impurities which alter its physical characteristics, as in tin-bronze production, or work hardened through the action of cold-hammering (forging). Cold-hammering alters the elastic properties of the metal so that the harder it becomes, the less malleable it is, until it will eventually crack. This may be reversed by annealing: heating the metal in a fire for a short time to a moderate temperature (400°-500°C) and then cooling it. This returns the metal to a soft and workable condition. The prehistoric smith is likely to have discovered this before either melting (which follows naturally from it) or smelting.

(a) Native Copper:

Although the linear evolutionary development of ancient metallurgy in which native copper played a key, initial role, is no longer regarded as axiomatic (cf. Berthoud, 1979; Maddin et al., 1980), the particular problems presented by this metal are of considerable interest. Native copper was available to pioneer metal workers in Iran, notably in the Talmessi-Anarak region (Berthoud, 1979), in Anatolia, and perhaps also in Cyprus; it has not been reported at Timna (cf. Patterson, 1971). To what extent it was used for the earliest reported artefacts from the Near East is debatable. Native copper may be cold hammered into artefacts, though there is some disagreement over how effective this is (Coghlan, 1975, 76ff; Smith, 1967); shaped by periodic annealing; or melted and cast. The basic problem for archaeometallurgists is the confident isolation of such artefacts. At present no adequate criteria exist for distinguishing artefacts of native copper from those of worked or recrystallized smelted copper of high purity, even when metallographic studies have been made (Maddin et al., 1980). It is generally assumed that all the earliest metal artefacts in the Near East were of worked native copper; but this is not certain. Native copper may contain significant quantities of arsenic, as at Talmessi in Iran (Berthoud et al., 1982), so "arsenical copper" objects could have been manufactured from melted native copper.

(b) Arsenical Copper:

General understanding of early metallurgy in the Near East has been slowly modified in the last thirty years by the demonstration that arsenical copper was the dominant copper alloy, either natural or artificial, from at least the fourth to well into the second millennium B.C. (particularly Charles, 1967; cf. Selimkhanov, 1977). Natural arsenical coppers are known in various parts of the region including Anatolia (de Jesus, 1980, 91), Iran (Heskel and Lamberg-Karlovksy, 1980, 233) and Cyprus (Panayiotu, 1979). Coghlan (1975, 79) has succinctly characterized these arsenical coppers:

> "The alloy may occur naturally through the smelting of arsenic-bearing ores, or the alloy may be intentionally made by simultaneous smelting of copper and arsenic minerals. There is also the

> possibility that the high arsenical coppers, in which the arsenic content may reach 10 per cent or more, could have been produced by a crucible process in which an arsenical mineral (orpiment or realgar), was added to molten copper under reducing conditions with the crucible well covered with charcoal...this process, if used, was probably a late development."

Maréchal (1958; cf. Böhne, 1965) showed that arsenical coppers were not inferior to tin bronzes for hardness or malleability in the annealed state, and were well-suited for the hot and cold forging methods used in antiquity. Their mechanical properties closely approach, if they do not actually match, those of normal early tin bronzes. There seems little doubt that they were adequate to meet the mechanical strength requirements for the tools and weapons generally used. As arsenic, like tin, also acts as a deoxidizing element in an alloy, it would facilitate a casting operation. Analyses increasingly show that much Sumerian and later metal statuary was made, quite successfully, of copper or arsenical copper not of tin bronze, as had previously been supposed. Although the toxic gases may be controlled by smelting in a closed reducing atmosphere (McKerrell and Tylecote, 1972, 209ff.), ancient smiths may have been inhibited in their use of arsenical coppers by the lethal effects of the industrial processes involved (cf. Charles, 1967).

In the absence of any direct information as to how these arsenical coppers were made, most authorities are at present sceptical of alloying in a direct sense; but, in certain areas, the use of native arsenic may not be ruled out. Experiments (Eaton and McKerrell, 1976; Charles, 1980, 168ff.) have demonstrated how partially roasted copper and arsenic sulphide ores might have been smelted together. So long as the atmosphere is kept partially reducing, co-smelting of any arsenic mineral (oxidized or sulphidic), with properly roasted copper ores, results in an almost complete recovery of arsenic. Roasting is an essential prerequisite to avoid any related iron content producing a brittle, useless metal. A high arsenic copper has another advantage, for it has a silvery colour, which tarnishes less readily than silver itself. On casting, even a relatively low arsenic copper exhibits inverse segregation: a small quantity of arsenical copper (c.15 to 20% As) is forced to the surface forming a complete outer skin of silvery metal. Although this was exploited in the manufacture of mirrors in Egypt and statuary in Anatolia by the third millennium B.C. (cf. Smith, 1973), it has not yet been certainly reported on a Sumerian object.

The argument (Eaton and McKerrell, 1976) that annaku was an arsenic-rich master alloy of copper to be used in the production of arsenical copper, not "tin" as it is usually translated (but see p.124 here), is not sustained by the published analyses said to demonstrate the case for this modification in translation. Currently available analyses, varying greatly in arsenic percentages, give little indication that metalsmiths could control arsenic content in the artefacts they were manufacturing. What they do show is that the transition from copper arsenic "alloys" to tin bronzes in the Near East was a far lengthier, more irregular, and more complex process than had previously been supposed. In certain regions, notably

Anatolia, from about 3000 to 1700 B.C., both arsenical coppers and tin bronzes were being used in what appear, on present evidence, to be separate but contemporary workshop traditions (cf. analyses in Esin, 1969; de Jesus, 1980, 94-5). At Ur, in Early Dynastic III, although tin and arsenic, in significant quantities, are found in the same object, the high arsenic contents are not found in tin bronzes, suggesting controlled alloying (Craddock, 1984; Sumerian Metals Project, Philadelphia).

Analyses of later fourth and third millennium base metal objects from Mesopotamia and Khuzistan (Moorey and Schweizer, 1972; Craddock, 1984; Berthoud et al., 1982) indicate the currency of arsenical coppers there, with over 1% arsenic and only traces of tin. As arsenic does not appear in significant quantities in the tin bronzes made from sometime in the middle of the third millennium, it is likely that by then, at least, the arsenic was a deliberate addition to the copper. In publishing their analyses of objects from Susiana Berthoud et al. (1980) argue that intentional arsenical copper alloys appear in Susa C, later fourth millennium, (the earlier Susa B is interpreted as a transitional period) and begin to decline by about 2000 B.C. Craddock (1984) has suggested that suitable arsenic ores may have been reduced in contact with molten copper in crucibles. Significantly, in working with material from Iran, Berthoud et al., (1980) have found that arsenic only appears in slags from urban metalworking sites, where alloying is likely, not in slags found at smelting sites in mining regions. Only with more analyses will it be possible to describe the eclipse of arsenical copper as the bronze Age passed into the Iron Age. They may have endured sporadically as long as the use of unalloyed copper (see p.20).

(c) <u>Bronze</u>:

The term <u>bronze</u> is taken here to embrace a range of binary alloys consisting of copper and tin, though commonly with a variety of other ingredients in small proportions. It is not used to describe alloys of arsenic and copper, whether natural or contrived (see above). It is assumed that in the manufacture of Mesopotamian bronzes the primary constituents were copper and tin (see pp.127ff. here), and that traces of other ingredients detected by modern analysis just happened to be present in the raw materials involved. Only lead is likely to have been introduced intentionally as an additive. A standard modern bronze has about one part tin to nine parts copper; but in antiquity the proportions, since less easily controlled, varied more. For instance, texts of the late third millennium B.C. from Ur give proportions of tin and copper that would yield alloys with between 9% and as much as 17% tin (Limet, 1960, 58ff.). Lower percentages of tin are more problematic, since they involve the distinction between a natural and an artificial alloy. This limit has been much debated and is, anyway, difficult to define with any confidence in isolation from specific groups of metalwork.

Some copper slags contain tin in amounts from 0.1 to 0.5% (Bachmann, 1982, 26; cf. comments of Craddock, 1980a). Maréchal (1962) has shown that, when copper ores with noticeable tin contents are smelted, reduction of tin is promoted in the presence of copper due to intermetallic affinity. Thus low tin bronzes might have been manufactured earlier than the reduction of tin ores to tin metal. Cleuziou and Berthoud (1982, 15), who have recently studied much Iranian metalwork of the later prehistoric and early historic periods, argue that "the definition of a minimum tin content is strongly dependant on local conditions... We have therefore concluded that tin values above 0.5% indicate deliberate combination of copper and tin at some point during the process which transformed ores into metallic bronze, though this is not to imply that the operation was totally controlled and that the results were standard". Such small percentages might well be deliberate additions in periods when contemporary bronzes have up to 10% tin; but earlier it is debatable and is probably more often accidental. Coghlan (1975, 35) "tentatively" suggested a deliberate copper-tin alloy as one in which tin exceeds 2% to 3%. The addition of tin to copper hardens the metal so long as it is not excessive; it lowers the melting point from 1,083°C to 1,050°C (5% Sn), to 1,005°C (10% Sn) to 960°C (15% Sn) (Lucas, 1962, 217); and much increases the liquidity of the metal.

The recognition of the role of arsenical coppers in the early history of metallurgy has modified views about the first emergence of copper-tin alloy in the Near East, as elsewhere. The distinction between copper and bronze is first made in the "archaic" texts from Ur (Early Dynastic I) (Burrows, 1935, no.373) and later in texts from Tell Fara (Early Dynastic IIIA), at the time of the earliest archaeologically attested tin bronze objects in Mesopotamia (Elam, 1932; Levey and Burke, 1959; Limet, 1960, 66). Wertime (1980) and Charles (1978; 1980) have both associated the exploitation of arsenical ores of copper and the subsequent introduction of tin ores by calling attention to the polymetallic character of many ore-bodies (Wertime) and to the community of colour between some tin ores and arsenical copper ores (Charles). The tin ore stannite, for example, resembles in appearance the copper-arsenical mineral enargite, so anyone used to enriching copper with enargite would be happy to use stannite for the same purpose. Both scholars have suggested that another context in which miners might inadvertently have come into contact with tin is offered by the stannic ferrous gossans that may cap sulphide ore bodies, especially if such gossans were used as fluxes in copper-smelting.

There is little certain information on early methods for manufacturing bronze. The key question is how the alloy was produced. Did the smith work with metallic copper and metallic tin or was cassiterite (SNO_2) added to the copper in smelting or melting? Both Charles (1978) and Wertime (1978) have seen the addition of cassiterite to molten copper as the most common means of bronze production in early times; but this may only have produced a low-quality bronze (Maddin et al., 1977, 41) and may anyway not have been a method much favoured at a distance from the source of the cassiterite. The oldest known Sumerian text relevant to the question is from Telloh, dating to Early Dynastic III (Limet, 1960, 67; with important emendation in Hallo, 1963, 139a). This document is the first to offer the standard formula for proportions, in this case 13⅓ shekels of tin, which is exactly one-sixth of the 80 shekels of copper which precedes it in the text. Sargonic and Neo-Sumerian texts use the phrase "seven parts bronze" i.e. in the proportion of six parts copper to one part tin. As appropriate weights are given, these, and later recipes (Dossin, 1970, 21; C.A.D. s.v. balalu 1c), indicate that metallic tin was used for bronze making from at least the middle of the third millennium B.C. in Mesopotamia, though, as both texts and analyses show, proportions varied. Since bronze is not known much before this time, it is probable that tin always reached Mesopotamia ready processed as ingots. In a country to which metals were brought from considerable distances, presumably much more often than not in a processed or semi-processed form, this is not surprising. Bronze as a raw material seems rarely to have been the object of commercial transactions in Mesopotamia, as it was customarily prepared on the spot from copper and tin by the manufacturer of bronze objects. In a Sumerian hymn to the god Ninurta a bronze producing country is mentioned, but its name is lost in surviving copies (Cohen, 1973, 28, lines 146-7).

The introduction of tin bronze as a commodity into Mesopotamia sometime in the first half of the third millennium B.C., and its slow adoption in the following centuries, remains to be charted systematically through a wide range of analyses. If, as now seems likely, tin was reaching Mesopotamia from as far away as Afghanistan, apparently at first by-passing workshops in many parts of Iran (Moorey, 1982a, 88; Berthoud et al., 1982), its advantages must have been known after long experiment closer to the source zone. Cleuziou and Berthoud (1982) have argued that tin, before it was widely used for alloying, was added in casting to facilitate the processing of copper by lowering the melting point and increasing fluidity. If this were so, it is surprising it was not used more consistently in this respect on arrival in Mesopotamia, for instance, in the production of metal statuary. Present analyses of early tin-bronzes in Mesopotamia show no correllation between tin levels and types of object; vessels may contain as much as weapons. Its use for decoration and for soldering, however, seem to have been recognized there as early as its advantages in the production of fine cast weapons, like daggers and axeheads (see p.127).

If the tin was travelling as ingots, at a high price, as seems very probable, and if tin bronze has no very obvious advantage over a well-made arsenical copper weapon (see p.16), its adoption is not easily explained in purely economic terms. Charles (1967) has seen the toxic effects of arsenical copper as a disincentive to its use; but, in view of this alloy's prevalence in time and space, it does not appear to have been a crucial factor. For those who could afford it, the key was probably technological as "the hardness imparted to the cutting-edges of arsenical copper axes is quite appreciably below that of tin-bronzes" (Coghlan, 1975, 83). In general the effect of tin upon copper in the cast or annealed state was more marked than in arsenical coppers and skilled smiths no doubt soon recognized this.

If status value was also involved, appearance or decorative factors may have had as much influence as efficiency. Tin-copper alloys are distinctive in colour. By the fourth century B.C. in Greece, if not earlier elsewhere, quality controls were exercised over copper-tin alloys through distinctions of surface colour, using a set of well-polished copper-tin alloys of known composition (Varoufakis, 1974, 95). So it is possible that initially the "golden" colour of some copper-tin alloys increased their status value.

The continuing "Sumerian Metals Project" in the University of Pennsylvania has analysed 32 Early Dynastic III and 27 Akkadian objects from Ur (Stech: personal communication 1983 gratefully acknowledged; cf. Woolley, 1934, 290; Moorey and Schweizer, 1972; Craddock 1984). There is no marked contrast between the two periods. More or less "pure" copper is relatively rare; arsenical copper is the most common metal, with arsenic percentages ranging up to 4.05%; bronzes, about a third of the total, have tin from 0.012 to 15.9%. A rather different pattern has been revealed for Tepe Gawra by the Philadelphia Project. 47 artefacts, ranging in date through levels XI (Late Prehistoric) to IV (Old Assyrian), but mostly from VI (Akkadian and post-Akkadian), were analysed and studied metallographically. No bronze was detected before level VI. "Pure" copper and copper with arsenic levels usually lower than 1% were common. Even in levels VI and V bronze is rare. These results isolate even more the old analysis of a pin from level VIII said to contain 5.62% tin (Speiser, 1935, 102). It may no longer be identified to check this analysis. Although, at both Gawra and Ur, arsenical copper was the metal in greatest use, Ur yielded relatively greater numbers of arsenical coppers than Gawra in both periods and the objects from Ur reveal a greater range of arsenic percentages.

It is already clear that well into the second millennium B.C. there was an extensive use of unalloyed and arsenical copper (McKerrell, 1977; 1978). It is possible that it was not until the move to sulphide copper ores in all the source zones that the demand for another additive was sufficiently stimulated to create a regular bronze industry throughout Mesopotamia.

In Neo-Assyrian texts of the earlier first millennium B.C. the logograms for "copper" and for "bronze" were both used to describe "bronze" (Zaccagnini, 1971, 123ff; Postgate, 1973a, 25, n.48). As the two metals cannot be distinguished by eye this is a perennial source of confusion and probably applies equally, if not so obviously, to the textual record in other periods. The role of bronze in the metal industries of the Iron Age has hardly been explored through analysis. In western Iran it appears to have been readily available (Moorey, 1971, 296), as also in Urartu (Hughes et al., 1981); analyses of bronzes of Eastern origin found in the Greek world (Filippakis et al., 1983) have been shown to contain significantly greater quantities of tin (8 to 12%) than does local work before about 750 B.C., whilst even thereafter, when Greek smiths introduced more tin into their alloys (perhaps under direct Oriental instruction), it is up to, rather than above, the 7 to 8% range. This has been taken to indicate that a tin source in the Near East, rather than in Europe or the West Mediterranean, was serving Greece at this time.

(d) Brass:

The early history of brass has long been confused by the use of the name in so fundamental a source as the King James Bible of 1611 to describe all copper alloys irrespective of composition; indeed Johnson defined the new word "bronze" in his Dictionary (1755-73) as "brass". In strict modern usage brass is a yellow-coloured alloy of copper and zinc, usually containing about a third of its weight of zinc. Zinc is found in nature as the sulphide blende or sphalarite; the carbonate calamine or smithsonite, and hemimorphite, once also called calamine. These ores are found together with copper and lead ores. Sulphide copper ores contain some zinc, but this is lost during the smelting of the copper and only in exceptional cases would any be retained in the copper to appear as small traces in artefacts. The earliest alloys in which zinc replaces other metals as the main constituent with copper appear in Anatolia in the 8th to 7th centuries B.C.; suggested earlier appearances are spurious (Craddock, 1980). Fibulae and metal vessels from the Phrygian capital of Gordion (Craddock, 1978, 3; Steinberg, 1981, 287-9) and at Urartian sites such as Altintepe and Toprak Kale (Hughes et al., 1981; Meyer, 1968) are the first properly documented "brasses". An Urartian bull's head examined by Gettens (in Hanfmann and Hansen, 1956) is made of tin-bronze with at least one horn of brass, originally providing a distinctive colour contrast. These are all low zinc brasses, with zinc below ten per cent.

To what extent such alloys were produced in Mesopotamia about this time is obscured by the absence of analyses. Two bronze bowls in the British Museum from Layard's finds at Nimrud contain 6.39% (N.94) and 5.51% (N.617) zinc respectively, but they may not be local products (Hughes et al., 1981, 144). Significantly, when Greek authors refer to <u>oreichalkos</u> it is as a rare, exotic and expensive material, manufactured in Anatolia. How it was made is still an open question. Craddock (1978, 9) suggested the cementation process: "fragments of copper were packed in a sealed crucible with charcoal and zinc oxide and heated to a temperature of between 900° and 1000°C; that is, hot enough for the zinc to have been vaporized but not so hot that the copper melted and ran to the bottom of the vessel. The zinc vapour in the sealed vessel readily dissolved in the widely dispersed copper and brass was formed. At the end of the process the temperature was raised and the then molten brass was stirred to form a uniform alloy" (cf. also Werner, 1970).

The reference to a "brass ring open at its ends" from the very early level XI at Yarim Tepe I is an error of some kind, presumably in translation (Merpert et al., 1977, 82).

D: <u>The Repertory of objects:</u>

The enormous range of objects for which copper and its alloys were used in ancient Mesopotamia is neatly epitomized by the despair of an eighth century clerk listing spoils for the renowned account of Sargon's eighth campaign. In reference to 120 objects of bronze he remarked "the setting down of whose names for the purpose of writing is not easy" (Thureau-Dangin, 1912, 567: line 364). This is a philological problem, unlikely to be much eased by the archaeological data (cf. the list in Limet, 1960, 195ff.). Archaeology has retrieved only a tiny fraction of the metal repertory for any period, usually the personal ornaments and weapons, with a scatter of tools and other isolated implements and vessels. The main source of such evidence is graves; if a period is not well represented in the archaeological record by graves, its metalwork is likely to be correspondingly little known. Only in exceptional cases are settlement sites rich in metal finds. This is to be explained by the systematic recycling of metal objects, since the practice of melting down and re-using base metal tools and weapons was a commonplace of daily life in Mesopotamia (Moorey, 1971a). No less devastating was the rigorous plundering by victors, as is explicitly recorded in a Hittite text: "The Araunna-people have looted the bronze implements of the town as is customary" (Kempinski and Košak, 1977, 87).

Two volumes in the series Prähistorische Bronzefunde, under the supervision of Müller-Karpe, have begun the long process of creating a corpus of metalwork from Mesopotamia (Braun-Holzinger, 1984; Rashid, 1983).

E: An Historical Survey:

(a) The Prehistoric Period:

The history of copper metallurgy in Mesopotamia is normally pushed back to the oval pendant from the Shanidar Cave, probably dating to the ninth millennium B.C. (Solecki, 1969, 311ff.). As it is now completely mineralized, it cannot be accurately described and may do no more than illustrate the use of malachite for ornamental purposes, as is evident elsewhere in the prehistoric period. For some years the rolled bead of worked native copper from an aceramic context at Tell Ramad on the Euphrates, and comparable objects from Cäyonü Tepesi in Anatolia and Ali Kosh in southwest Iran, provided the only indicators of the earliest copper repertory (Cambel and Braidwood, 1983, 157; Smith, 1969; France-Lanord and de Contenson, 1973, 109ff.). Recently Russian excavations on a series of sites in the Sinjar Plain of northern Iraq, extending in time from the pre-Hassuna to the fully mature Hassuna period, have revealed significant evidence for a pioneer copper metallurgy in some of the first village settlements of the region (Merpert et al., 1976, 41; 1977, 82; 1978, 36). Pieces of copper ore, as well as simple copper trinkets and tools, have been reported down to the bottom levels of Yarim Tepe I. From one of the lowest levels is reported a small copper ring of "flattened tight plait" (Ibid., 1978, 36); earlier still was a lead bracelet (see p.122). Low in Tell Sotto, of the pre-Hussuna period, copper beads were associated with an inhumation burial in a pot; contemporary pieces of copper have been reported from Kültepe and at Tell Magzalia "a hammered copper awl has been found on the floor of one of the houses" (Bader et al., 1981, 62). Some level of copper working was also evident at Tellul eth-Thalathat II, northeast of the sites excavated by the Russians (Fukai et al., 1981, 65); but none was recognized at Hassuna itself. Down the line of the Tigris at Tell es-Sawwan the excavator reported "three beads of copper...on the floor of building No.3 level II, and a very small knife with a boring at one end...in level I inside a grave..." (Al-A'dami, 1968, 59). Analyses of the early copper work from Yarim Tepe I and Sawwan have revealed copper with high levels of iron (1-10%), for which an iron flux may have been responsible, indicating that already copper, like lead, was being smelted in the source area, presumably in Anatolia (Merpert et al., 1977a).

An equivocal position is held by the copper spatula, pin and two simple sheetmetal bowls from Samarra (Herzfeld, 1930, pp.4, 5, figs. c, e), from the layer containing prehistoric graves. Initially Herzfeld accepted the metal as contemporary, but three years later (1933, 29) he rejected it as intrusive from Islamic levels and said there was no metal contemporary with the graves. With informed hindsight this might be an unnecessary revision since, typologically, a Samarra/Halaf date is not impossible (cf. iron fragments, p.100).

The important cultural complex named after Tell Halaf in Syria is still singularly devoid of metal finds in Mesopotamia. Yarim Tepe II, the Halaf period settlement, was far less productive in this respect than I (Merpert et al., 1977, 95). Various copper implements and fragments at Arpachiyah were tentatively attributed to this period (Mallowan, 1935, 104). A fragment from Tell Shemshara proved on analysis to be a weathered piece of copper pyrite, with no trace of human working (Smith, 1970, 123-4). A bead from Chagar Bazar 12 was analysed by Desch: "No trace whatever could be found of arsenic, nickel, tin, zinc or sulphur. This is the purest specimen of ancient copper that we have had", (Mallowan, 1936, 26-7). By approximately this period in Anatolia, to judge by the almost pure copper macehead from Can Hasan, shaft-hole castings were already current (de Jesus, 1980, 41). It should perhaps still be emphazised in passing that a copper or bronze axehead, a spearhead, a dagger and an arrowhead from Tell Halaf itself have nothing to do with this stage in the history of metallurgy (Schmidt, H, 1943, pl.CXIV).

Nor is the evidence for the subsequent Ubaid period in Mesopotamia much more revealing. Gawra XVII, of the later Ubaid, yielded a copper ring and a rectangular-sectioned awl; level XIII a copper awl; level XII a hemispherical copper button and a flat copper blade with a splayed cutting edge (Tobler, 1950, 213, pl.XCVIIIa, 5-6). Analysis of the latter indicated a relatively pure copper with only 0.05% arsenic. A flat copper adze or axe blade from Arpachiyah falls late in the Ubaid period (Mallowan, 1935, 104, pl.XI; Curtis, 1982, fig.22 (BM 127757)). Analysis has shown it to be 99% copper, 0.1% arsenic; no tin was detected (down to 0.002%) (de Jesus, 1980, II, 207). It may be the product of a simple smelting technique (Craddock, pers. comm.). At Nineveh a copper pin was attributed to "Nineveh 3" (Campbell Thompson and Mallowan, 1933, 145, p.LXVIII, 5). In the south, at Ur, in a grave placed by Woolley in the Ubaid III horizon (but possibly of the later Protoliterate period) was a solid, tanged spearhead of very pure copper (Woolley, 1934, 291; 1956, 21, pl.30). It is the much more common base metalwork from Susa in Iran which indicates how inadequate is this body of evidence from Mesopotamia for assessing the state of the Mesopotamian metal industry by the end of the Ubaid period (cf. Berthoud and Francaix, 1980; Amiet, 1979). It is not certain how many of these objects were of melted copper, how many of smelted; but casting is still only in simple open moulds and working methods are unadventurous.

The status of copper technology in southern Mesopotamia at this time is commonly assessed in the light of the argument that the baked clay shaft-holes axeheads and hammer axes characteristic of the period on many sites, though persisting later, are faithful copies of metal forms (Childe, 1952, 116). No metal exemplar has yet been found and the hypothesis may be redundant. The much more common baked clay sickles of this and subsequent periods in the same area were definitely working tools. Their sizes, contexts and the condition of their working edges, make this clear (Adams and Nissen, 1972, 205ff.; Anderson-Gerfaud, 1983). With exactly the same criteria it may be shown that such was also the case with full-sized shafthole axes and hammers in baked clay, (Lloyd and Safar, 1943, pl.XVIb, XXIXa), though what they were used for is still an open question.

There certainly were such model tools; but they are clearly distinguished by size and condition. For instance, in grave 21 at Tell al-Ubaid a clay model of a flint hoe was found beside a real one, (Hall and Woolley, 1927, pl.XLVI T.O. 40). The flared blades of baked clay shaft-hole axeheads may be explained by reference to stone tools, for there are pierced stone axeheads of this period with finely ground and flared cutting edges (Tobler, 1950, pl.XCIVd, e). It may then not be assumed from the shaft-hole baked clay tools that bivalve casting of putative copper counterparts was practised at this time. It may have been; but more convincing evidence, of the kind now available from Iran, is still needed in Mesopotamia. It may be, as Deshayes (1960, 408-9) argued, that in this period Sumerian metal-working was as much indebted to Iran for technical stimuli as it was for raw materials.

By the early fourth millennium B.C., the coppersmith's repertory in Mesopotamia appears to extend little beyond simple personal ornaments, such as pins, rings and beads, flat tools like awls, adzes or chisels, and metal stamp seals; by the end the range of techniques and artefacts is considerable. Many copper artefacts were probably by now of smelted copper, imported ready processed, for casting in open and bivalve moulds, then hot or cold worked after casting. Sheet-metal working had been mastered in all its simplest forms. Small-scale lost-wax castings appear early enough in this period to indicate some kind of previous development in Ubaid contexts, either in Mesopotamia or closer to a supply zone. One major archaeological discovery outside Mesopotamia is very relevant to this question, since it shows a remarkable mastery of the art of lost-wax casting, both for relatively plain, if intricate objects and for free-standing, three-dimensional ornaments, long before any such thing is evident in Mesopotamia. The enormous hoard of copper and arsenical copper tools, weapons and "ritual" objects, from a cave in Nahal Mishmar on the western shores of the Dead Sea in Israel is dated within the second quarter of the fourth millennium (c.3750-3500 B.C.) (Bar-Adon, 1980). It is a product of the Chalcolithic Beersheba Culture of southern Palestine and, even on relatively conservative dating, would be earlier than Uruk IV in Mesopotamia. Nothing to compare with it has yet been found elsewhere.

A free-standing lion (4.7x3.4x1.1cm) amulet from Uruk is of copper with 9% lead (Heinrich, 1936, 25, 47, pl.13a). This high lead content suggests that the craftsman was already aware that its presence facilitated casting (cf. Braun-Holzinger, 1984, 3ff.). However, it is not clear how much control was exercised at this time over copper alloys. Berthoud and his colleagues (1982, 45-6) have shown how, in the period on either side of 3000 B.C., there was a degree of lead-copper alloying not encountered later. Relatively high levels of lead characterize objects from Susa at this time. Analyses have now also revealed that a group of metal cylinder seals of late Prehistoric type have copper as the main constituent, but with relatively high lead percentages (up to 20%) (cf. Buchanan, 1981, XIII; Gorelick, pers.comm.).

Among the many small finds from the "Riemchengebäude", in the western corner of the Eanna sanctuary, built soon after the destruction of the "Stone-cone Temple" in Uruk IVA, were various copper vessels and spearheads, as well as animal horns (Lenzen, 1958, 24-5;

1959, 10, pl.17, 39b). The Uruk excavation reports rarely detail more mundane copper finds. Between levels "C" and "D" of the Anu ziggurat were an "astonishing" number of completely oxidized lumps of copper, some as large as a man's fist (Nöldeke, 1938, 25). Worked copper was also noted in levels of Eanna XI, (Jordan, 1932, 30). Objects of copper are noted in levels of "D" and "E" of the Anu ziggurat, (Nöldeke, 1937, 53). No details are given of a spear point analysed as Cu: 99.13%; Fe: 0.27%; Ni: 0.14%; As: 0.17%; Zn: trace (Heinrich, 1936, 47). Survey in the Uruk region offered some evidence for sheet metal vessels in the Uruk period; but the various copper tools and weapons recovered might have been made at any time down to Early Dynastic I (Adams and Nissen, 1972, 205-6).

Excavations in levels of the second half of the fourth millennium B.C. at Tepe Gawra in the north do not substantially alter the picture offered by Uruk. It is only with Stratum VIII, at the outset of the third millennium B.C., that an appreciable rise was observed in the number of copper objects retrieved. Thereafter they increased markedly with 22 copper specimens in VIII, 42 in VII and 334 in VI, towards the end of the third millennium B.C. (Speiser, 1935, 103). An adze blade from XI was primarily copper with 1.63% arsenic and 3.34% nickel, an arsenical copper typical of this phase over much of the Near East (see p.15 here). By this time smiths clearly had considerable knowledge of the results produced by a manipulation of various ores, selected by colour and sometimes by form, to improve the hardness of the tools and weapons. So far there is no clear evidence in Mesopotamia, as there is slightly later from Egypt, Anatolia and Iran, for the use of arsenic compounds to give objects a surface colour like silver. At Gawra the repertory in base metals was standard and simple, save for a cast copper object which, to judge from bone parallels, may be a kohl tube. It looks like the "mouthpiece of a musical instrument" with ribbed sides and was in an Uruk period archaeological context. A pin (no longer identifiable) from level VIII, usually dated to the early third millennium B.C., contained 5.62% tin; if correct, this would be the earliest tin bronze yet reported from Mesopotamia (Speiser, 1935, 102). Arsenical coppers have been reported from Habuba Kabira on the Upper Euphrates, a settlement of the Uruk IV period closely linked to Sumer (Riederer, 1976), whilst recent work in Philadelphia on objects from Gawra indicates a predominance there of copper and arsenical copper until the end of the third millennium B.C. (see pp.54-5).

For some guide to the routine production of the later Protoliterate and Early Dynastic I periods recourse must be made to Ur. There the evidence comes from a number of graves and levels of occupational debris in Pits W and X. Woolley divided the graves into three groups, early (A), middle (B), and late (C) (for recent complex and unresolved chronological discussions see Gockel, 1982, 1983; Kolbus, 1982; Korbel, 1984). Of the 54 graves assigned to "A", 15 contained metalwork: of lead, 11 tumblers and 1 dish; of copper, 5 bowls, 1 vessel and 1 ladle. Of the 130 graves attributed to "B", 26 contained metal objects: of lead, 13 tumblers; of copper, 14 bowls, 5 pins, 1 razor, 1 fishhook, 1 harpoon. Of 148 graves classified in "C", 22 contained metal artefacts: of lead, 12 tumblers and 1 bowl; of copper, 3 pins, 1 chisel (?), 5 bowls and 1 tumbler, 1 needle, 1 ladle and 1 mirror; of silver, 2 pairs of earrings (Woolley, 1956, 104ff.). The copper vessels, allowing for heavy corrosion, had percentages of arsenic varying from 0.3 to 1.3. There was no trace of tin, and nickel readings were low; a copper rod gave the same result (Desch in Woolley, 1934, 165). Predictably, it is tools which predominate in the occupational debris: fishhooks, projectile points and spatulae; no vessels were reported there (Woolley, 1956, passim). Such a pattern is evident elsewhere in this and the subsequent period. Techniques remain simple; arsenical coppers are current, but still, surprisingly, there is no published artefactual evidence for bivalve casting. Vessels were hammered, usually in one piece. Metalwork from Jamdat Nasr was meagre, adding nothing to the evidence of Ur (Mackay, 1931; Moorey and Schweizer, 1972, 180).

The excavations of de Genouillac at Telloh yielded a little metalwork of the Protoliterate period, though its identification is not without problems. The original publication is very confused and chronological attributions are not always certain. Buchanan provided a critical review of this material, which sifts it into groups (de Genouillac, 1934, 44ff; Buchanan, 1967, 535ff.). Apart from a few gold beads, the finds are predominantly copper pins, mirrors and sheet metal vessels, some spouted; but a pin topped by a pair of nude female "dancers" is an important early example of lost-wax casting (de Genouillac, 1934, 46, 50, pl.10.2-5; cf. Braun-Holzinger, 1984, 7ff.).

Metal temple fittings rarely survive from any period in Mesopotamia so the fragments of copper panelling found in the Late Prehistoric Temple at Tell Brak are all the more remarkable. Some pieces were fluted, some decorated with a linear eye design. They were originally fixed to a wooden backing with small copper nails, (Mallowan, 1947, 32, pl.VIII.2, n.1., pl.XXX.8).

(b) <u>The Early Historic Period, c.3000-2300 B.C.</u>:

In the "archaic" texts of Early Dynastic I from Ur, there is the earliest certain mention of <u>ZABAR</u> (bronze) distinguished from <u>URUDU</u> (copper), although analytical evidence of tin copper alloys in Sumer is not available until Early Dynastic III. Although arsenical coppers were then slowly superseded, copper was long to remain the staple metal in Mesopotamia. Supplies of tin were always unpredictable and its use in tools and weapons by no means constant. The extent to which Mesopotamia used tin at this time is a complex question for which far more analyses are needed before sound conclusions may be reached.

In the south, in the cemetery at ʿUbaid, a few of the graves are late prehistoric but the majority are of early Dynastic II-III, (Martin, 1982; Hall and Woolley, 1927). The metal tally contrasts with the "Jamdat Nasr graves" at Ur. There are no lead vessels; only 8 graves had copper vessels, (Hall and Woolley, 1927, c.11, 27, 37, 46, 63, 76-7, 92), 2, copper daggers or knives, (Ibid. c.28, 91), 4, copper pins or kohl sticks, (Ibid. c.3 (E.D.I.), 27, 65, 73), 1, a copper harpoon head, (Ibid. c.77, (E.D.I.)), and 4, twisted copper wire earrings, (Ibid. c.4 (E.D.I.), 4, 59, 73). A copper flat axe, an adze blade and a fishhook loose in the soil do little to alter the overall sample, (Ibid. pl.XLVI.2).

In Central Iraq the earliest phases of the Early Dynastic period are covered by the ill-excavated, poorly recorded graves of the "Y" sounding on Tell Ingharra at Kish, (Moorey, 1978, Fiches E09-G03). They are, for these reasons, best seen in comparison with the grave sequence at Khafajah in the Diyala region. Here, of 53 graves attributed to the end of the Protoliterate period, only 5 yielded metal: simple sheet copper vessels and pins as well as 2 lead vessels and a lead ornament (Delougaz, 1967, 59ff.; graves 1, 4, 5, 8, 18). Of about 30 graves attributed to Early Dynastic I only 3 yielded metal: 2 copper mirrors and a pin, (Ibid. graves 66-7, 83). Slightly over 30 graves were allocated to Early Dynastic II, 13 produced metalwork, with sheet copper vessels predominating. Isolated mirrors, pins and a razor appear; but the only clues to a more sophisticated range of metallurgical skills are 2 vessel-stands, cast and welded (Ibid. graves 86-7, 89, 91-2, 94, 97, 108-110, 114-4, 119A). It is only with Early Dynastic III that quantity markedly increases and the repertory in graves extends to include tools and weapons. But here the arbitrary witness of the archaeological record is all to clear when attention is switched, as in a moment it will be, from private graves to temple furnishings. At Kish Early Dynastic I-II private graves are furnished exactly as at Khafajah, though 2 cast copper vessel-stands, 1 set on a frog, the other with elaborate interlaced openwork sides,are more elaborate than those in graves at Khafajah, (Watelin, 1934, pl.XX. 1-2). It is the furnishings of cart burials at Kish that indicate the wider range of metalwork now available to those of wealth and social standing: tools (saws, goads and awls), weapons (fine daggers and shaft-hole axeheads), still of copper or arsenical copper, and finely cast zoomorphic rein-rings (not yet analysed; cf. Braun-Holzinger, 1984, 33ff.).

Metal temple furnishings certainly of Early Dynastic I are still unknown; only in the Diyala Valley are such objects available for Early Dynastic II, with one possible exception. In pit W at Ur, at the lower end of the S.I.S. 4-5 were found 4 hooves of oxen, (Woolley, 1956, 38, pl.29.77), made of thin sheet copper hammered over a wooden core, originally fixed to a stand. Woolley believed them to be much earlier than the comparable Tell al-Ubaid temple metal sculptures. The context, if undisturbed, certainly indicates an Early Dynastic I-II dating; but Frankfort (1932, 16, note) did not accept it. Three vessel-stands cast as bearded male figures, nude save for a girdle, of almost pure copper, with minor casting faults filled in with lead, were found in a context attributed to Early Dynastic III in the Temple Oval at Khafajah; but Frankfort suggested they might have been made earlier (Frankfort, 1939, 38ff. pl.98-103). In the Shara Temple at Tell Agrab, in an early Dynastic II context, was a badly corroded anthropomorphic stand, just like those from Khafajah, (Ibid. 11, pl.55). Here also were 3 solid cast copper statuettes, 2 male, 1 female, of worshippers, (Ibid. 11-12, pl.56); technically more outstanding is the three-dimensional cast copper model of a one-man chariot, drawn by four equids (Ibid. 12-13, pl.58). These objects, and the vessel-support modelled as two wrestlers wearing only girdles, from Nintu Temple V at Khafajah, illustrate the great skill of the Sumerians in making elaborate cast copper statuettes by the lost-wax method by at least Diyala Early Dynastic II (Ibid. 12, pl.54). In the highest surviving levels of the Shara Temple at Tell Agrab, in Early Dynastic III, were found the fragments of an anthopomorphic stand, considerably larger than those from Khafajah, and the toes of a human foot three-quarters life size.

Other body fragments are hollow cast with clay cores and look like debris from a faulty casting on a comparable scale, (Frankfort, 1943, 11, pl.61, 311-12; in general, cf. Braun-Holzinger, 1984, 10ff.).

Although these fragments suggest that complex castings of some size were not beyond the skill and ingenuity of Early Dynastic craftsmen, an elaborate programme of temple decoration in cast statuary of any size would have been costly in terms of metal and craftsmanship. This probably explains why, at sites like Tell al-Ubaid, large free-standing animals and relief panels were made mainly of hammered sheet copper fitted over roughly shaped bitumen or bitumen-coated wooden cores and secured with rivets, (see p.49), with separately fitted cast bronze heads (Hall and Woolley, 1927; Braun-Holzinger 1984, 26h; J.C. Curtis, pers.comm.). A number of cast copper (or bronze) animal heads have survived from Early Dynastic III, though only at Tell al-Ubaid is their original destination known, (cf. Moortgat, 1969, pl.53; de Sarzec and Heuzey, 1884-1912, pl.5 ter.; Parrot, 1960, 186; Glob, 1955, 178, fig.I; Hilprecht, 1904, 539, fig. on p.540; Frankfort, 1943, pl.29B; Braun-Holzinger, 1984, 31ff.). At Ur five copper bulls heads were found together in a hole below a plano-convex brick wall alongside the Neo-Sumerian grave 1850 (Woolley, 1934, 212, 594, pl.143).

The base metal equivalents of the well-known Early Dynastic II/III votive stone human statuettes (cf. Braun-Holzinger, 1984, 10ff.) have also still to be studied primarily through a few examples from unknown sources. Three are particularly important examples of this aspect of the sculptor's art in metal; all are lost-wax castings. The Louvre has a "bronze" bearded male figure in a kilt standing on a small base plate; in the Metropolitan Museum, New York, is a copper statuette of a man wearing only a girdle, who carries on his head a box or pile of bricks; and the Schimmel collection contains a copper male worshipper, best paralleled by a stone statue from Mari, set on an open-work base, inscribed by a scribe with a dedication to the goddess Ninegal, (Contenau, 1940, 37ff., pl.IV; Crawford, 1960, 264ff., fig.6; Muscarella, 1974, no.106). Closely related to these figures is a whole group of cast copper foundation figurines, plain, anthropomorphic and zoomorphic, many of known source and date, (Ellis, 1968, 46ff; Rashid, 1983). One text of this period relevant to the history of metal statuary speaks of a silver statue of the deity Šag-šag made in the second year of Urukagina of Lagash and the craftsmen involved in its production: the master-sculptor, foundry-men, a worker in stone and jewellers, (Spycket, 1968, 34-5). This assembly suggests that the sulptor created the original wax model for casting, that its eyes at least were inlaid with stone, and that its surface was further ornamented by jewellers. The size of the statue is not given.

The "Royal Cemetery" at Ur continues to dominate discussions of metalworking in third millennium Mesopotamia and beyond. In base metals it provides a comprehensive repertory of weapons, tools, vessels, personal ornaments and miscellaneous fittings across the whole period from Early Dynastic IIIA to Ur III, (Woolley, 1934; Nissen, 1966, passim). Here for the first time in Mesopotamia tin-bronzes are evident, (Woolley, 1934, 290; Craddock, 1984; for chronological comments see p.26), as they are in Early Dynastic IIIA/B at Kish and elsewhere (Moorey and Schweizer, 1972). A comprehensive analytical programme of the copper and bronze objects from this cemetery would give a vital indication of the relative place of bronze at this time. The Sumerians primarily used gold, for its beauty, and copper, for its utility, and bronze is likely to have remained a "luxury" product until the second millennium B.C. (see pp.127ff. on tin).

(c) The Akkadian Period, c.2300-2100 B.C.:

Limet has provided a useful study of the textual evidence for metals and metalworking in the Akkadian period, (Limet, 1972), emphasizing its limitations. Generally speaking the archaeological record for the routine repertory of metalsmiths is equally meagre outside the Ur graves, since there is still so little published excavated material from occupation levels securely dated to this period. There is nothing to suggest a radical break in the evolution of metallurgy, indeed the contrary, nor any certain indication of a widespread break in the lines of metal supply, as had been suggested on the basis of restricted evidence from Ur. Tin continues to be used at Ur and its use appears to be carefully regulated (cf. Craddock, 1984). There is one remarkable aspect of metalworking in this period, only obliquely indicated in Early Dynastic III by the fragments of large-scale bronze castings from the Diyala sites. This is the unequivocal evidence for large scale copper castings of the very highest quality (Braun-Holzinger, 1984, no.49, 54, 61).

The first of these, the well-known metal male head from Nineveh, (Campbell Thompson and Mallowan, 1932, 72), now known to be of cast copper not bronze (Nylander, 1980), has long been known, though never published in full metallurgical detail. The casting core has largely gone, though chaplets survive, leaving the head at least partially hollow, though it is sealed across the base of the neck. A tiny hole here, which may not be ancient, leads into the cavity and there is the "ghost" of a square peg on the base plate, either produced by the original casting process or else the remains of the fixture which secured the head to the body, probably in other materials. The detail of the head is remarkably fine and, although some of it may have been improved after casting with chasing tools, it looks as if it was predominantly cut on the original hard wax model. The head had been intentionally damaged at some point in its history, perhaps when Nineveh was sacked in 612 B.C. (Nylander, 1980).

Diakonov (1947) argued, from its hair and beard styling, that the life-size cast head of a bearded man, also of copper, now in the Metropolitan Museum, New York, once in the Brummer Collection and said to be from western Iran, should also be dated to the Akkadian period, though subsequent commentators prefer a dating in the early second millennium B.C. (Porada, 1965, 62, fig.38, n.34; Schlossman, 1981/2, 156, figs.22-4). It was cast solid with a cavity in the neck to take a rectangular dowel for fitting to a body; the eyes were originally inlaid as on the Nineveh head. The reported companion piece to the New York head, now in Cincinnati, has not yet been analysed (Pope, 1939, IV, pl.107). It is smaller, has eyes modelled in one with the face, a short beard, and the casting extends down onto the shoulders in a way neither of the other heads do. Its date is a much more open question and the original association of the heads on the art market may mean no more than a common geographical source in modern times.

The achievement of the Akkadian craftsmen in casting large-scale human statuary was confirmed in 1975 when, at Bassetki near Dohak in north-east Iraq, a bulldozer turned up the lower part of a cast copper crouching nude man, clasping the base of a standard or flag pole, set on a circular podium bearing an inscription of Naram-Sin. It may originally have been one of a pair of "guardians of the gate" in a temple, (Al-Fouadi, 1976, 63ff.; Oates, J, 1979, pl.17). It is a massive, hollow copper casting which still retains some of its clay core with the chaplets still in place. The whole figure, in so far as it survives, is modelled with striking realism and sensitivity, exactly placed within the circle provided by the circumference of the base. The right leg, at a right angle, frames the flag-pole socket, the left is drawn back more sharply to provide room for the panel of inscription. A tasselled girdle provides the only ornament. Like the Nineveh head it would be regarded as a masterpiece of the sculptors' art in metal in any age or place, epitomizing the level of skill available to the Akkadian kings towards the end of the dynasty. Copper is particularly difficult to handle in large castings of this kind, probably done with multiple piece-moulds. The technology has not yet been expertly studied; it might be direct casting in divisible refractory piece-moulds.

The relative roles of metals and stone in the later third millennium may be gauged, at least for weapons, by the persisting use of flint arrowheads. At Brak they were found in considerable numbers, though none of metal was recovered by the excavator (Mallowan, 1947, 180, pl.XXXVII.). For agricultural tools wood, stone and baked clay continued to play a crucial role and were only slowly superseded by metal (Moorey, 1982, 32).

(d) Ur III to the Neo-Assyrian Period: c.2100-1000 B.C.

The Ur III period marks the point from which textual information on copper and bronze-working in Mesopotamia steadily supersedes the evidence of archaeology. In his fundamental textual study of the Ur III metal industry Limet (1960, 198ff.), provided a useful list of the objects made of metal by the Sumerians. Many may not be identified with certainty, but the list demonstrates how restricted the range of surviving material evidence is. This point is again well made three centuries later within the context of an operating metal industry, by Rouault's (1977) study of the activities of Mukannišum, one of the high officials of the Mari palace administration at the time of Zimri-Lim. Fortunately, the surviving documents for his role as controller of metals and metalworkers are the most important and most detailed in his archive, (Rouault, 1977, 162ff.). He was the middleman, the senior official who received and stored consignments of metal, who executed royal orders for their manufacture or disbursement, and who ordered and stored the artefacts, depending upon expert advice on technical matters. This valuable documentation is complemented by more scattered information on metals and the repertory of objects in the Mari archives (Bottero, 1957, 293ff.; Birot, 1960, 310ff.; Durand, 1983). Against this background it is easy to appreciate why, in a region where the archaeological evidence for metalworking is predominantly mortuary, a sharp decline in the amount of metalwork placed in graves increasingly reduces the data-base for study. When, at the same time, the number and range of excavated graves is also more restricted than before the problem is still more acute. At present, evidence is virtually confined to Ur in the Ur III period; is sparse and scattered for the Isin-Larsa and Old Babylonian periods; and is negligible for the Kassite and Middle Assyrian periods in publications.

The nature of the contrast in mortuary evidence between the third and second millennia B.C. is best illustrated at Ur where the graves of the Isin-Larsa period may be compared to those of the earlier "Royal Cemetery". 168 graves are listed in detail by Woolley (1976, 195ff.). Apart from personal ornaments, the following copper or bronze items were recorded: 21 graves produced metal vessels, (24, 27, 33, 41, 44, 45, 59, 106, 131, 142, 145, 148, 152, 157, 164, 179, 182, 184, 186, 189, 196); three had pins or kohl sticks (5, 33, 61); 4 held razors, (27, 142, 175, 187); one had what may be an arrowhead (43); one contained a "chisel" blade, (double-ended, 45); three, significantly, were equipped with sets of miniature tools, (27, 38, 44), and 3 had sets of scale-pans, (23, 38, 170). This sample of the graves of an urban community, largely small merchants and craftsmen, without any conspicuous signs of wealth or status, (only 12 graves contained pieces of gold or silver jewellery) is, where it can be checked on other sites, typical of the period. The situation does not substantially change thereafter. The range of personal ornaments, weapons and vessels from graves of the Ur III period at Assur complements the Ur evidence, (Haller, 1954, 6ff.).

Jewellery is the only aspect of metalwork, for which a continuous, if at times thin, thread of material evidence extends through the second millennium B.C. (Maxwell-Hyslop, 1971, 64-93, 163-179); but, when personal ornaments are in base metal, they do little more than copy, as far as possible, the forms and techniques of gold and silver jewellery. The development of weapons at this time in Mesopotamia is ill-documented from controlled excavations. It has been largely reconstructed from scattered artefacts (spearheads, daggers and axeheads) assembled typologically. In general the shaft-hole axeheads show increasing variety of form (Maxwell-Hyslop, 1946, 1949; Deshayes, 1960; Curtis, 1983), though Mesopotamian craftsmen do not seem to have favoured relief or incised decoration as much as their colleagues to the east. Daggers, or short swords (the true sword is not evident), developed flanged hilts to replace the tangs of the third millennium B.C., perhaps under western influence, and by the end of the millennium had evolved into very efficient bronze weapons, notably some with royal inscriptions (Maxwell-Hyslop, 1946; Calmeyer, 1969, 59ff.; Moorey, 1971, 71ff.; Boehmer, 1983; Curtis, 1983). The evidence for spearheads (Maigret, 1976), now generally socketed, and arrowheads (Medvedskaya, 1982), flat and tanged, is even more inadequate in Mesopotamia; but what there is, does not demonstrate any marked divergence from the mainstream Mesopotamian tradition nor any significant modification across the millennium. Metal maceheads are known, but have not yet been systematically studied. Nor, for this long period, do pictorial illustrations, save very rarely, (cf. Maxwell-Hyslop, 1970a), offer any new information.

It is surprising that excavations on town sites in second millennium Mesopotamia have not yet revealed more information on the tools of the period. Scattered examples are published in excavation reports, at least showing that metal was now regularly used for agricultural tools; but it is only by association and context that identification of specialist craft tools is possible. It is unfortunate that neither of the two sets of miniature tools from Ur may be confidently attributed. The first (Woolley, 1976, pl.99: U.16699(1-8)) is of 5 miniature spearheads, 1 miniature axehead with bone handle, what might be a miniature ingot, and an enigmatic piece of folded copper. The second (Woolley, 1976, pl.99: U.16773(1-5)) appears to consist more certainly of model specialist tools, "spatulae", and "knives". Deshayes (1960) assembled and typologically classified the base-metal tools of Mesopotamia, but only in one case, that of the Old Babylonian hoard from Tell Sifr (Moorey, 1971a), is it possible to identify what has every appearance of being an assembly of agricultural tools, predominantly of copper, from which only plough-shares and hoes are absent; axeheads, mattock-heads, adze-blades, hammers, spade-blades, double-edged and single-edged knives, tanged sickle-blades, "bill-hook" blades, chains, various ring fittings, perhaps clamps for assembling a wooden plough, nose rings for animals, and vessels, are all there.

Copper or bronze statuary is only known from small-scale pieces, with the exception of some bulls' hooves of copper hammered over wood from the Shulgi "Mausoleum" at Ur (Woolley, 1974, 14). Metal peg figurines in foundation deposits are known until the time of Rim-Sin of Larsa (c.1822-1763 B.C.); thereafter only inscribed stone and metal tablets are reported from such contexts (Ellis, 1968; Rashid, 1983). Spycket (1981, 227ff.; Braun-Holzinger 1984, 43ff.) have assembled the votive statuary in metal for this period, none of which shows any marked technical improvement over earlier work, though some examples are rather more elaborate than the small statuary known from the third millennium. Sheet gold continued to be used to plate at least the faces of human figures and animals (see p.79). Neither from the Old Babylonian, nor from the Kassite, period in Mesopotamia is there any piece of cast copper or bronze to compare with the metal temple furnishings of the Middle Elamite period recovered from Susa (Amiet, 1966, figs.280, 291, 297, 305). Some elaborate cast objects of uncertain attribution, but probably Old Babylonian, are discussed here on p.43.

In some instances, new aspects of the base metal repertory may be identified. Although there are isolated examples of what may be sheet metal plaques for body armour before this period (cf. Boehmer, 1972, 102ff. for a review), it was not until the earlier second millennium B.C. in Mesopotamia, in texts and isolated finds, that it seems to have become a regular part of armoury for men and horses (C.A.D. s.v. <u>gurpisu</u>; Starr, 1939, 475ff. pl.126, A.L.; Mahmoud, 1970, pl.24). It is not until the second half of the millennium that there is material evidence for metal horse-bits with snaffle action; previously organic materials or nose-rings had been used (Littauer and Crouwel, 1979). Examples of these early bronze bits are reported from Assur (Potratz, 1966, 105, pl.105-6); but there is, as yet, no archaeological evidence from Mesopotamia for the metal chariot fittings that would also have been a major innovation in the second millennium (Kendall, 1975, 230ff.).

In major buildings of the second millennium B.C. metal fittings have rarely survived. The eighteenth century palace of Mari provided a set of metal fittings for a door (Parrot, 1959, 87, fig.66, pl. XXXIII). At Nuzi four centuries later there were metal door pivots (Starr, 1939, 471, pl.124B), sheet copper wall "knobs" (Ibid., pl.127E), and door studs of bronze, their heads covered with sheet silver (Ibid., 142, 473, pl.123I). Some dome-shaped bronze nails were given gilded heads (Ibid., 473). Fifteenth century Nuzi also yielded other uncommon base metal objects. One of the rooms adjacent to the palace chapel contained a cylindrical cast bronze stand with excised sides and with lions couchant on the upper edge (Starr, 1939, 439ff.). A cache of bronze objects concealed in a pot consisted of: "two sun disks (pl.127B), two crescents (pl.127C), nine sickles (pl.123J, 124C,D), six pins (pl.125R-T), a bell (pl.126Y), a concave plate pierced at the edge and two bracelets (pl.126N)." A bitumen mask of a bull was though to be the core for a statuette of precious metal (Starr, 1939, 136, pl.103G). Among a conventional repertory of tools and weapons, two spoons were unusual: "too small for food, too large for an ear spoon", (Ibid., 485, pl.126U).

Analyses of second millennium base-metal objects are still extremely rare. What information is available indicates the continuing use of unalloyed copper for a large proportion of the metal repertory. Thirty objects from excavations on Bahrein, said to date c.2300-1800 B.C., were reported to be of unalloyed copper without a trace of tin (McKerrell, 1977, 167). Surface analyses, again by McKerrell, of a selection of objects in the Tell Sifr Old Babylonian hoard are instructive (Moorey, 1971a; cf. Iraq, XXXIV (1972), 2). Axeheads, mattock-heads and adze-blades (Moorey, 1971a, nos. 1, 2, 5 and 8) alone proved to be good tin-bronzes. One mattock-head (no.4), one spade-blade (no.13), one knife-blade (no.28), one flat ring (no.81) had some tin in them; but all the other objects tested, notably all the sickle-blades, were made of unalloyed copper. Fourteen objects from excavations at Tell Rimah and four from Kar-Tukulti-Ninurta, all of the Middle Assyrian period were, with one exception, also of unalloyed copper (McKerrell, 1977, 167ff., table 13; 1978, 21, table 10). There is no evidence as yet, either in Mesopotamia or in western Iran (Moorey, 1971, 297ff.), for leaded bronzes, which are a significant feature of Late Bronze Age metallurgy in certain parts of Europe. As they are associated with industries where tin may, at the time, have been in short supply and a degree of "built-in obsolescence" was customary, this is not surprising.

(e) Neo-Assyrian and Neo-Babylonian Periods, c.1000-550 B.C.

It was not until the Neo-Babylonian period in the earlier sixth century B.C. that iron became cheaper than copper/bronze, though long before that it had been adopted as the primary metal, at least for the working parts of tools and weapons (see pp.102ff. here). The sources of archaeological evidence for metalwork at this time are surprisingly restricted. The range of evidence from graves is confined to a few sites, such as Assur, Babylon, Kish and Nippur, and even then only personal ornaments and vessels are commonly represented. The situation is somewhat improved by the vivid presentation of Neo-Assyrian metalwork on sculptured palace reliefs (cf. Hrouda, 1965; Madhlum, 1970), and by significant quantities of metalwork excavated from storerooms in the palaces themselves. But although the sequence of reliefs begins about the middle of the ninth century

B.C., the majority of surviving metalwork belongs to the last century or so of the Assyrian Empire. This means that it is not yet possible to write a history of Neo-Assyrian copper or bronze working in any meaningful way, since developments through time are obscured by the nature of the surviving information.

The paucity of surviving copper and bronzework of this period has led to a persistent underestimation of the Neo-Assyrian metal industries, particularly by contrast with those of Urartu for which evidence is more extensive and varied (cf. Van Loon, 1966; Azarpay, 1968; Vanden Berghe and de Meyer, 1982). On the evidence of the texts and sculptured reliefs, Assyrian metalworkers were no less active than their colleagues elsewhere at the time. The requirements of their army alone, not to mention the demands of the kings constantly building and rebuilding, decorating and redecorating, palaces and temples, were a recurrent stimulus to production. Much of the drive behind Assyrian imperialism came from the quest for guaranteed lines of supply for vital raw materials, metals conspicuous amongst them (Jankowska, 1969). But, with so much booty and tribute entering royal palaces, and the well-documented presence of foreign artisans in major Assyrian workshops, caution is needed in the identification of Assyrian, as distinct from foreign, metalwork, when so much of the known repertory is from palace storerooms. Unless compelling arguments to the contrary can be sustained, architectural fittings in metal, particularly when they follow long established Mesopotamian traditions, metal elements from furniture of types regularly shown on Assyrian reliefs, and standard items of military equipment for men and horses may safely be taken as local products. Where decoration is present such distinctions are clearer, as among the imported bowls from Nimrud, especially when re-inforced by inscriptions, (Barnett, 1967).

Curtis (1979) has surveyed the whole range of Neo-Assyrian metalworking, with particular reference to finds at Nimrud. To judge from developments elsewhere in the Near East a major iron industry in Assyro-Babylonia is unlikely to have been established before the first centuries of the first millennium B.C.; during that time a full repertory of iron tools and weapons emerged (see pp.102ff. here). Since iron could not be cast, though it might be decorated by swaging and repoussé work, copper and bronze long remained pre-eminent for all forms of decorative casting and for all objects or applications requiring sheet-metal. The American excavators at Khorsabad (Loud and Altman, 1938, 15-16) reported that bronze was found much more often than any other metal. It was extensively used for tools and implements; architecturally for door pivots and nails, some with heads covered in sheet silver. Wooden columns were encased in sheet bronze, scaled to resemble a palm tree (Place, 1867-70, I, 120-2; III, pl.73; Loud, 1936, 97ff.) and embossed plaques decorated doors. Sheet metal overlays for doors, more fully described here on pp.50ff., are one of the better known aspects of Neo-Assyrian metalworking, notably as exemplified by the sets of decorated doors from the Temple of Mamu at Balawat set up by Assur-nasir-pal II and Shalmaneser III (King, 1915; Barnett, 1973). Wall plaques (cf. BM 22488 from Nimrud), overlays for "Hands of Ishtar" (see p.180 here), and brick-stamps (cf. BM 124598; 135465-7) further illustrate the use of bronze in architectural contexts.

Elements of cast and hammered bronze formed an important part of furniture construction and decoration (cf. Salonen, 1963, 247-8). The evidence of palace sculptures is particularly valuable in this instance in offering the basis for the reconstruction of furniture from which only displaced metal fittings have survived (cf. Hrouda, 1965, pl.13-16; Dentzer, 1982, 48ff.; Born, 1984). Layard (1853, 198ff. with figures) described his discovery of a royal throne at Nimrud:

"With the exception of the legs, which appear to have been partly of ivory, it was of wood, cased or overlaid with bronze... The metal was most elaborately engraved and embossed with symbolical figures and ornaments... As the woodwork over which the bronze was fastened by means of small nails of the same material, had rotted away, the throne fell to pieces, but the metal casing was partly preserved. Numerous fragments of it are now in the British Museum, including the joints from the arms, and legs; the rams' or bulls' heads, which adorned the ends of the arms (some still retaining the clay and bitumen with the impression of the carving, showing the substance upon which the embossing had been hammered out), and the ornamental scroll-work of the cross bars, in the form of the Ionic volute. The legs were adorned with lion's paws resting on a pine-shaped ornament, like the thrones of the later Assyrian sculptures, and stood on a bronze base...

In front of the throne was the foot-stool, also of wood overlaid with embossed metal, and adorned with the heads of ram and bulls. The feet ended in lion's paws and pine cones, like those of the throne..."

Sheet bronze was also used for cauldrons, cauldron-stands and a wide range of vessels, some of which may also have been cast (cf. Hrouda, 1965, pl.18-19; Luschey, 1939; Layard, 1853, 176ff.); for shields, for helmets, and for body armour (cf. Hrouda, 1965, pl.23). Cast bronze was used for the decorative parts of weapons (cf. Hrouda, 1965, pl.21-2) and in the harness-trappings of horses (cf. Hrouda, 1965, pl.26-30). Horses on the reliefs also have elaborate sheet metal trappings but few such objects have been reported from Assyria itself, though they are well represented in some neighbouring countries (cf. Winter, 1980). Some fine cast bells found by Layard at Nimrud (1853, 177, figure) are probably from horse-harnessing. Some of the finest surviving representatives of the Neo-Assyrian cast metal tradition are lion-shaped weights (Layard, 1849, I, 128; BM 91220-35); but they do little more than bear witness to much now lost, judging from the evidence of texts (cf. Braun-Holzinger, 1984, 111ff.). Of large-scale statuary virtually none has survived (see p.45 here) and the surviving corpus of small-scale Neo-Assyrian statuary is meagre; heads and statuettes of the demon Pazuzu are the commonest (Spycket, 1981, 372ff.; Braun-Holzinger, 1984, 74ff.).

At present the published repertory of Neo-Babylonian bronzework is too meagre to offer any contrast to that recorded from Assyria. There is no reason, however, to suppose the two traditions would have diverged much, save perhaps in styles of decoration. One rare glimpse re-inforces this view. On a relief from Sennacherib's Palace at Nineveh, probably carved very late in Neo-Assyrian times (c.630-620 B.C.), scribes are shown recording booty from a Chaldaean settlement (BM 124955-6: Reade, 1983, pl.61 (colour)). In this scene metal bowls and cauldrons are evident, two on conical stands, as are weapons and furniture, which appears to have structural elements of metal as did Neo-Assyrian furniture. Most individual is a stand supporting a model fortress which, on Urartian analogies, would have been of metal. Substantial pieces of Neo-Babylonian bronze-work are almost unknown archaeologically, though constantly evident in texts. An isolated example is provided by a bronze "door-step" from Birs Nimrud, inscribed for Nebuchadnezzar. It is a massive casting (164x53x95cm.) decorated with moulded floral designs (British Museum Guide, 1922, pl.XXIX).

Copper and its alloys continued to be used in Mesopotamia through the Achaemenid Period with no evidence for significant innovations in the repertory of forms or techniques.

F. Workshops and Manufacturing Equipment:

The confident identification of metal workshops on excavated sites in Mesopotamia is by no means as easy as might be supposed and the literature already contains a number of debatable or demonstrably erroneous attributions. The best known is the case of the palace at Mari, where archaeological evidence of secondary metalworking, at least, is to be expected (Muhly, 1973, 212; Rouault, 1977, 162ff.). Here a confusion over the translation of <u>suripum</u> ("ice" not "copper-ore") led Dossin and Forbes to unwarranted conclusions about smelting, and other metalworking, in specific parts of the Mari palace, (cf. AHw.s.v. <u>suripum</u>; Muhly, 1973, 211; cf. Margueron, 1979, 14ff.). Nor may a case be sustained for a copper workshop in the AH site, of the earlier second millennium B.C., at Ur, cited by Leemans (1960, 55), and Muhly (1973, 211), on the basis of a preliminary report by Woolley. In his final excavation report of this part of the site Woolley (1976, 33) wrote: "In no.1 Baker's Square the entire house had been remodelled and turned into a workshop and one old room served as a stoke-hole and in another room and in the courtyard were furnaces (pl.50); judging from a miniature set of model tools in the owner's grave he was a working copper-smith". This is a reference to grave LG144, found in the "chapel" of 1B Baker's Square (Woolley, 1976, 158, 199), which contained this set of models (U.16773: Woolley, 1976, 244, pl.98). But as their appearance (and Woolley's own catalogue entry) indicates they look more like a leatherworker's tools than a coppersmith's, and there is no reported trace of metal-working debris from the building itself. Indeed the "furnaces" are more like ovens, perhaps for baking bread (cf. Barrelet, 1974, 270ff., fig.15: Mari). The same objections (the absence of diagnostic industrial debris, of working tools, and of characteristic installations) apply to the proposed identification of metal workshops in later prehistoric levels at Uruk (Lenzen, 1960, 10, pl.39, 3b, 4a-d; Nissen, 1970, 110ff., pl.VI) and in Early Dynastic Khafajah (Delougaz, 1940, 33; 1967, 176) and in Palace "A" at Kish (Mackay, 1929, 87).

Delougaz identified a "smithy" in the Akkadian to Ur III house levels at Tell Asmar. This conclusion was based on the fact that "identical fireplaces are used by metalworkers today in the bazaars of small Iraqi towns", though even here, the associated small finds do not make the comparison absolutely compelling (J.18.1; Delougaz,

1967, 228); no more convincing are workshop identifications in the "Northern Palace" at Tell Asmar (Delougaz, 1967, 198; cf. Margueron, 1979).

By far the best identification of a "workshop" is in a town of the Isin-Larsa period at Tell edh-Dhiba´i (al-Gailani, 1965), which yielded a whole series of small finds indisputably demonstrating its purpose. Sadly, these were recovered under weather conditions that prevented proper recording. Winton, on the basis of photographs only, (al-Gailani, 1965, 37-8, pl.6), presented a preliminary identification of the finds. More recently Davey (1983) has offered a valuable first-hand study of them with good illustrations. They include the following items:

1. Two handmade baked-clay pot-bellows (Davey, 1983, fig.2, pl. Ia), about 0.4m. in diameter and 0.16m. high; probably operated, by the feet, as a pair, on analogy with illustrations in the New Kingdom tomb of Rekhmire at Thebes in Egypt (cf. Davey, 1979).

2. Five crucibles and a fragment of another (Davey, 1983, figs. 2, 3, pl.Ib, c), made of a coarse fabric containing silica and traces of straw. X-ray fluorescence of debris in crucible 4 revealed copper, tin, arsenic, with traces of nickel and iron, indicating that bronze had been melted in it. These crucibles are the only examples of their type so far reported from the Near East; their capacity has not been reported. Davey (1983) reconstructed their use with the assistance of reliefs showing metalworkers in Egyptian tombs of the fifth Dynasty: "the reliefs reveal that melting is performed in two crucibles placed back to back and with six men using blow-pipes aimed at the charcoal placed in front of each crucible. After the metal was melted the crucible was taken to the mould, held with the assistance of two stones or lumps of clay and then a blockage in the mouth of the crucible was removed to allow the metal to discharge into the mould" (Davey, 1984, 174) (cf. also the Egyptian hieroglyphs for copper: Gardiner, 1957, 490, 529: this shape is not an "ingot" but a "crucible").

3. A number of fragmentary baked clay open moulds (Davey, 1983, figs.3 and 4, pl.Id) were recovered. Two, Davey nos.9 and 14, were probably for casting bar ingots, perhaps then bent round to form rings for ease of handling. Mould 10 is similar, but is larger and might be a blank to be hammered into a knife or dagger blade. Mould 11 produced rectangular plaques of metal, perhaps for hammering into sheet. Mould 18, with thin sides, for a rectangular bar, might be for a metal of lower melting point than copper and its alloys. The use of ingot-moulds is referred to in texts, both practical and literary, where it is a matter of recycling metal by melting it down for re-use: "like statues poured into an ingot-mould we will be killed" (Cooper and Heimpel, 1983, 81-2).

4. One baked clay mould cover (Davey, 1983, fig.4, pl.IIa) and the fragments of another; such covers inhibited oxidization of the surface of the casting and, by allowing the mould to be heated in the furnace, retarded the solidification of the metal, allowing it to flow more freely into the mould.

5. One mould for casting a pin by the lost-wax method (Davey,

1983, fig.5, pl.IIb); made of two different types of clay, the inner of fine yellow clay, the outer shell of coarser buff coloured clay.

6. A broken clay model axehead (Davey, 1983, fig.5, pl.IIc, d), probably used as a pattern in mould making, and the core for creating a shaft-hole in an axehead. Davey (1983 178-9) suggested that this was evidence for "sand casting". This is an open question, as the model's purpose is not self-evident.

7. A baked clay ladle (Winton in al-Gailani, 1965, 38), or possibly the clay nozzle of a blow pipe (Davey, 1983, no.13).

8. At least one fragment of a tuyère (cf. Tylecote, 1981).

9. A number of small flat round dishes, some with green and brown slag indicating they had been subjected to considerable heat.

It is important to notice the absence of stone moulds here (cf. Davey, 1977), the use of baked clay instead and possibly earth or sand, and the presence of closed crucibles as in Egypt, rather than the open ones of Anatolia, the Levant and Iran (cf. Davey, 1983, 183; Tylecote, 1976, 20).

There is no known illustration from Mesopotamia of copper-working. Limet's view (1960, 121) that a scene on an archaic sealing from Susa shows metalworking is mistaken. It is just one in a whole series of contemporary sealings which show pots and potting (Baudot, 1979).

Scattered evidence for metalworking is no more common in Mesopotamian excavations; nor have any sets of metal worker's tools been identified. Before bellows, smiths probably used blow pipes for annealing, brazing and other low temperature operations. Sufficiently high temperatures (1063°C. to melt gold) can be obtained locally with a brazier-type hearth and a blow pipe blown by human lungs. "A pair of human lungs can produce an intermittent flow of 401/min. but only 10-201/min. on a continuous basis. This is more than enough to superheat a hearth to 1000°C over a small area, already brought to a high temperature by a natural draught" (Tylecote, 1981, 108). Such pipes, perhaps reeds tipped with clay, are unlikely to leave recognizable archaeological traces. Pot bellows are not evident in Mesopotamia before the earlier second millennium B.C., when examples are known from Tell edh-Dhiba' and Tell Asmar (Davey, 1979, 106-7). The earliest stone moulds, significantly from northern Iraq at sites like Chagar Bazar (Mallowan, 1937, 160, pl.18B) and Tepe Gawra (Speiser, 1935, 104, pl.47), appear in the later third millennium; a Middle Assyrian group, some for shaft-hole axeheads, have recently been found at Tell Mohammed Arab, in the Eski Mosul region (M.Roaf, pers.comm.). In central and southern Iraq stone moulds are always likely to have been exceptional, for baked clay and possibly sand would have been the standard medium there, as in the Tell edh-Dhiba'i workshop. Forbes cites, without reference, tuyères "found near a furnace at Telloh (Ur III period, c.2300 B.C.)". I have not traced this in the published reports.

In Level IV, of the earlier second millennium B.C., at Tepe Gawra there was an object reported as "a handmade lamp of greenish grey ware with two legs on one side balancing with the base" (Speiser, 1935, 59, pl.LXXIV, 201). It was associated with a number of storage jars. In "ambiente 5" of period IVD at Arslantepe in eastern Anatolia exactly comparable objects used as crucibles were found in direct association with stone moulds for casting copper or bronze artefacts (Palmieri, 1973, 109-110, pl.45, 1-3, 46, 1-3,5) dated to Early Bronze IIIB, late in the third millennium B.C. (Palmieri, 1981, 102). This is one of the best published examples of a metal workshop in the Near East (Palmieri, 1981, figs.34, 37; cf. "Ozgüc, 1955). The room also contained pestles and mortars and had benches running round the sides. There may be no doubt of the function of these crucibles, for there was metal debris in at least one. A crucible from Amuq "G" in Syria had been used for bronze (Braidwood et al., 1951); another, of the later third millennium B.C., from Tell Sweyhat on the Euphrates, attested to local metalworking (Hedges, 1976a, 66-7). An isolated crucible was reported from the early second millennium B.C. levels at Nuzi (Starr, 1939, pl.57); but there is no clear indication that it was for metal.

The evidence both of texts, as with the gate and the quarter named after metalworkers at Assur (Andrae, 1913, 18ff.), and archaeological surface surveys (cf. Mariani, 1984), is that specific craft activities were concentrated within settlements, not scattered at random in them.

G. Techniques:

"The nobler metals provide a fruitful source of evidence for technical methods of manufacture. This, unfortunately, is often denied to us in the case of copper and its alloys, which are ravaged by corrosion to such an extent as to hide all traces of tool marks unless in exceptional cases" (Plenderleith in Woolley, 1934, 295).

(a) Annealing:

There is some uncertainty about the earliest use of annealing (see p.22). To those who argue that native copper is extremely difficult to cold hammer successfully (Coghlan, 1975, 26ff; 76ff.), it is believed to have been an early innovation in the appearance of native copper artefacts in the Near East. But Smith demonstrated (1967, 28) that "native copper can be worked almost infinitely without cracking" and that the earliest known copper artefacts from Cayönü in Anatolia, and from Ali Kosh and Sialk I in Iran, "are unmistakably hammered copper made without annealing" (cf. his more cautious statements in Smith, 1969, 427). By Sialk II annealing is evident (Smith, 1967, 28). This is the later Ubaid period in Mesopotamia; but no relevant metallographic information is available there to indicate the transition from unannealed to annealed native copper. Thereafter it is a crucial technique, for it alone allows for the continuous workability of copper. For planishing, a light hammering to toughen and polish the metal, polishing stones may have been used. Recent work by the "Sumerian Metals Project" in Philadelphia has shown that in both Early Dynastic III and the Akkadian Period at Ur some tools were left as cast, others were annealed, as were contemporary pins; but not to a consistent extent. By contrast, Gawra in the third millennium showed a much more standard, if less enterprising, tradition of manufacturing.

(b) Casting:

1. Tools and weapons:

Casting involves, at its simplest, pouring liquid metal into a suitably shaped mould of baked clay, stone, metal or sand. The earliest moulds to survive in archaeological contexts are one-piece, of clay or stone. They remained usual for the manufacture of simple tools, flat weapons such as tanged arrowheads, bar ingots (cf. Levey, 1959, 207, fig.46), and jewellery. Simple jewellery moulds of stone are more common in excavations than their more complex relatives used for tools and weapons. Wartke (1980) has catalogued and discussed the stone jewellery moulds from Assur, Sinjirli and Uruk, now in East Berlin; isolated examples have been published from other sites (Opitz, 1933, 208ff.; see p.123 here). Even a one-piece mould would have been covered with a flat stone to prevent oxidization.

Two-piece (bivalve) moulds were introduced sometime in the fourth millennium, if not before, with core pieces for sockets when required, as on axe, adze and hammer heads. A few tools and weapons, of the later third millennium B.C. from Tepes Gawra and Billa, have been the subject of metallurgical study (Levey, 1959, 201ff.). Three tin bronze axeheads (Sn:6%; 15.7%; 5%) were the only objects among the ten listed specimens that had apparently not been worked after casting. Of the specimens tested, the flat "axes" and chisels, all of unalloyed copper, "show microstructures indicative of cold working and annealing or of hot-working subsequent to casting" (Levey, 1959, 204). It was probably common practice to cast the simple tools in open moulds and subsequently hammer them to the desired shape. Two shaft-hole adze-heads, neither with appreciable amounts of tin, had been "worked to some extent subsequent to casting although the amount of working is slight". There appeared to be "no indication that specimens were heavily cold worked near the edge to harden them. While the smiths must have been aware of the hardening effect of working the metal while cold, they apparently made no attempt to utilize this, in giving a cutting edge that could last longer".

The only elaborate moulds made of metal so far reported from Mesopotamia are for casting socketed arrowheads of the so-called "Scythian type" in the seventh to sixth centuries B.C.; one, said to be from Mosul, is for producing three at a time (Coghlan, 1952); the other is from Woolley´s excavations at Carchemish (Woolley, 1921, pl.23) and casts singly. The triple mould consisting of six separate pieces, probably of the seventh century B.C., is a beautiful and masterly construction and an isolated indicator of the sophistication of casting technology by this time. Coghlan (1952) argued that it was for the production of actual arrowheads; Underwood (1958) was of the opinion that it was for producing wax arrowheads, then reproduced in metal by lost-wax casting. It has yet to be demonstrated that bronze moulds can be repeatedly used. British Prehistoric metal moulds have lead on their inside surface indicating use in a process of "lost-lead" casting, so Underwood´s suggestion may be more likely for the "Mosul" mould. By the Neo-Assyrian period parts of metal vessels, if not entire vessels, were being cast; but no metallographic work has yet been done on this problem. It is impossible to tell by eye which of the richly embossed phiale (Luschey, 1939) are cast, which hammered. Only cast handles may be so identified (cf.

Woolley, 1962, pl.24, top). Finger-rings from the Achaemenid period onwards were cast in moulds of baked clay, stone and metal (Wartke, 1980).

Davey (1983, 178-9) recently suggested that a fragmentary baked clay model axehead and a shaft-hole core, included among workshop debris from Tell edh-Dhiba´i (see p.38), might have been used for sand-casting. This technique is rarely mentioned in the current literature on ancient metalworking, since it seems generally to be regarded as an aspect of modern rather than of ancient metallurgy. This may be a matter of definition rather than a reflection of the actual situation. Contemporary practice among Bedouin, cited by Davey (1983, 80), indicates that the use of loose earth for casting in antiquity is probable. Although Coghlan (1975, 50-1; cf. Wulff, 1966, 18-19), in his comprehensive discussion of casting techniques, included a full description of sand-casting: "the most simple method by which non-ferrous castings are sometimes still made today...", he did not expand the discussion, as he believed the technique to have been largely unknown in antiquity. Textual references do not necessarily elucidate such matters, but a key phrase in the famous passage in I Kings 7:46, where the manufacture of bronze objects for Solomon´s Temple is described, is rendered in the Jerusalem Bible, following the Greek, as "sand-casting". Even the more conservative renderings speak of "in the compactness of the soil" or "in the clay ground". The archaeological problem is obvious. A moulding box filled with sand or loam, or just a negative made in the ground, is not going to survive for recognition in an excavation. It is only an accumulation of stray indicators that will reveal such a procedure. Patterns and cores are one aspect of the matter; objects not cleaned down may be another. Some of the distinctive bar ingots of the later third millennium B.C. in Palestine (cf. Tufnell, 1958, pl.21, 11-15), for example, look as if they might possibly be sand-castings.

2. Elaborate or decorative fittings:

Lost-wax casting:

Investment casting, the cire perdue or lost-wax method, was used for small scale work by the fourth millennium B.C. in Mesopotamia. First a model of the object to be cast was made in beeswax, then covered with clay or a clay mixture (the investment material to form a mould, see C.A.D. under "iškuru" (wax), usage "b"). This would then be supported in some way and heated. The wax was thus melted and either burnt away or ran out through holes provided in the mould. When the mould was sufficiently baked, metal was poured into the holes whence the wax had escaped. It was allowed to cool, the mould broken away and the casting cleaned and surface treated as required (cf. Coghlan, 1975, 143ff., pl.XI ff. for illustration of the method). Smith (1981, 134) has commented generally on this remarkable technique: "There is, of course, no evidence whatever of what "wax" was used. Any easily shapable material that was fusible or volatile, or that left a light removable ash on combustion, would serve - for example, a hard animal fat, beeswax, resin, some of the harder forms of bitumen, and, on occasion, the bodies of insects or a light low-ash wood. With this technique the artist is free from any restraint arising in the negative nature of the mold, for the pattern does not have to be withdrawn and the mold surfaces, however contorted, reflect all the details of the wax and do not have to be accessible to further tooling. The sculptor could, indeed, model things as freely as he had done in clay for millennia. At first these castings were solid, but later the art of molding the wax over a sandy clay core was introduced".

The earliest castings of this type known from Mesopotamia are tiny, solid-cast animal amulets, parts of composite statuettes of animals, and zoomorphic handles for seals, from the "Sammelfund" found in the latest part of Eanna, level 3, at Uruk (Heinrich, 1936, pl.13a, 14a, 17a-b). Although buried at the very end of the prehistoric period, this hoard is believed to contain objects manufactured earlier in the Uruk IV period, (Goff, 1963, 265ff.). In the case of one amulet, a copper lion, analysis revealed 9% lead, indicating that the maker already realised the value of this additive in casting. By at least Early Dynastic I-II more elaborate, larger anthropomorphic and zoomorphic copper artefacts were cast, notably a series of stands for vessels found in temples in the Diyala Valley (Frankfort, 1939, 39ff.; 1943, 11ff.) and in graves of the "Y" sounding at Kish (Watelin, 1934, pl.XXI.1-2; for dating Moorey, 1978, pp.112-11), where they also decorated rein-rings (Watelin, 1934, pl.XXV). This work already shows remarkable mastery of the technique, particularly when it seems the medium was copper or arsenical copper rather than leaded tin bronze.

The use of divisible piece-moulds for such work is best shown by lost-wax castings in precious metals (p.85). The most remarkable single work in this category, in base metal, remains the copper quadriga from the Shara Temple at Tell Agrab (Frankfort, 1943, 12-13, pl.58ff.). The only object of this type to receive a detailed examination is an unprovenanced, wild-goat stand in the Metropolitan Museum, New York (Moorey, 1982, pl.1b), probably dating to Early Dynastic III. Its structure elucidates many points which may only be surmised about the excavated pieces yet to be studied. This stand is made of three separate units: a foot comparable to those on the Kish and Diyala stands, a central standing wild goat, and an upper bracket set on its back. The animal is arsenical copper (Cu: 94.0%; As: 2.6%; Zn: 0.9%; Sn: 0.2%; Pb: 0.3%) cast round a clay core, held in place by chaplets through the shoulders and haunches. The hollow central, circular strut passes through the animal's body and must have been cast around the core. The solid animal head was cast separately and fitted by a "tongue in groove" device to the neck, which is tapered and reached into the base of the head, where the junction is secured by a pin. Radiographs indicate that the base was a single casting. Tangs below the animal's feet fit into cavities in the base and are secured by "collars" of extra metal, perhaps cast on by puddling, i.e. by pouring molten metal into a fired clay dam at the junctions, probably with the stand inverted. The four-ring upper bracket is a single casting, possibly made in an open mould. The three vertical struts were cast on separately and the whole joined to the animal (Lynda Aussenberg and Pieters Meyers report; cited with permission).

That lost-wax casting was early exploited for relatively routine objects is evident from a cast open-work handle for a copper dagger from a Kish grave of Early Dynastic I (Watelin, 1934, pl.XVIII.1). It was to be used throughout for small-scale statuary, predominantly solid-cast, examples of which survive only sporadically (Spycket, 1981, passim, pl.23, 40, 42, 52, 62, 152, 153, 156-9, 171, 191, 242-5, 252; Braun-Holzinger, 1984), and foundation figurines (Ellis, 1968; Rashid, 1983). Decorative cast fittings of one sort or another are equally elusive in the archaeological record. To the earlier second millennium B.C. belong an enigmatic series of rollers and frames (Moorey, 1977; Calmeyer, 1969, 177ff.) and some standards (Calmeyer, 1969, 50ff.) that may be Babylonian rather than "Elamite". In contrast to Iran and Anatolia, Mesopotamia seems never to have developed a taste for weapons with cast decoration.

Whether or not cast metal furniture fittings had been used for furniture before the Neo-Assyrian period is not clear, but by then both reliefs and isolated finds bear ample witness to their currency (Kyrieleis, 1969, 6ff; Dentzer, 1982). At Nimrud Layard (1853, 198ff.) found the bronze metal fittings from an Assyrian royal throne and foot stool, some of which are cast (see p.35). It may be compared with the even more elaborate and various cast fittings to have survived from the throne and footstool of the god in the Urartian temple at Toprak Kale, near Van (Barnett, 1950). A fine hollow-cast bronze "dragon's" head in the Louvre was, perhaps, once the head of a sceptre sacred to Marduk (Encyl. Photo, II, 1936, pl.39). Fine zoomorphic castings on a small scale are illustrated in the same period by the bronze lion weights found in Assyrian palaces at Nimrud and Khorsabad (Spycket, 1981, 433-4, pl.284).

For substantially larger castings, and even for some smaller ones, it was important to save metal and for this cored lost-wax castings were devised. For this a core of clay or clay mixture suitable for rough moulding was shaped to the desired form and covered with wax to an appropriate thickness and consistency, so that all ornament and detail might be cut into it. An outer coating of clay was then applied and the same process of baking and metal pouring used as with smaller work. Most such castings would have needed projecting spikes of copper, bronze or iron ("chaplets") to secure the core in position. By this method, particularly when skilfully executed, there can be great economy in the use of metal. It is also extremely versatile. Fine detail can be worked on the wax model and exactly reproduced in the casting. Any shape may be produced since, unlike moulds (even piece moulds) the pattern does not have to be extracted from the mould before casting.

There is already some evidence in the later Early Dynastic period for cast metal statuary of considerable size (Frankfort, 1943, 11, pl.61) in which the surviving limbs and feet are hollow cast: "It may be, of course, that the statue was of some other material with only the face, hands, and the feet cast of copper" (Frankfort, 1954, 26). Both the Akkadian copper head of a king from Nineveh (Nylander, 1980) and the fragmentary copper statue inscribed for Naram-Sin from near Dohak in northern Mesopotamia (Al-Fouadi, 1976) are hollow cast. Chaplets and much of the core survive in the Dohak statue. The latter is interesting as the earliest surviving statue of this size certainly made completely of metal, as was its base. The Nineveh head well illustrates the extent to which craftsmen had appreciated the potential of cutting the fine details of hair and beard in the wax for reproduction in casting, though the head is heavier than it need be, suggesting rather less control. Yet both sculptures would be regarded as masterly achievements in any time or place, so far unmatched by anything else to have survived from Mesopotamia.

For the second millennium B.C. it is only at Susa that excavators have found large-scale castings of the type that texts indicate would also have been current in other major Mesopotamian cities. The famous bronze barrières inscribed for King Shilhak-in-shushinak (c.1130-1100 B.C.) are 20′ long, 8" in diameter and about ½" thick; perhaps cast by pouring simultaneously from numerous crucibles through many funnels (Maryon and Plenderleith, 1954, 632). The now headless bronze statue of Napir-aṣu, wife of Untash-Gal (c.1250 B.C.) is almost bell-shaped; it consists of an outer shell of metal usually about an inch thick, though more at the hands. The core is a solid mass of less compact metal, probably from successive pourings from small crucibles (Amiet, 1966, no.280, pl.372; 1.29m. high; 1750 kg. weight). Maryon and Plenderleith (1954, 632) commented that: "perhaps the additional metal was added to strengthen a crack that formed in the casting when it cooled. It is to avoid this mischance that modern bell founders rake out the core of a large bell as soon as the metal has set, and before it has time to contract strongly."

The most famous text from Mesopotamia relating to large-scale casting forms part of Sennacherib′s description of a "Palace without Rival" and has often been commented on. The key portions, as rendered by Luckenbill (1927, 162ff.), are as follows:

> "Eight lions, open at the knee (i.e. free-standing), in the posture of advance, which were cast of 11,400 talents of bright copper... With two colossal pillars whose copperwork came to 6,000 talents... Four mountain sheep, as protecting deities, of silver and copper... Following the advice of my head (will) and the prompting of my heart, I fashioned a work of bronze and cunningly wrought it. Over great posts and cross bars of wood, 12 fierce lion-colossi together with 12 mightly bull-colossi, complete in form, 22 cow-colossi clothed with exuberant strength and with abundance and splendour heaped upon them, at the command of the god, I built a form of clay and poured bronze into it, as in making half-shekel pieces, and finished their construction".

The key final sentence might be rendered: "I made their figures perfectly, as (easily as if they had been) objects weighing half a shekel each" (S. Dalley: pers.comm.).

Some commentators (e.g. Underwood, 1958) have been misled by the reference to shekels into proposing methods based on dies and on coin striking, when the reference is to three-dimensional solid bronze weights on the shekel standard made by long-established, if small-scale, techniques of casting in moulds. Maryon and Plenderleith (1954, 632-3) offered a plausible reconstruction of Sennacherib′s procedures:

> "From this description it is clear that the legs

> of these figures were in the round, and sufficiently strong in themselves to support the body. Sennacherib's moulds were of clay; they would be piece-moulds, built up individually against the original models. When removed from the model they would be fitted with a core, and the necessary jets and vents for casting would be provided. Next, the core and mould would be set up near the furnace and fired to enable them to withstand the pressure of the molten metal. Then brick walls would be built round them, strengthened by the 'great posts and crossbars of wood'. These could not have been placed within the moulds themselves, for the heat of the molten metal might have set them alight, with disastrous results; nor could the mould have been baked if they were within."

It may be that these were not hollow lost-wax castings, as had been customary before Sennarcherib's time, but enormous free-standing solid metal figures. Most surviving Assyrian weights whether zoomorphic, notably a fine series in bronze (see p.35) or small, plain ones in geometric shapes, were solid cast in multiple clay moulds. Perhaps it was this type of casting that Sennacherib, at least for the first time in living memory, exploited on a monumental scale. It is possible that a relief, from Ashurbanipal's palace at Nineveh, carved about 645 B.C., shows the facade of Sennacherib's palace, with its columns standing on cast bronze bases set on the backs of lions passant, just as Sennacherib's inscription describes them (Reade, 1983, pl.56).

Very little Neo-Assyrian metal sculpture has survived, although the archive of photographs left by Place includes one captioned "debris d'une statue de bronze de Kalah-Chergat" (Börker-Klahn, 1973, fig.1). This appears to be a metal version of a type of male caryatid stone statue, of a deity carrying a vase with flowing streams, known from Khorsabad (Strommenger, 1970, 23ff.). Such a column-shaped casting would not have presented the same problems as Sennacherib's free-standing colossi. It is in the tradition of the Queen Napir-aṣu statue. A striding cast bronze bull-man, in an "atlas" pose, was found by chance near Kirkuk some years ago. It seems to be a solid casting made in two-piece or multiple moulds. It stands about 35cm. high and may have been an element in a substantial piece of Assyrian furniture upon which similar "supporters" are shown when represented on palace reliefs (J. Curtis; pers.comm.; Hrouda, 1965, pl.15:1). Assyrian soldiers are shown breaking up a statue, for its metal, on Sargon's depiction of his sack of Muṣaṣir, in the later eighth century B.C., on the reliefs at Khorsabad (Botta and Flandin, 1849-50, pls.140-2; Mayer, 1979).

The Louvre in Paris has a damaged, headless inscribed bronze statue of King Assur-dan II (c.934-12 B.C.), now 30cm. high, weighing 2kg.; it is of simple, cylindrical form (Spycket, 1981, 372).

Equally rare are surviving fragments of cast-bronze relief sculpture comparable to the familiar stone palace reliefs of the Neo-Assyrian period (Braun-Holzinger, 1984, 105ff.). The Louvre possesses a very small piece of inscribed bronze relief (AO 20.185; Parrot, 1956a, 0.33m. high; 0.31m. wide; 13kg. 900 in weight) as does Baghdad (Basmachi, 1962). They show royal figures in low relief, perhaps as a part of a seventh century B.C. altar or comparable furnishing. Such work is known earlier at Susa (Amiet, 1966, no.305, p.404), but not in Mesopotamia, save in one highly dubious instance. Without a thorough metallurgical examination it is hard to assess a large copper or bronze casting, perhaps a piece of temple furniture, with a frieze in relief, now in the Museum of Canakkale/Dardanelles in Turkey (Börker-Klahn, 1974). The relief is very damaged, but Börker-Klahn has identified it as Sumerian, stylistically of Early Dynastic III. If this is correct, and that remains open to question, it would be a remarkable addition to the cast metal repertory for the middle of the third millennium B.C.

(c) Sheetmetal working:

1. Vessels and miscellaneous fittings:

Maryon and Plenderleith (1954, 635) resolved the question of whether hot or cold forging was used at Ur in antiquity on the basis of the existence, or absence, of efficient tools for holding the work for hammering. For hot-working the hands must be about 8" or more from the metal. No hinged tongs are known from the ancient Near East before the mature Iron Age and it is assumed, from Egyptian practices shown on reliefs (Newberry, 1900, pl.XVIII), that crossed strips of young or damp wood were used, insulated as necessary with clay. The grip would not have been very secure. "This beaten work, such as that from Ur, must have been done mostly by cold-forging, a process still employed wherever hand-working is practised. To the craftsman, it has the immense advantage that by holding the work in his hand he had complete control over it" (Maryon and Plenderleith, 1954, 636). It is likely that the metal vessels were raised, by means of blows on the outside, rather than by sinking into a suitably cut hollow in a wood or stone block. Maryon (1949, 95ff. fig.10) describes the procedure in detail, with particular attention to a range of standard vessel shapes from the "Royal Cemetery" in which he isolated "the shapes which the vase would be likely to have taken while passing, under the hammer of the craftsman, from the original flat disc of metal to its completed form". These techniques had been developed by the fourth millennium and were to undergo no appreciable change (cf. Woolley, 1976, 185, pl.100; 1962, 104, pl.24, 32). They included tricks for including handles and spouts: "If desired, the material to produce a lip or spout, or a handle, may be left on the disc from which a jug or saucepan is to be raised. Such material does not get in the way very much during the raising of the body of the work. When that is completed the shaping of the handle or spout may be taken in hand" (Maryon, 1949, 99).

Maryon believed, great though the competence of Sumerian sheet-metal workers was, that they lacked the "snarling iron" or its bronze equivalent. Hodges (1964, 78) describes this tool as follows: "a Z-shaped rod, one end of which is held in a clamp and the other, free end of which is rounded. This end is passed through the mouth of the vessel and held against the area to be raised. The middle section of the rod is then struck with a hammer, and as the whole free end of the rod rebounds from the blow so the rounded end in contact with the metal forces it out". In the British Museum collection from Ur, said to be from the suburb of Diqdiqqeh is a broken bronze tool, originally identified as a "steelyard" (Woolley, 1976, 86, 233), that might have served a comparable purpose (cf. B.M.: 122186 = U. 12683; Maryon, 1949, fig.12, for illustration of use). Its date is unknown, but the majority of the objects from this area are dated to the first half of the second millennium B.C. There is no evidence before the Achaemenid period, and then it is equivocal, for the use of a lathe to turn or spin metal vessels (Cooney, 1965).

Most of the sheetmetal copper or bronze vessels at Ur were beaten up from one piece of sheet metal, though rivetting was used for larger pieces, notably type 77 (Woolley, 1934, 302, pl.237), "which was made in three pieces, the edge of the base being bent so as to fit round the bottom of the sides". In making raised bases "the general method has been to bend the bottom edge of the walls inwards and upwards and then to splay it out so as to make a ring-ledge to which the base, a disk cut separately, is fixed by soldering: in some cases, e.g. Type 9 a disk has been soldered or sweated directly on to the hammered base so as to strengthen it where the metal was thinnest." With one exception (Type 17; Woolley, 1934, pl.233), handles were separately fixed on, as occasionally were legs (Type 24; Woolley, 1934, pl.233). Collon (1982) has examined the distinctive and unusual double-tubes of metal, fixed to the vessels by rivets, through which wire loop-handles were passed.

More sophisticated joining methods had already been developed by the middle of the third millennium B.C. If copper or any of its alloys is heated a film of copper oxides forms on the surface so that neither a flux nor hammering will unite two surfaces. Unlike iron, copper cannot be welded together by heating or hammering; a solder is required (see pp.87ff. for distinction of "soft" and "hard" solder). The stag's antlers on the Ubaid copper relief are square copper bars and the joints between the main stem and the tines were made with soft-solder, though of what type is not specified (Maryon, 1949, 114). Tin was used as a solder at Ur on a copper bowl with silver lugs from the "Royal Cemetery" (Craddock in Collon, 1982, 101; Craddock, 1984). Hard soldering (brazing) with a low melting-point silver-copper-tin-alloy has been recognized on an Early Dynastic III "bronze" axehead from Ur (BM 121574: Roberts, 1974, figs.12-13).

None of the base metal vessels from the "Royal Cemetery" at Ur was, apparently, decorated though some were inscribed. The main copper object to exhibit a developed mastery of _repoussé_ decoration is a fragmentary plaque from RT 789, which Woolley believed might have decorated a shield (Woolley, 1934, 69, pl.169b.: U. 10475). This is a rectangular strip, decorated with an addorsed pair of lions passant above recumbent men, probably corpses, in some way associated with a rosette. It is generally argued that a true cutting tool (a graver or scorper) of bronze could not be used on bronze (Maryon, 1949, 117ff; Lowery _et al._, 1971) so all linear work before an advanced date is likely to have been chased rather than engraved. Caution, however, remains in order. "We have found a bronze tracer very much less convenient to use on bronze than a steel one, frequently needing re-shaping of the edge, though hammers and styles of use different from those customary with steel tracers reduce the differences...we are not yet certain how far" (Lowery _et al._, 1971,

170).

From subsequent periods embossed work in copper or bronze has very rarely survived. In the Neo-Assyrian period, apart from the work on a larger scale considered below, high repoussé is known from an isolated piece, in the Boston Museum of Fine Arts, which may again have been from a shield (Porada, 1950), and on fragments from what may be overlays for boxes, some from Nimrud (Organ, 1963, 129, fig.2). Some fragments without known source, now in the Louvre (Parrot, 1958) were published as Neo-Assyrian, but may be from further afield; all are decorated with animals or with scenes of animal combat and hunting. The most remarkable examples of the technique at this time, sheet-metal vessels in the shape of animal heads, were made in Mesopotamia and in neighbouring countries (Calmeyer, 1979), following a tradition that went back at least a millennium. It is possible that a series of drinking goblets, decorated with repoussé and chased scenes, dated mainly to the tenth century B.C., illustrate an aspect of "Middle Babylonian" metalwork. They are reported predominantly from clandestine excavations in Western Iran; but a few inscribed examples bear witness to a strong Babylonian connection (Calmeyer, 1973).

2. Overlays:

(i) Statuary:

As elaborate programmes of temple decoration in cast statuary of any size would have placed a heavy burden on Mesopotamian metal resources, it is hardly surprising that the Sumerians developed methods of producing substitutes. It may not have been until the Neo-Assyrian period, when copper came so readily as tribute or booty, that rulers were able to contemplate and execute schemes like that undertaken in cast metal by Sennacherib (cf. pp.44). Statues of hammered sheet copper or bronze fixed to roughly shaped bitumen or bitumen-coated wooden cores of the shapes desired were current by the earlier third millennium B.C. Composite statuary, combining parts of stone, wood, and metal, though evident from textual sources, (Spycket, 1968, passim), is not so readily illustrated from the archaeological record, where the original destination of fragments is not certain. That the Sumerians had a taste for such creations is clear from the outset of the sculptural record, notably where animals are concerned (Spycket, 1981, pl.31). The two rampant he-goat supports from RT 1237 at Ur remain the best examples of the genre (Woolley, 1934, pl.87), though in them precious materials are the visible components, and the scale is small.

The outstanding archaeological instance of wrought statuary from Mesopotamia is that from the Early Dynastic III temple at Tell al-ʿUbaid, excavated in turn by Hall and Woolley (1927). This temple had been dismantled, possibly in the Ur III period, for extensive restoration. The fittings had been removed, and assembled at the base of one of the access ramps, for burial in the new terrace or platform. This had preserved them remarkably well, though with little indication of their original disposition, presumably within the temple building. This metal-work has never received the detailed technical study it deserves, as year by year, the unique character of the find is more clearly evident.* Various early analyses (Hall and Woolley, 1927, 36ff.) showed that the sheetmetal of the bodies was copper, with no clear evidence of alloying: "all being wonderfully pure copper", as was said at the time (Ibid., 38). Maryon (Maryon and Plenderleith, 1954, 639-40) has given the most concise description of the technique used to make one of the standing bulls at Ubaid:

> "Their construction is worthy of consideration. A piece of wood, measuring about 18x6x3", was roughly shaped with an axe to the general form of the bull's body. To it were fitted, by mortise and tenon joints, further pieces for the head and legs. The whole wooden core was then covered with sheets of thin copper: head and legs first, then body, horns etc. each piece being fixed in place by large-headed nails. Each leg was constructed from one piece of metal. Now, though a piece of copper could be bent round the core to form a kind of leg, much work would be needed before it assumed its final shape. Thus the formation of the concave anterior border of a hind leg would entail much hammering near the back edges of the folded sheet to stretch the metal over the hock, and this demanded frequent annealings. Then the smith, after a final annealing, tacked the long edge onto a bitumen-covered core. He would warm the metal so that the bitumen would flow closely against it to form a support during the chasing of the surface modelling. The last task would be to fold over the free edge so that it overlapped that already nailed to the core and to drive in a row of large-headed nails, which held all together."

Detail was worked on the surface of the metal with tracers and some of the design is in <u>repoussé</u>. Maryon may not be correct in all his deductions. The heads of the lions and leopards appear to be of cast tin bronze, not of hammered sheet copper, with bitumen poured in after casting to re-inforce them and support the eye inlays and the stone tongue-tips, which project from the open mouths. The few cast heads that have been properly examined are said to show traces of the multiple piece-moulds in which they were cast after an original wax model; "On some of them, between the horns, down the front of the head, and under the lower jaw there remain traces of casting webs." (Maryon, 1949, 99). In some cases the horns were made separately and fitted on; the joints between the main antlers and the tines on the stag's head were soft soldered (Maryon, 1949, 114).

Later examples of the same technique were found in clearing the temple of Dagan at Mari, where the terrace had been guarded by lions, or the foreparts of lions, made of copper or bronze plaques hammered

---

* A detailed investigation is now in progress initiated by Dr. John Curtis.

over wooden cores, probably in the reign of Zimri-Lim (c.1800 B.C.). Little save inlaid eyes remained of the external ones, but two from inside had survived intact (Spycket, 1981, 290-1).

(ii) Furniture:

Some types of religious furniture were made of sheet metal over wood. A "ringed gate-post", 3.27m. high, from one of the Early Dynastic III temples at Telloh (Girsu) was of hammered copper over a wooden core, with a bitumen terminal on the top, perhaps also originally plated (de Sarzec and Heuzey, 1884-1912, pl.57:1; Parrot, 1948, 106 fig.26c). That this tradition endured is evident at Assur, where in one of the rooms of a substantial Neo-Assyrian house was found the copper or bronze cladding for a standard of some kind, in this case a radiate rosette set on a shaft (Preusser, 1954, 41, 44ff., pl.19). Hammered metal fittings, and silhouettes cut out of bronze sheet were important components of elaborate pieces of ceremonial Neo-Assyrian furniture, as on the throne found at Nimrud by Layard (1853, 198ff.; see p.35 here) and earlier evident in texts (Salonen, 1963).

At Ur a pair of bronze coffins, made in the late Neo-Assyrian period, are equipped with panels on the outside with a design of wild goats standing on rosettes (Woolley, 1962, pl.18; Curtis, 1983a).

(iii) Architectural:

By far the best known instance of sheetmetal decoration on monumental scale is provided by the Assyrian examples of decorated bronze overlays for palace and temple doors, first referred to under Adad-Nirari I (1305-1274 B.C.), sporadically mentioned down to Ashurnasirpal II, and thereafter more regularly. Layard found fragments in chamber I of the North-West palace at Nimrud; Place found pieces in the Adad Temple at Khorsabad; Loud recovered some from the Nabu Temple there, and further pieces were found in the throne room. At Assur the door of the Anu-Adad Temple and the Tabira Gate offered some examples, as did the temple of Nergal at Tell Hadad (Sirara), in the Hamrin, built or restored by Ashurbanipal (see Curtis, 1982, 118). But the best known and most elaborate are the gate coverings set up by Shalmaneser III, discovered by Rassam and Mallowan at Balawat. They are of sheet-bronze, with about 10% tin. Maryon and Plenderleith (1954, 646) report that "the bands (i.e. of decoration) were in repoussé, worked mainly from the front, where the tilting downwards of the background toward the outlines of the figures, and the marks of chasing tools and tracers, are clearly visible in the original". The elaborate scenes, miniature renderings of the activities more familiar from monumental Neo-Assyrian palace stone sculptures, were first traced in outline on the surface with a graver. Not sufficient work has been done to define the earliest appearance of engraving in this period, but Maryon and Plenderleith (1949, 649), believed that the inscriptions on the Balawat Gates "are probably the earliest examples of true engraving on metal, the technique of cutting inscriptions on stone with steel tools having been transplanted to a new material - bronze".

What appear to be bronze facings for Neo-Assyrian terracotta wall plaques were found at Tell Deim in northern Iraq (al-Tikrit, 1960, 107, fig.9 (in Arabic section).

H: The use of metal tools:

In recent years increasing attention has been paid by archaeologists to the manner in which ancient tools were used. Two distinct approaches have emerged, the one based on the study of wear on stone tools (Semenov, 1964), the other on the idea that it is possible to tell the tool from surviving tool marks on stone artefacts. Since corrosion destroys the surface of metal work, the second approach is the one most likely to yield information on how and when tools of copper and its alloys superseded those of stone, and how and when in turn iron tools (cf. Nylander, 1970, 50ff.) were widely preferred. Such research on Mesopotamian artefacts is in its infancy. Gwinnett and Gorelick (1979) have made a start with cylinder seals, seeking to establish the relative use of wood, stone and metal tools in their manufacture, on the assumptions that "standards" for ancient tool marks may be developed and the marks themselves faithfully duplicated by experiment. Their work has done much to confirm the existence of rotating cutting-disks by 3000 B.C. They are also able to recognize the emergence of metal tool marks, as distinct from those of wood or stone, in the seal cutter´s repertory in the third millennium B.C.; indeed for hard stones drills of copper may already have been used in the previous millennium B.C.

I. Copper and its Alloys: Notes on Published Analyses

A widely scattered and heterogeneous series of analyses of copper and alloyed copper objects is now available in the literature of Mesopotamian archaeology over the past century and a half. Disparities of analytical methods, often inadequate or unreliable documentation, and the absence of any general system for the presentation of results, renders tabulation impossible. What information may be drawn directly from these analyses has been absorbed into the main text here. The following bibliographical survey, by site, does not attempt a critical review of the kind so successfully undertaken by Caley (1962) for pioneer analyses of glass. It seeks merely to indicate the range and nature of the data presently available, and to facilitate rapid reference to it. Here, as in the main text, Mesopotamia is broadly defined. Susa is not given a specific reference, since it is the only archaeological site for which a comprehensive modern analytical programme has been undertaken (Berthoud, 1979; see also pioneer analyses in Berthelot, 1906, 83ff.).

1. ARPACHIYAH:
   Late ´Ubaid "chisel"
   Mallowan, 1935, 194, pl.X; analysis by Craddock in de Jesus, 1980, 207
   Cu: 90%; Sn: undetected (down to 0.002%); As: 0.1%.

2. ASHERAH:
   Uncontexted flat axe or adze blade.
   Th.Dangin et al., 1924, 292
   Cu: 91.159%; Sn: 6.941%; Pb: 0.132%; Fe: 0.159%;
   Ni: 0.532%; As: 0.071%; Sb: 0.020% (sulphur: 0.019%;
   oxygen: 0.967%).

3. TELL ASMAR:
   Openwork hilt of a dagger (As 32:1080), late Early Dynastic

III/Akkadian, with traces of an iron blade. Frankfort, 1934, 58-9; Delougaz et al., 1967, pl.74, 184-5; Moorey and Schweizer, 1972, 185
Sn: 7.6%
Frankfort, 1934, also refers to a "knife blade from one of the private houses, however, contained 2.8% tin".

Hemispherical bowl (A.11294; Oriental Institute Chicago, 32.1082) Dayton, 1978, 159
Sn: 10%; Ag: 1%; As: 1%; Pb: 0.5%; Ni: 0.4%; trace: cobalt.

"Rim-Sin" type axehead (A.9446) Dayton, 1978, 159
Sn: 10%, Ag: 1%; Pb: 0.5%; As: 0.3%; trace: cobalt.

"Miniature axehead of the Ur type" Dayton, 1978, 159
Sn: 10%; Pb: 0.3%; As: 0.5%; traces: silver, cobalt.

4. BALAWAT:
Analysis of the metal overlays in the Neo-Assyrian gates (King, 1915). Reported by Sebelien (1924, 13-14) as tin-bronze.

5. BASSETKI:
Copper statue fragment with inscription of Naram-Sin
Al-Fouadi, 1976, 66
Cu: 98.2%; Ni: 0.32%; As: 0.80%; Fe: 0.41%; Co: 0.04%
(Analysis by Laboratory for Archaeology and the History of Art, Oxford University).

6. TELL BILLAH:
1. Shaft-hole adze-head: Philadelphia: 31-51-468: level 5; late 3rd millennium. Levey, 1959, 209, fig.39
Sn: 0.3%; Pb: 0.1%; As: 0.01-0.1%; Ni: 0.01-0.1%; Fe: 0.1%.
2. Analyses of objects from Billah is proceeding in Philadelphia ("Sumerian Metals `roject" 1983-4; Muhly, pers.comm.).

7. BRAK:
A: Mallowan, 1947, 258 (analyses by Desch and Barr; not fully quantified; all objects Late Prehistoric):
1a. Fragment of copper panel from "Eye-Temple" Sanctuary.
1b. Pin from "Eye-Temple".
2. Fragment, as 1.
3. Nail from "Eye-Temple" Sanctuary.
(4. Silver object)
5. Copper pin from mudbrick wall behind "Eye-Temple" podium.

On these Desch commented "There are no bronzes among the specimens, tin being completely absent...Of the coppers, nos.1a and 5 are of high purity, whilst nos.1b, 2 and 3 contain both arsenic (below 0.5%) and nickel...Silver is usually present in small quantities".

B: Moorey and Schweizer, 1972, 186: Akkad to Ur III

Ashmolean Museum 1939.295: sickle-blade (F.434), Mallowan, 1947, pl.XXXI.1 As: 0.6%; Sn: n.d.

1939.294: knife-blade (F.435), Mallowan, 1947, pl.XXXI.2 As: 2.8%; Sn: n.d.

1939.281: small wedge (or chisel), context uncertain, perhaps later than Ur III (cf. Speiser, 1935, pl.XLVIII.6) As: 0.4%; Sn: 9.3%; Sb: 0.6%.

1939.276: fragment of rough bar casting (f.437) As: 2.0%; Sn: 9.4%.

1939.275: folded arrowhead (F.402) ("Rimush Treasure Vase": Mallowan, 1947, pl.XXXIV, p.177) As: 0.7%; Sn: n.d.

1939.296: folded piece of sheet-metal (F.402); as 1939.275. As: 1.2%; Sn: 0.5%.

1939.277: Awl (F.500?) Mallowan, 1947, pl.XXXI.7 As: 0.4%; Sn: n.d.

1939.280: Awl, Mallowan, 1947, pl.XXXI.8. As: 1.2%; Sn: n.d.

1939.279: Unworked casting for a pin-shaft or awl (F.438) As: 1.1%; Sn: 0.9%.

1939.285: Pin (F.405): Mallowan, 1947, pl.XXXI.5. As: 1.2%; Sn: 2.2%.

1939.287: Pin with coiled head (F.444), Mallowan, 1947, pl.XXXI.6: As: 4.5%; Sn: n.d.

1939.243: Pin with coiled head (F.450), JNP-Ziggarat. As: 1.4%; Sn: n.d.

1939.286: Pin with domed head (F.429), Mallowan, 1947, pl.XXXI.3: As: 0.2%; Sn: 1.4%.

First half of second millennium B.C.

1938.51: Pin with moulded shank (B.537), cf. Mallowan, 1937, pl.XVIb: As: 1.1%; Sn: 1.4%.

1938.47: Needle or bodkin (B.535), JNP: top. As: 0.2%; Sn: n.d.

1938.46: Needle or bodkin (B.528): As: 2.0%; Sn: n.d.

1938.52: Pin with coiled head (B.538), cf. 1939.287 above. As: 1.3%; Sn: n.d.

1938.50: Pin with flat head (B.515), JNP:
As: 1.6%; Sn: n.d.

8. CHAGAR-BAZAR:

A: Level 12 (Halaf): Mallowan, 1936, 27: bead: "purest specimen of ancient copper that we have had" (Desch).

B: Second Millennium B.C.: Moorey and Schweizer, 1972, 187:

Ashmolean Museum 1937.231: knife-blade, Mallowan, 1937, fig.13.7, c.1700 B.C.:
As: 0.6%; Sn: 1.4%.

1938.39: Belt or pulley fitting: near surface:
As: 1.4%; Sn: 13.7%.

1937.221: socketed awl (A.528), cf. Mallowan, 1947, pl.XLI.2:
As: 0.4%; Sn: n.d.

1936.103: pin with flat head (Me 616; grave 66M, level 3, c.2000 B.C.)
As: 1.4%; Sn: n.d.

1938.42: pin with pierced shank and large flat nail-like head:
As: 0.6%; Sn: n.d.; Sb: 0.8%

9. ERIDU (Abu Shahrein):

Objects undated: Campbell Thompson, 1920, 144;
Woolley and Hall, 1927, 39:
"It may be added that the analyses made by Dr. Scott and his assistant Mr. E.C. Padgham of copper nails and a dagger(?) from Shahrain, found by Mr. R.C. Thompson, show metal of the same purity as that from al-ʿUbaid, without any trace of tin, antimony or arsenic."

9A. FARA: objects in Philadelphia ("Sumer Metals Project", Philadelphia; analyses, 1983-4; Muhly, pers.comm.).

10. TEPE GAWRA:

A: Pin (level VIII)
Speiser, 1935, 102:
Cu: 93.22%; Sn: 5.62%; As: 0.67%; Pb: 0.24%; Zn: 0.05%; Fe: 0.04%; Al: 0.02%; Sb: 0.09%.
Chisel (level VI)
Speiser, 1935, 102:
Cu: 97.01%; Sn: 0.37%; As: 1.59%; Pb: 0.04%; Zn: 0.10%; Fe: 0.49%; Al: 0.06%; Sb: 0.04%.

B: Flat adze (Philadelphia: 35-10-246: level XII)
Tobler, 1950, 212:
Cu: 95.36%; As: 0.05%; Fe: 0.19%.
Flat adze (Philadelphia: 35-10-101a: level XI)
Tobler, 1950, 212:

cu: 91.80%; As: 1.63%; Ni: 3.49%; Fe: 0.02%.

C: Levey, 1959, 209:

	Flat axe/adze	(38-13-38: level XII):	Sn: 0.3%
	Flat axe-adze	(31-52-231: level VI):	Sn <0.3%
	Chisel	(31-52-233: level VI):	Sn <0.3%
	Shaft-hole axe	(31-52-249: level VI):	Sn: 6%
	Shaft-hole axe	(31-52-251: level VI):	Sn: 15.7%
fig.39	Shaft-hole axe	(32-21-93: level VI):	Sn: 5.7% As: 0.1% Ni: 0.1% Fe: 0.1%
fig.39	Shaft-hole adze	(32-21-94: level VI):	Sn: 0.3% Pb: 0.1% As: 0.01-0.1% Ni: 0.01-0.1% Fe: 0.1%
	Flat axe/adze	(31-52-6: level IV):	Sn: 1.7% As: 0.1% Zn: 0.1% Ni: 0.1% Fe: 0.1%
	Chisel	(32-21-20: level IV):	Sn: 0.3%

D: A series of analyses have been undertaken by the "Sumerian Metals Project", University Museum, Philadelphia (Stech; pers.comm., 1983).

11. JAMDAT NASR:

Ashmolean Museum 1926.449: flat adze-blade; late Prehistoric; Mackay, 1931, pl.LXXI.32.LXXV; analysis: As: 0.2%; Sn: 0.5% Moorey and Schweizer, 1972, 180.

1926.455: fish hook; late Prehistoric; Mackay, 1931, pl.LXXV.2481; analysis: As: 3.3%; Sn: n.d. Moorey and Schweizer, 1972, 180.

12. KAR-TUKULTI-NINURTA:

"Holdfasts"; Middle Assyrian.

British Museum: 1922.8.12.142: Sn: 0.5; As: -; Pb: -; Ag: -

141: Sn: 0.5; As: -; Pb: -; Ag: -

143: Sn: 0.5; As: 0.5; Pb: -; Ag: -

140: Sn: 0.5; As: 0.5; Pb: -; Ag: -

McKerrell, 1977, 167, table 13 (not Assur as published).

13. KHAFAJAH:

Copper Statuette: Early Dynastic III (sample from Kh.I 351c, analysed by Desch; Temple Oval III)
Cu: 99%; Sn: 0.63% (Analysis A).
Frankfort, 1932, 78; Woolley, 1934, 291; Frankfort, 1939, no.183, pl.102D-F; Delougaz, 1940, 151.
Pin (Kh.III 688): Sn: 0.32%; As: 0.83%
Pin (Kh.III 729): Sn: 3.33%; As: -
Pin (Kh.III 850): Sn: 10.64%; As: 0.68%
Early Dynastic IIIA; Delougaz, 1940, 152; Moorey and Schweizer, 1972, 185.
Pin (Kh.III 215): Sn: 3.44%; As: 1.06%
Dagger-blade (Kh.III 44): Sn: 2.98%; As: 0.94%
Arrowhead (Kh.II 75): Sn: n.d. As: 0.47%
Early Dynastic IIIB; Delougaz, 1940, 152; Moorey and Schweizer, 1972, 185.
Blade (Kh.III 904): Sn: 9.82%; As: 0.23%
Early Dynastic III; Delougaz, 1940, 152, Moorey and Schweizer, 1972, 185.

14. KHORSABAD:

Foundation tablets of gold, silver, bronze and lead (cf. Ellis, 1968, 194).
M. Berthelot, Introduction à la Chimie des anciens (1906), 219-228.

15. KISH:

A: Peake, 1928, 453-4: eight undated specimens (six from 1928 season); all save one (from mound "A") were tin bronzes; published in a slightly variant way in Woolley, 1934, 291 (table II).

B: Moorey and Schweizer, 1972, 182-4:
(1) From graves in the "Y" Sounding: E.D.I.

Ashmolean Museum 1929.376:	Fragment of a conical stand (gr.479) cf. Watelin, 1934, pl.XXI.1; Moorey, 1978, Fiche 2: F05: As <0.1%; Sn: c.1%?
1929.331:	disk mirror (gr.524) cf. Watelin, 1934, pl.XIX.1-3; Moorey, 1978, Fiche 2; F05: As: 0.6%; Sn: n.d.
1931.174:	hemispherical bowl (gr.687), cf. Watelin, 1934, pl.XX, 2-3; Moorey, 1978, fiche 2:F09: As: 4.8%; Sn. c.1%.

(2) From "Y" Sounding, without exact context: E.D.I.

Ashmolean Museum 1928.482:	Adze-blade, Moorey, 1978, fiche 2:D05; As: 0.6%; Sn: n.d.
1928.483:	"Harpoon head", Moorey, 1978, fiche 2:D05: As: n.d.; Sn: n.d.
1931.189:	"Harpoon head", Moorey, 1978,

fiche 2:D05:
As: 3.1%; Sn: 1%

1929.229: Needle or bodkin, Moorey, 1978, fiche 2:D05:
As: 0.9%; Sn: n.d.

1931.478: Spouted vase with cast spout, Moorey, 1978, fiche 2: D05:
Spout: As: 4.3%; Sn: c.1%?
Body: As: 2.6%; Sn: n.d.

1931.479: Disk mirror, Moorey, 1978, fiche 2:D05;
As: 3.1%; Sn: c.1%?

(3) Cemetery A: E.D.IIIb - Early Akkadian

Objects from graves:

Ashmolean Museum 1924.292: Bowl (gr.16), cf. Moorey, 1978, fiche 1: D14;
As: 2.5%; Sn: n.d.

1926.525: Dagger-blade (gr.20), Mackay, 1925, pl.XVII.11;
As: 2.3%; Sn: n.d.

1926.526: Dagger-blade (gr.23), Mackay, 1925, pl.XVII.14;
As: 2.8%; Sn: n.d.

1925.174: Dagger-blade (gr.47), Mackay, 1929, pl.LXII;
As: 4.3%; Sn: 3.0%.

1925.173: Pin (gr.47), Mackay, 1929, pl.LVIII.19;
As: 6.5%; Sn: n.d.

1925.176: Pin (gr.63), Mackay, 1929, pl.LVIII.6;
As: 1.5%; Sn: n.d.

1925.177: Knife-blade (gr.67), Mackay, 1929, pl.LXI.22;
As: 0.5%; Sn: n.d.

1925.178: Axe-head (gr.80), Mackay, 1925, p.159;
As: 0.3%; Sn: n.d.

1925.186: Pin (gr.104), Mackay, 1929, pl.XL.4(1);
As: 3.1%; Sn: n.d.

1925.160: Axe-head (gr.104), Mackay, 1929, pl.XXXIX.7, lower;
As: <0.2%; Sn: 0.5%.

1925.183: Flat triangular blade (gr.105), Mackay, 1929, pl.LXI.20;
As: 0.3%; Sn: n.d.

1925.250: Axe-head (gr.135), Mackay, 1929, pl.XLIII.11;
As: 1.9%; Sn: 15.5%.

1925.245: Handled bowl (gr.135), Moorey, 1978, fiche 1: F10;
As: 1.2%; Sn: 11.6%.

1925.246: Cosmetic dish (gr.135), Moorey, 1978, fiche 1: F10;
As: 0.2%; Sn: 2.2%.

1925.193: Bowl (gr.139), Mackay, 1929, pl.LVII.10;
As: 0.6%; Sn: n.d.

1925.197: Bowl (gr.144), Mackay, 1929, pl.LVII.2;
As: 3.3%; Sn: c.1%?

Objects without grave context: E.D.III to Akkadian:

1925.528: Pinhead, Moorey, 1978, fiche 1: A09:
As: 5.6%; Sn: n.d.

1925.182: Axehead, Mackay, 1929, 158, pl.LXII.3:
As: 1.4%; Sn: n.d.

(4) Objects from Tell Ingharra: E.D.III to Akkadian

1926.452: Pin, Moorey, 1978, fiche 1:C09:
As: 1.0%; Sn: 3.5%.

1926.451: Small wedge, Moorey, 1978, fiche 1:C09:
As <0.2%; Sn: 6%.

1926.454: Pin with ram modelled on top, Moorey, 1978, fiche 1:C08-9:
As: 0.3%; Sn: n.d.

1928.481: Bowl, Za at 3m.; Moorey, 1978, fiche 2:C11.
As: 0.2%; Sn: n.d.

(5) "Sumerian Metals Project", University Museum, Philadelphia: analyses of objects from Kish in the Field Museum, Chicago (1983-4; Muhly, pers. comm.).

16. NIMRUD:

A: British Museum:

1. Vessel, Neo-Assyrian; Cu: 89.51%; Sn: 10.63%.
2. Hook-shaped object, Neo-Assyrian; Cu: 89.85%; Sn: 9.78%.
3. Bronze bull's foot from tripod; Neo-Assyrian; Cu: 88.37%; Sn: 11.33%.
4. Bell Neo-Assyrian; Cu: 84.79%; Sn: 14.10%.

Layard, 1853, 670ff.: "Notes of the specimens from Nineveh (i.e. Nimrud), kindly favoured to the author by Dr. Percy, of the School of Mines".

B: British Museum: wall-peg; Neo-Assyrian:
Sn: 6.4%; Pb: 13.4%.
Hughes et al., 1981, 143, n.5.

C: "Nimrud Bowls" (British Museum): Hughes et al., 1981, 144, n.7, reports existence of unpublished analyses:

"More than 200 analyses have been included in a study by Curtis (1979; unpublished) and the range of tin percentages for Assyrian metalwork broadly matches that for the present Urartian objects, as does also the typical scatter of values for lead, with many objects containing under 2% lead, and a few (mostly bells) higher values of 10-20%".

17. NINEVEH:

A: 1. Thick grey rod; ?Neo-Assyrian:
Cu: 88.03%; Sn: 0.11%; Pb: 3.26%; Fe: 4.06%; Sb: 3.92%; As: 0.60%.

2. Bent rod; ?Neo-Assyrian:
Cu: 86.84%; Sn: 12.70%; Pb: 0.28%; Ni: 0.18%.

3. Furniture decoration; ?Neo-Assyrian:
cu: 86.99%; Sn: 12.33%; Pb: 0.38%; Fe: tr; ni: 0.30%.

4. Rim of a dish; ?Neo-Assyrian:
Cu: 80.84%; Sn: 18.37%; Pb: 0.43%; Fe: 0.16%; Ni: 0.20%.

Objects from Layard´s excavations on Kuyunjik analysed by L.R. von Fellenberg given by E. von Bibra, Die Bronzen und Kupferlegirungen der alten und ältesten Völker (Erlangen, 1869), 94ff., 154; cited by Partington, 1935, 250.

B: Royal statue head; Akkadian period; Campbell-Thompson and Mallowan, 1933, 72, n.2:

"Not yet tested whether copper or bronze (i.e. the head): but the spearhead found with it is certainly copper, as Mr. Padgham of the British Museum was so good as to test for us".

Nylander, 1980, 329, n.1, refers to an unpublished technical and metallurgical analysis undertaken in Berlin. He describes the head as "copper" not "bronze".

C: Wall-peg (Kuyunjiq: BM 83-1-18, 669):
Cu: 87.8%; Sn: 6.7%; Pb: 1.4%; As: 0.36%; Sb: 0.19%; Bi: 0.01%; Fe: 0.19%; Ni: 0.09%; Ag: 0.06%.

18. NIPPUR:

1. Partington (1935, 247) cites Helm´s (1901) analyses of four objects from Hilprecht´s excavations: "fragment of a sword; stylus; copper nail; dish"; only the stylus contained any tin (Sn: 5.45%). There is no way now of dating these objects.
"Large pin of Ur III period"
Cu: 86.9%; As: 52%; Pb: 3.2%; Sn: 8.3%; Ni: 0.13%; Bi: 0.028%; Ag: 0.54%; Fe: 0.39%. Dayton, 1978, 158.

2. Objects in Philadelphia and Chicago ("Sumerian Metals Project", Philadelphia; analyses, 1983-4; Muhly, pers.comm.).

19. TELL RIMAH:

All Middle Assyrian period:

		Sn.	As.	Pb.	Ag.
Knife	(TR 4502)	0.5	--	--	--
Sickle	(TR 45a)	--	--	--	--

		Sn.	As.	Pb.	Ag.
Sickle	(TR 4505)	--	0.5	--	--
Sickle	(TR 4551)	--	--	--	--
Disc	(TR 4541)	5.10	0.5	--	--
Statuette	(TR 3512)	--	--	--	--
Cymbal	(TR 5306)	1.5	--	--	--
Bowl	(TR 5379)	--	0.5	0.5	--
Awl	(TR 4512)	--	1.5	--	--
Strainer	(TR 5386)	--	0.5	--	--
Arrowhead	(TR 5341)	--	--	--	--
Arrowhead	(TR 5359)	--	--	--	--
Arrowhead	(TR 5342)	--	0.5	--	--
Arrowhead	(TR 5389)	--	--	--	--

Non-dispersive XRF surface analyses: McKerrell, 1977, 167, table 13.

20. TELL ES-SAWWAN: (see under Yarim Tepe)

21. TELL SIFR:

McKerrell ran a series of non-dispersive XRF surface analyses on the Old Babylonian agricultural tools in the "Tell Sifr Hoard" in the British Museum (Moorey, 1971a). These have only been briefly referred to in a publication (Iraq 34 (1972) 2). The axeheads, mattock-heads and adze-blades (Moorey, 1971a, nos.1, 2, 5 and 8 (7 not tested)), alone proved to be good tin bronzes; nos.4 (mattock-head), 13 (spade), 28 (knife), 66 (ring), and 81 (ring) had some tin in them; but all the other objects were of copper or arsenical copper, not tin bronze. Nor were the alloys necessarily consistent for any one type of artefact, as no.3, exactly comparable to nos.1 and 2 in form and function, did not contain tin.

22. TELLOH:

1. Inscribed spearhead (Early Dynastic III); de Sarzec-Heuzey, 1884-1912; Berthelot, 1906, 76; nearly pure copper with a trace of antimony only.

2. Shaft-hole axehead (? Early Dynastic); very pure copper; Berthelot, 1906, 77.

3. Flat axe or adze-blade (? Early Dynastic): pure copper with traces of arsenic: Berthelot, 1906, 78.

4. Rim-Sin foundation figurine (Louvre: Rashid, 1983, no.176): Cu: 92.9%; Berthelot, 1906, 78-80; he compared a figure of Amar-Sin in the Louvre (Rashid, 1983, no.170): Cu: 77.4%; Pb: 17.0%; similar statuettes of Ur-Bau and Gudea (cf. Rashid, 1981, 17ff.) were also said to be of copper.

Compare: (1) Entemena foundation figurine (BM 116685; Rashid, 1983, no.67) McKerrell surface analysis: copper with trace of arsenic.

(2) Gudea foundation figurines (BM 91056-8; Rashid,

1983, nos.99-101) McKerrell surface analysis: copper with traces of arsenic and lead.

(3) Ur-Nammu foundation figurine (Pierpont Morgan Library, New York): E. Porada (ed.), Ancient Mesopotamian Art and Selected Texts (1976),9ff., pl.5-6: Cu: 93.5%; Sn: 0.06%; Pb: <0.2%; Zn: 0:05%; As: 0.4%; Fe: 0.05%.

These figurines, acquired on the Art Market, are attributed to Telloh.

23. TELL al-UBAID:

Various analyses are given of relatively pure copper fragments from the bulls and lions found in the Early Dynastic III temple cache; but one fragment contained 2.81% tin. Recent examination of the figures, by eye, suggests that though the body panels were of hammered copper, some of the heads at least were cast tin bronze (J.E. Curtis, pers. comm.) (Hall and Woolley, 1927, 36-8; also Peake, 1928, 454; Woolley, 1934, 291: table II).

Scott reported in Hall and Woolley, 1927, 37, that the antlers of one of the stag's heads were of very pure copper. The "nails" were also copper (with, in one case, "a notable quantity of silver with only a trace of lead"). A small "ingot" is reported as Cu: 95.0%; Pb: 4.4%.

24. UR:

A: Three undated copper objects from trial trenches. Campbell Thompson, 1920, 144.

B: Three undated high-tin bronze objects (11.78/12.00/14.52% Sn) said to be from "the earliest graves at Ur, dating from before the first dynasty of that city" (see below, Woolley, 1934).
Peake, 1928, 453-4.

C: Spearhead (U.14992: Ubaid period): Cu: 99.69%. Woolley, 1934, 291; 1956, 164, PFG/G.pl.30. "several vessels...all may be described as arsenical copper" ("Jamdat Nasr" level) i.e. Late Prehistoric to Early Dynastic I (Woolley, 1934, 165). "a copper rod... marked PG, Pit W, JNG" was also regarded as an arsenical copper.
Woolley, 1956, 165.

Woolley, 1934, 290: Table II: must be re-assembled chronologically; the identity and date of the unnumbered objects may no longer be independently established. Even when numbered, the object type is rarely given "Ist grave" has not been identified; it may well have been Neo-Sumerian rather than Early Dynastic IIIA as given (all three analyses are high tin-bronzes (see above, Peake,

1928).

E.D.IIIA: U.11887 (PG 1054 D: "Royal Grave"); Sn: n.d.
U.12239 (PG 1305) Cu: 97.2%; Sn: 0.56%.
"Shubad Bowl" Sn: 8.3%.

E.D.IIIB: U.12229 (PG 1301) Sn: 1.45%.
U.12672 (PG 1382) Sn: 8.8%.

Late Akkadian/ Neo-Sumerian: U.11436 (PG 970; axe) Sn: n.d.
U.11475 (PG 1035) Sn: 1.64%.
U.12483 (PG 1422) Sn: 5.8%.

Date uncertain: U.12098 (PG/s) Sn: 11.65%.

D: Moorey and Schweizer, 1972, 184; Early Dynastic III
Ashmolean Museum 1935.870: shaft-hole axe-head; cf. Woolley, 1934, pl.223, Type A.3a
As: <0.2%; Sn: 1-1.2%.

1933.248: shaft-hole axe-head; cf. Woolley, 1934, pl.223, Type A.3a
As: 1.2%; Sn: n.d.

1935.869: shaft-hole adze-head; cf. Woolley, 1934, pl.229, Adzes, Type 1
As: 1.6%; Sn: 5.5%.

1933.247: spear-point (U.14062); cf. Woolley, 1934, pl.227, Spears, Type 1
As: 1.7%; Sn: n.d.

1933.245: dagger-blade; cf. Woolley, 1934, pl.228, Type 9a;
As: 1.1%; Sn: n.d.

E: University Museum, Philadelphia: Dayton, 1978, 158-9:

1. Pin (type 7: RT 777: U.9962): Early Dynastic IIIA.
Cu: strong; Pb: 1.0%; As: 1.5%; Co, Sb, Sh, traces; no Ag.

2. Object (no. and type not given) (PG 1589): Early dynastic IIIB.
No Sn or Pb; Sb: 0.2%; As: 2.5%; Co: 0.1%.

3. "Chisel" (PG 744: U.9865): Late Akkadian to Neo-Sumerian.
Cu: 99%; As: 0.3%; Co: 0.1%.

4. Pin (PG 1625: U.14061): Early Dynastic IIIA.
Pb: 3.0%; As: 5.0%; Sb, Ag, traces; Sn, Co: n.d.

5. Miniature Axehead (U. number uncertain? 8559 or 11932)

Sn: 9.0%; Ni: 0.1%; Fe: 0.1%; Ag, Pb, As, Zn, Co traces.

6. Pin (type 5a: U.14018): Early Dynastic IIIB.
sn: 11%; Ni, Fe: weak; As, Pb, Co: traces; no Ag, Sb.

7. Dagger (PG 1589: U.14021): Early Dynastic IIIB.
As: 2.5%; Sb: 0.2%; Co: 0.1%; No Sn or Pb.

8. Dagger (RT 789: U.10825): from body 46; Early Dynastic IIIA.
As: 1.0%; traces of Ag, Pb; no Sn or Sb.

F: (see table 2 overleaf) pp.66-68.
P. Craddock MASCA Journal 3 (1) (1984), 7ff.
In the accompanying table the analyses have been ordered chronologically and given their Ur excavation numbers (for which I am grateful to Dr. J.C. Curtis)

G: "Sumer Metals Project", Philadelphia; analyses of objects in the University Museum (in progress).

25. URUK:

(A) Lion pendant (W 14766d; heinrich, 1936, 47, pl.13a);
Uruk IV/III; Cu with 9% Pb.

(B) Spearhead (W 15068D): Uruk IV/III; Heinrich, 1936, 47;
Cu: 99.3%; Fe: 0.27%; Ni: 0.14%; As: 0.17%;
Zn: trace.

26. YARIM TEPE I and II:
The following information is tabulated in Merpert et al., 1977a, 160: table on pp.64-5.

27. ZERGHUL (Siraran):
Zoomorphic foundation figurine inscribed for Gudea (c.2150 B.C.):
"Essentially copper, with small amounts of silicon, iron and nickel, and traces of silver, arsenic, tin, lead, aluminium, magnesium and calcium" (Sollberger, 1975, 178, pl.IX: BM 135994).

Table 1
(After Merpert et al., 1977a)

Site	Object	Lab.nos.	Cu	Sn	Pb	Zn	Ag	Sb	As	Fe	Ni	Co	Mn
YT= Yarim Tepe													
YTI (Hassuna)	Ore	2483-p	Main Constituent	-	-	4	-	-	4	2	5	5	4
YTI (Hassuna)	Ore	2484-p	Main Constituent	-	-	4	-	-	4	2	4	5	4
YTI (Hassuna)	Bracelet	16623	Main Constituent	-	5	-	present	-		traces	-	-	-
YTI (Hassuna)	Ring	16624	Main Constituent	5	5	?	?	-	4	3	6	-	4
YTII (Halaf)	Bead	16625	Main Constituent	5	-	-	5	-	-	3	5	-	4
YTII (Halaf)	Pendant	16626	Main Constituent	4	4	3	5	-	-	2	3	-	4
YTII (Halaf)	Ore	16627	Main Constituent	-	4	3	3	4	4	3	-	-	3
YTI (Hassuna)	Ore	16628	Main Constituent	-	?	4	-	-	-	2	5	-	3
YTI (Hassuna)	Ore	16629	Main Constituent	?	4	4	5	-	4	2	4	5	3

Table 1 (continued)

YTI (Hassuna)	Ore	16630	Main Constituent	-	5	4	-	-	3	2	4	5	3
YTI (Hassuna)	Ore	16631	Main Constituent	-	-	5	?		3	2	5	?	3
YTI (Hassuna)	Ore	16632	Main Constituent	?	5	4	5	-	-	4	6	-	4
YTI (Hassuna)	Ore	16633	Main Constituent	-	-	4	?	-	4	4	4	3	4
YTI (Hassuna)	Ore	16634	Main Constituent	-	-	4	-	-	3	2	5	5	3
YTl (Hassuna)	Ore	16635	Main Constituent	-	-	3	-	-	4	2	4	5	3
Tell es-Sawwan	Bead	13616	Main Constituent	5	5	5	5	-	3	4	6	-	5
Tell es-Sawwan	Bead	13617	Main Constituent	6	-	4	-	-	-	3	6	-	4
Tell es-Sawwan	Ore	2482-p	1	5	-	4	6	-	-	2	5	-	4

Key: 1=10%, 2=1-10%, 3=0, 1-1%, 4=0. 01-0.1%, 5=0.001-0.01%, 6=0.001%.

Table 2

UR: Section F (after Craddock, 1984)*

	Period	Object	Ur No.	BM No.	Cu %	Pb %	Sn %	Ag %	Fe %	Sb %	Ni %	Co %	As %	Bi %
1.	E.D.III	Socketed axe	8148	121643	93.0	0.15	<0.005	0.15	0.015	0.11	2.8	0.005	3.7	0.003
2.	E.D.III	Socketed axe	12261	123680	96.0	0.37	2.0	0.18	0.33	0.75	0.16	0.015	1.8	0.007
3.	E.D.III	Socketed axe	12108	1929, 10-17,539	85.0	0.36	8.50	0.07	0.12	0.12	0.21	0.005	0.75	0.007
4.	E.D.III	Socketed axe	8107	120719	88.0	0.27	8.7	0.095	0.16	0.03	0.23	0.005	0.47	0.005
5.	E.D.III	Socketed axe	9972	121574	93.0	0.015	5.20	0.004	0.35	0.08	0.70	0.005	1.3	0.007
6.	E.D.III	Socketed axe	10173	121577	86.5	0.02	0.10	0.03	0.22	0.17	0.11	0.020	0.37	0.005
7.	E.D.III	Spearpoint	12274	1929, 10-17,504	96.5	0.21	0.05	0.14	0.19	0.11	1.42	0.015	2.4	0,007
8.	E.D.III	Spearpoint	9337	123635	90.0	0.40	9.00	0.03	0.31	0.05	0.095	0.002	0.26	0.005
9.	E.D.III	Spearhead	10825B	121591	99.0	0.09	0.005	0.012	0.30	0.035	0.035	1.05	1.05	0.005
10.	E.D.III	Socketed adze	10956	121668	98.0	0.06	0.1	0.1	0.5	0.06	0.17	0.035	1.5	0.005

Table 2 (continued)

	Period	Object	Ur No.	BM No.	Cu %	Pb %	Sn %	Ag %	Fe %	Sb %	Ni %	Co %	As %	Bi %
11.	E.D.III	Chisel	9106	120791	94.5	0.07	<0.005	0.03	1.0	0.05	0.80	0.110	3.0	0.005
12.	E.D.III	Chisel	10158	121579	84.0	0.13	9.70	0.03	0.25	0.06	0.40	0.035	1.1	0.003
13.	E.D.III	Pin	10826B	121608	97.0	0.10	0.10	0.055	0.50	0.035	0.085	0.003	0.80	0.005
14.	E.D.III	Staple	9916	121615	96.5	0.37	0.050	0.16	0.50	0.065	0.25	0.012	0.90	0.005
15.	E.D.III	Rod	9915	121618	98.0	0.005	0.070	0.06	0.70	0.02	0.57	0.002	0.80	0.005
16.	Akkad	Socketed axe	12653	1929, 10-17,536	90.0	0.85	7.90	0.055	0.19	0.06	0.33	<0.005	0.80	0.005
17.	Akkad	Socketed adze	12014	1929, 10-17,531	93.0	0.54	2.00	0.10	0.27	0.15	1.3	0.025	1.7	0.005
18.	Akkad	Dagger	9901	121580	94.5	0.48	0.20	0.035	0.25	0.47	1.7	0.090	3.2	0.005
19.	Akkad	Pin	12028	1929, 10-17,569	98.0	0.65	0.050	0.017	0.20	0.09	0.47	0.010	1.7	0.005
20.	Akkad	Socketed axe	9059	120776	99.5	0.025	<0.005	0.01	0.14	0.05	0.34	0.005	0.25	0.005
21.	Akkad	Socketed axe	9928	121607	75.5	2.0	6.80	0.035	0.14	0.05	0.065	<0.005	0.21	0.005
22.	Akkad	Harpoon	11750	1929, 10-17,611	97.5	0.05	<0.005	0.03	0.41	0.21	0.34	0.010	1.7	0.005

Table 2 (continued)

23.	Akkad	Chisel	9898	1928, 10-17,259	86.0	0.05	10.2	0.015	0.43	0.02	0.21	0.002	0.55	0.005
24.	Akkad	Staple	10355	121617	99.0	0.17	<0.005	0.17	0.42	0.11	0.02	<0.005	0.65	0.003
25.	Akkad/ Ur III	Chisel	11766	1929, 10-17,607	90.0	0.015	7.50	0.02	0.31	0.075	0.26	0.015	0.70	0.065

Data obtained by absorption analysis (see Hughes et al. 1976). major and trace element determinations have a precision of 2% and 30% respectively. Detection limits for trace elements included here are 0.0005%. The presence of zinc was also sought, but not found in concentrations greater than 0.010%.

Artifact form and sampling location:

4 (socketed axe): underside, at blade/socket junction
20 (socketed axe): socket
5 (socketed axe): underside, at blade/socket junction
6 (socketed axe): distal end of blade
1 (socketed axe): top, into socket/blade junction
2 (socketed axe): underside, at blade/socket junction
3 (socketed axe): underside of blade, near socket
8 (square section lance): centre
11 (chisel): centre
12 (flat chisel): centre
9 (spearhed): tang
13 (pin): centre
14 (staple): centre
24 (staple): centre
15 (rod): centre
10 (socketed adze: side of blade
7 (squared section spearhead): head/tang junction
16 (socketed axe): top, at socket/blade junction
23 (flat chisel): centre
18 (dagger): tang
21 (pin): head
17 (socketed adze): side of blade, near socket
19 (pin): head
22 (harpoon): centre of shank
25 (chisel): broken end

* After Craddock, 1984; some slight differences in dating; analyses tabulated differently to include Ur excavation numbers.

IV: ELECTRUM:

Electrum, which derives its name from the Greek word for amber, describes a mixture of gold and silver, pale yellow in colour. In mineralogy it is applied to native argentiferous gold containing 20% to 50% of silver; its definition when artificial is arbitrary. Pliny used the term to describe alloys of gold and silver with one part of silver to four of gold. Laboratory research has not proceeded far enough with ancient objects of precious metal from Mesopotamia to make separate treatment of this metal possible, since the distinction between natural and artificial electrum may not be confidently drawn. Many of the rare analyses of Mesopotamian "gold" show that it is in fact electrum, but whether a natural or a deliberate alloy is not invariably clear. As its history involves both that of gold and silver, electrum is, for the moment, best treated with them as and when the question arises. An alloy of gold with silver is harder than pure gold and will consequently stand wear better. Silver may only be separated from gold by a complicated process (see p.108); but base metals may be slagged off by repeated meltings of native gold in an oxidising atmosphere, routine in many goldsmithing techniques.

V: GOLD:

(a) The metal:

Gold, whose distribution is almost universal, occurs either in veins of quartz rock (*reef gold*) or in alluvial sands and gravel (*placer gold*); a distinction found in the textual sources (*ḫurāṣum ša abnišu; ḫurāṣum ša maʾēšu*). As it combines chemically with neither oxygen nor sulphur, it is present predominantly in the metallic state (Patterson, 1971). In its native state gold is always alloyed with silver in proportions that vary greatly; sometimes also with copper and traces of iron and other metals. A gold with only a trace of silver, or no silver at all, has been deliberately refined.

(b) Recovery:

As there are no gold deposits in Mesopotamia, the processes of recovery would have been remote from the experience of her craftsmen. They received the metal either in nuggets, or prepared or partially prepared as powder, or as ingots, taking that term to embrace both block and ring shapes (Limet, 1960, passim; Leemans, 1967-71, 505). Methods of recovery are rapidly reviewed here as a brief preliminary to understanding the manipulation of this metal.

Reef gold occurs as irregular masses in quartz veins or lodes. The mining process consists first of freeing the gold particles by crushing and sifting the ore, then of separating the gold by making use of the metal's higher density. The classic description of gold mining in antiquity is that by Agatharcides, preserved by Diodorus Siculus (iii, 12-14), who, in the second century B.C., visited and graphically described the rigours of gold mining in Egypt (cf. Lucas, 1962, 224ff.).

Placer gold is produced through the action of weathering, or other agents, by which gold deposits are broken down and carried into river beds as secondary, alluvial gold. As gold has a high specific gravity (19.3), it does not move so easily with the current as lighter matter, and so collects in pockets, commonly found concentrated in the middle of a river. To recover it the sand and gravel of the river bed are agitated with water in pans, troughs or cradles, so that the rocky matter floats off and the gold particles collect in the bottom as gold dust or nuggets. Maxwell-Hyslop (1970) sought to identify a type of metal pan found at Troy and in a grave at Assur as an instrument for panning alluvial gold. Further study of these vessels has indicated that a domestic role is more likely; such a pan was found at Sippar (Calmeyer, 1977).

(c) Colour variation:

Ancient terminology for gold (Leemans, 1957-71, 504ff.; Limet, 1960, 43-4; for Egypt, Harris, 1961, 32ff.) is notoriously complicated, probably arising from the different sets of circumstances in which it was employed (Ungnad, 1935; cf. also Eilers, 1954-9). There was first the technical terminology conveying, for the specialist administrator and craftsman, important information on source, appearance and degrees of purity; then a poetic or symbolic vocabulary; and finally what may best be termed the craftsman's slang or demotic phraseology. At present it is impossible to relate these terms directly to the material evidence in any meaningful or useful way. The quality of the gold from the "Royal Cemetery" at Ur, for instance, could not be assessed by eye from the colour alone, as some specimens of different colour had practically the same composition, whilst one example, with 10% copper, had the rich yellow colour of a purer metal (Plenderleith in Woolley, 1934, 292-3). Significant colour changes in native gold require a fairly substantial modification in the alloy and insufficient laboratory work has been done to allow for detailed comments on this phenomenom in Mesopotamia, where much of the little "gold" so far sampled is, in fact, "electrum" (Plenderleith in Woolley, 1934, 294; Partington, 1935, 232). Substantial additions of copper would tend to turn the gold red, perhaps illustrated by a text from Ur in which copper is alloyed with gold in making of earrings (Limet, 1960, 43-5: UET III, no.452). Substantial additions of silver would make it green or grey in tone. It is not only the composition of the alloy which modifies the colour; since surface colour may also be changed. If the surface treatment of the metal were varied, the surface would reflect light differently and give the impression of different colours, even in the same alloy. Surface colour may as easily be modified accidentally, for instance when organic matter stains it, or when a film of iron oxide, natural or contrived, turns it rose-pink in colour (cf. Lucas, 1962, 233-4), or when there is selective enrichment of the surface by contact with a salty conglomerate of irregular activity (Maryon, 1949, 106, n.14).

In the graves of the "Royal Cemetery" at Ur colour contrasts were directly contrived in more obvious ways, although Plenderleith (in Woolley, 1934, 294) noted how rarely gold and silver were directly juxtaposed. Striking exceptions are the rampant he-goats from RT 1237 with their gold-covered legs and genitals and their silver bellies (Woolley, 1934, 582, pl.77, 87-90). A most unusual patterning technique is illustrated by a silver bowl (U.10891) from RT 800 (Woolley, 1934, pl.173): "the bowl itself is of silver, there are very slight vertical grooves in its sides, and in them strips of electrum have been fixed either by sweating or by lapping the silver over the edge of the inlaid material" (Plenderleith in Woolley, 1934, 297). The famous rein-ring from Pu-abi´s land-sledge (Woolley, 1934, pl.166) has an electrum animal on a silver ring and such silver vessels as U.11821 from PG 1133 have tubular handles of electrum (Woolley, 1934, pl.173b, right). A brooch from PG 55 (U.8007: Woolley, 1934, pl.219, lowest right) has a background of filigree silver wire contrasting with its electrum boss. A roundel of gold from the Ur III "Royal Mausolea" was decorated with a whorl pattern whose outlines were inlaid with silver (Woolley, 1974, 32, 102: U.17429). The decorative potential of variously coloured gold combined with siver was exploited throughout down into the Achaemenid period (cf. C.A.D. s.v. sariru).

Even with the handful of analyses at present available for the "Royal Cemetery" caution is necessary, since alloying of precious metals is already evident. A spear-head (U.9122) has 30.30% gold, 59.37% silver and 10.35% copper (Woolley, 1934, 294: table III). The famous rein-ring from RT 800 (Pu-abi) has an equid of electrum (whether natural or artificial is an open question), comprising 65.60% gold, 31.45% silver and 2.65% copper, whilst the ring on which it stands is virtually sterling silver (93.5% silver, 6.10% copper and 0.08% gold). Bowls from PG 755 are also gold-silver-copper alloys (Woolley, 1934, 294).

(d) Refining and assaying:

The date of the introduction of refining in the Near East, as in Egypt, is still an open question. Cupellation will remove metals from silver or gold, but by itself will not remove silver, adding salt (see below) enables silver to be removed from gold (cementation). de Jesus (1980, 89) is cautious with reference to Anatolia, arguing that silver separation was not known and only native gold was used. For Egypt opinions vary from dates as early as the Old Kingdom (Notton, 1974) to the conservative position of Lucas (1962, 229), who saw no clear evidence for it before the Achaemenid period. This is certainly the period to which the best archaeological evidence now belongs, for at Sardis the American excavators found convincing indications of both processes operative in the same industrial area in the sixth century B.C. (Hanfmann and Waldbaum,1970, 310ff.; Ramage, 1970; Goldstein, 1970; Waldbaum, 1983). To this century also belong the texts of Nabonidus cited by Fossey (1935) as the clearest documentary evidence for refining in Mesopotamia (see below).

For cupellation the gold is alloyed with lead in a special clay vessel, known as a cupel, and the product oxidized by a strong current of air blown into the surface of the molten metal. The base metals are consumed, or drossed, as the oxides formed are absorbed by the porous walls of the cupel, whilst gold and silver survive. Precisely what form cementation took before the Classical authors offer descriptions of it (Diodorus Siculus iii, 14, 1ff. (after Agatharcides); Strabo iii, 2, 8) is a matter for continuing debate. Notton (1974) experimented with the method described by Diodorus Siculus and it proved very successful. A series of tests were conducted on a nine-carat alloy (in the absence of a suitable gold ore). In the first salt and the gold alloy were smelted under various circumstances. After five days heating at 800°C, gold recovery was high.

It was less so when salt, brick dust and the alloy were used and also when tin and lead were added in two different ways. The presence of tin to a greater extent, lead to a lesser, inhibited the recovery rate.

On the evidence of these experiments earlier suggestions about ancient methods of cementation gain extra credence. The Sardis excavators argued that "the gold is hammered into thin sheets, which are then stacked in a vessel with layers of dry "pickling mixture" like common salt or alum, and heated for a long time, c.700°C. Silver especially combines with salts and the gold is left pure" (Ramage, 1970, 22-3).

Textual references to "refined gold" (Limet, 1960, 43; cf. C.A.D. s.v. _ṣirpu_) in texts of the Neo-Sumerian period indicate the practice of refining by at least the end of the third millennium B.C. After a review of the evidence for gold prices Young (1979) concluded that in the Ur III period it was the purity of the gold which determined its price, and not the availability of either gold or silver. A few Sargonic texts permit a determination of gold prices, calculated on the basis of the silver equivalence; but before then there is nothing to indicate gold purity or price. Until a much greater corpus of analyses is available, it may only be surmised that the native alloy alone was used at this time and that refinement of gold was a development of the mid. to later third millennium B.C. Even then there is no evidence as to how the various purities were established. A common cited point-of-reference is provided by two letters from the fourteenth century B.C. Tell el-Amarna archives (Fossey, 1935; Levey, 1959, 193); but they are not particularly illuminating. In these letters the Babylonian king complains to Amenophis III of Egypt about the quality of the gold sent to him by the Pharaoh (Knudtzon, 1915, 70-71, 92-3). In one case only 25% of the metal was gold. From both letters it is clear that the Babylonians had a means of refining the metal; but the reference is simply to putting it into a furnace, without further comment (cf. C.A.D. s.v. _ṣâdu_).

The role of _assaying_, in which a sample is removed for analysis, is no clearer. Assay by fire, used in a qualitative not a quantative way, is described in the Leyden Papyrus (X, no.3; Oddy, 1983):

"If the gold is pure, it keeps the same colour and remains pure like coinage after heating. If it appears whiter it contains silver; if rougher and harder, it contains copper and tin; if black and soft, it contains lead".

The use of a touchstone (_lapis Lydius_ i.e. black chert) to test the quality of gold is attested by Classical authors from at least the sixth century B.C. (Theognis, 417; Pindar, _Pythian_, X, 67). Theophrastus (_de Lapidus_, 45) attributes them to the river Tmolus. In this case it is siliceous schist (flinty slate), black in colour, usually for this purpose deep black, and fine grained; but other black stones may be used. The surface has to be ground perfectly flat and left matt for efficient assaying by streak. Streaks are taken from alloys of known composition for comparison with streaks taken from the metal to be assayed. If the identifications are correct, touchstones were used in Mesopotamia by at least the later third millennium B.C. A tomb of the Ur III period at Telloh has yielded a scale and weights, with a piece of haematite, taken by Arnaud et al. (1979, 23) to be a touchstone by analogy with another such piece found in a jeweller's hoard of the Old Babylonian period at Larsa (Arnaud et al., 1979, 20-1, fig.8).

Gold, like silver, was regularly recycled in Mesopotamia and texts make clear the care taken by administrators to ensure its honest use (cf. Durand, 1983).

(e) Sources:

(i) Ancient textual references:

The evidence currently available for the sources of the gold used in Mesopotamia is uneven and far from precise (Leemans, 1957-71, 505:2). There is also a need, in using the documentary sources, to beware (if no reference is made to mines) of confusing transit zones or ports-of-trade with source areas. Sumerian literary texts refer to gold from Aratta (Pettinato, 1972, 79). Gudea records receiving gold from the mountain of Ḫaḫ(ḫ)um (Statue B, col.VI.33-5; Falkenstein, 1966), generally taken to lie in that part of modern Turkey northeast of Malatya, on the route from Diyarbakir, via Elazig, to Kültepe (Karum Kanesh) (Groneberg, 1980, 85-6; Yakar and Gürsan-Salzmann, 1979, 37-8) and from Melluḫḫa. In far less explicit terms the names of various mountains reputed to have gold sources are also recorded (Limet, 1960, 94). Various texts refer to the almost mythical land of gold known as (Ḫ)arallu, perhaps somewhere in the Iranian hinterland (Komoróczy, 1972; Groneberg, 1980, 20). Shu-Sin (c.2037-29 B.C.) refers to gold from "Su-land", probably in western Iran, though its location is still open (Edzard, 1959-60, 16-18) and from (Mar)daman, possibly identical with Mardin in southeast Turkey (ibid., 7; Edzard and Farber, 1974, 118). Documents relevant to the Dilmun trade in the later third and earlier second millennium B.C. indicate that some gold was still reaching Ur up the Gulf at this time, but whence it came, perhaps Meluḫḫa, is not stated (Oppenheim, 1954, 7; Leemans, 1960, 120-1, is more cautious). The renowned resources of Egypt and Nubia (cf. Lucas, 1962, 224-8) contributed most certainly in the fourteenth century B.C. to Mesopotamian royal gold holdings (Edzard, 1960; Wilhelm, 1974). Then for the first and only time in its history Babylonia adopted the gold standard and large amounts of gold came from Egypt to assist Kurigalzu I in his major building projects (Brinkman, 1972, 274-5; Müller, 1982). Hittite inventory texts of the thirteenth century B.C. record gold from Babylon and Lukka (Lycia) (Košak, 1982, 195). Jankowska's study (1969, 254-5) of the booty recorded in Neo-Assyrian texts contributes little to isolating the origin of gold, since what is recorded is either the local treasuries or centres of manufacture. The most considerable amounts of gold were held by Tyre and Muṣaṣir, smaller quantities by Carchemish, Damascus and Jerusalem, and the towns of Tabal, still less by Bit-Zamani and Zamua. The tribes of Hama were richer in gold than those of Harran. An Iranian source, reminiscent of earlier references, is recorded by Tiglath-Pileser III (c.744-27 B.C.): "...the land of Rû(a), as far as the alkali desert, the lands of Ushkakkâna, Shikraki (the land) of gold, provinces of the mighty Medes..." (Luckenbill, 1926, 285:795). According to the "Foundation Charter" of Darius I, the gold used at Susa in his time was from

Sardis and Bactria (Vallat, 1971).

(ii) Modern source regions:

So far as is known there were no sources of gold exploited in antiquity in Mesopotamia, Syria or Palestine; but the metal is widely reported on the periphery of this region (Maxwell-Hyslop, 1977): in Turkey, where there is a preponderance of sources in the west and south-west (de Jesus, 1980, 82ff.); in Egypt and Nubia (Lucas, 1962, 224-8); gold existed in considerable quantities in western and southern Arabia, not least in Ophir (I Chron.29:4; I Kings 10; Job 22:24; cf. Maisler, 1951), being well known both to Biblical and Classical authors (Diodorus Siculus iii, 45, 6ff.; Strabo XVI, 4, 18, 22; Periplus 36 etc.); in greater Iran, notably in the north of the country and eastwards into Transoxiana and the region of modern Afghanistan, where there are substantial vein and placer deposits (Dunlop, 1957; Allan, 1979, 3ff., 111-114; Chmyriov et al., 1973). Diodorus Siculus (ii, 36, 2) and Pliny (NH xxiii.66) refer to rich sources in India, though traces are no longer easy to detect (Allchin, 1962; Ratnagar, 1981, 106ff.). With so wide a range of potential sources within her orbit, many of them in regions whence she is known to have received other metals, Mesopotamia is likely to have had a variety of choices if and when supplies were for some reason blocked in any particular locality. It will be immediately clear that the ancient texts already considered indicate import at various times from all the potential source zones listed here.

(iii) Scientific source detection:

The wide distribution of potential sources does not ease the quest for a scientific method through which Mesopotamian gold sources might be "finger-printed". It is known that platinum-group elements (hereafter PGE), such as ruthenium, rhodium, palladium, osmium, iridium and platinum, occur in gold objects from the ancient Near East both in solid solution and as inclusions. As such an association is rare in primary gold deposits, this has been taken to indicate the use of secondary, or placer, deposits, in which gold and PGE became associated by fluvial transport. On the evidence of the inclusions of the platiniridium alloy in a number of Sumerian and other ancient gold objects, notably Lydian gold coins supposedly of gold from the Pactolus valley in western Turkey, Young (1972; cf. Whitmore and Young, 1973) argued that this region had been an important source of Sumerian gold. Within a few years Ogden (1977), after a comprehensive review, concluded that direct correlation of a gold object and its metal source, through the PGE inclusions, was not feasible. Maxwell-Hyslop (1977) also questioned the Whitmore and Young hypothesis, pointing to other more likely and closer sources for Sumerian gold, both in eastern Turkey and in Iran. Meeks and Tite (1980) investigated the problem through a wide range of objects in the British Museum. They found PGE inclusions in more than half the Early Dynastic III to Larsa Period jewellery they tested from Ur and in the majority of the Sargonid jewellery from Brak. They also identified them in two-thirds of the Achaemenid jewellery in the "Oxus Treasure". In contrast, they were rarely found in second millennium jewellery from Tell Ajjul in Israel, Tell Atshana in Turkey, Enkomi in Cyprus, Ialysus in Greece, and Crete. Research on the evidence from Sardis now reveals a situation more complex than

Young supposed (cf. Waldbaum, 1983, 3ff).

Although absence of inclusions may be indicative of source, other possible explanations at present preclude categorical statements. Certainly, since the rare alloy type platiniridium may not generally be used to characterize or identify the placer deposits used in antiquity, the Whitmore and Young hypothesis is not tenable. Nor do the iridium-osmium-ruthenium alloy inclusions necessarily provide a basis for characterizing sources; but there was enough consistency, in some cases, for Meeks and Tite (1980, 273) to conclude:

"It is therefore possible that, following considerably more work on PGE inclusions in gold from placer deposits of known location, some limited source characterization of the gold used in antiquity might, in favourable circumstances, be possible on the basis of the compositions of the inclusions".

(f) The repertory of objects and uses:

The surviving material evidence for the uses of gold and electrum in ancient Mesopotamia, except for small pieces of jewellery and miscellaneous fragments of sheetmetal, is to all intents and purposes concentrated into the few Early Dynastic III "royal" graves at Ur. Had comparable graves for other periods been discovered, no doubt the picture would be different; but for the present it is extremely restricted in range. The varied repertory of the goldsmith evident from texts has left virtually no trace in the archaeological record, apart from this single outstanding instance. The reason is obvious, and applies almost equally to silver. The precious metals were constantly recycled; they were always carefully plundered from any accessible place when cities were sacked; and they were always sparingly placed in graves, where they were also subject to plunderers. All the literary and documentary sources (Leemans, 1957-71, 505ff.) indicate that gold was pre-eminently for ceremonial and prestige purposes. It was amassed in temple and royal treasuries, where it was largely reserved for the decoration of their buildings and furniture, for the adornment of high ranking people, or for gifts and offerings in both religious and secular contexts. The archaeological evidence (save through intensive analysis of what little has survived to resolve questions of purity) does not offer answers to such key questions as the relative roles of gold for bullion or for manufacture; or as currency in "private hands"; nor whether the recorded range of gold prices indicates the action of the laws of supply and demand or the metal's relative purity (Leemans, 1957-71, 512ff.; Young, 1979). It is significant that gold was generally treated like other commodities and its value was reckoned against a silver standard, save briefly in the Kassite period (Müller, 1982).

The only gold objects for which some kind of systematic survey may be offered have been assembled and fully discussed by Maxwell-Hyslop (1971; cf. Spycket, 1973) in her study of jewellery. Consequently no historical survey is offered here, as has been attempted for the base metals. Instead attention has been focussed on categories of artefacts, so assembling under a single head whatever significant material evidence has survived.

(i) Personal ornaments and amulets:

As this section does not repeat information given, and well illustrated, in Maxwell-Hyslop (1971), it is inevitably brief.

Gold jewellery of the later prehistoric period was not included in Maxwell-Hyslop´s book. The evidence concentrates at present in two places: in the temple deposits of Uruk in the south, and in one or two tombs at Tepe Gawra in the north. The two earliest occurrences of gold are a piece of gold wire in what was said to be a late Ubaid period context at Ur (Woolley, 1956, 185: U.16981) and a few fluted beads in level XII at Tepe Gawra of approximately comparable date (Tobler, 1950, 193), sometime in the late fifth or early fourth millennium B.C. They indicate that manipulation of the metal was already varied. Gold metallurgy is evident earlier than this in Bulgaria, where in graves near the Lake of Varna, close to the Black Sea coast, dated by calibrated radiocarbon to the third quarter of the fifth millennium B.C., hammered gold ornaments were remarkably numerous. It has been argued, on the basis of their platinum content (Hartmann, 1978), that they were made of metal from an Anatolian source. In Anatolia itself the earliest evidence for goldwork may only be traced back to the much later Troy I (de Jesus, 1980, 88), whilst in Iran no goldwork has yet been reported before the later fourth millennium (Moorey, 1982a).

In the tombs of levels XI-X at Gawra, of the later prehistoric period, the gold jewellery included beads, pendants of insects, combining gold and lapis lazuli (Tobler, 1950, pl.CVI:46), hoof-shaped and spatulate pendants, and gold spirals (ibid., pl.LVIb, CVII:47-8). Gold bands decorated hair-pins and spatulate objects (ibid., pl.CIV:20; CV:29, 31). Rosettes, from levels X-VIII, cut from gold foil, some with pendant ribbons, some with insets of stone secured with bitumen, may have been made to sew onto clothes (ibid., pl.LVIII; CVII:53-7). If so, this is the earliest instance of an enduringly popular means of decorating fine garments in Mesopotamia (Oppenheim, 1949). At Gawra studs of gold sometimes had inlays of semi-precious stones (ibid., pl.LIX:1-5; CVIII:59-63). By level VIII (Late Prehistoric to Early Dynastic I), gold ornaments and pendants have a bitumen core. A sheet gold tube from tomb 109 may have been a cap or ferrule for some kind of wooden handle (Tobler, 1950, pl.LIXa; CVIII:64). Undoubtedly the most significant object, for the extent to which it reveals technical mastery, is an electrum wolf´s head terminal from tomb 114 attributed to level X (ibid., pl.LIXb; CVIII:65). It is 30mm. long and 25mm. high; described as follows:

"The entire head is a single piece of metal with the exception of the ears, the lower jaw, and the teeth. The ears were attached by means of copper pins, long since decomposed into copper-oxide. The lower jaw was carefully jointed into the rest of the head, and held in position by an electrum pin leading from the roof of the mouth, which passed through the lower jaw, where it was bent back. The eye sockets contain bitumen, which presumably held eyeballs of some kind of stone... the mouth is open, showing the teeth which were made of electrum wire finely drawn and sharp. The inside of the head was filled with bitumen, which served together with the socketed neck, and the two pins through the neck, to hold the objects in position" (Tobler, 1950, 92).

Later prehistoric Uruk yielded a sheet gold amulet in the shape of a goat with a bitumen core (Nöldeke, 1936, 14, pl.231).

To Maxwell-Hyslop's comprehensive survey of the historic period little may usefully be added. Although its origin is unknown, a gold pendant of "Gilgamesh" flanked by rampant bulls, probably of the Akkadian period, illustrates a use of the metal, not otherwise well attested, for elaborate ornaments (Margueron, 1965, pl.3: rear view in colour; Contenau, II, 1931, 695-6, fig.486-7). A recent discovery in the temple area at Larsa (Arnaud et al., 1979) is of much greater importance. Here a hoard of gold jewellery, fragments of silver, beads and weights, and tools from a jeweller's workshop, were found packed into a jar. They could be attributed, through associated epigraphic evidence, to the reign of Samsuiluna (c.1749-12 B.C.). This has added considerably to knowledge of jewellery in an important period for which the gold "Dilbat" necklace, in the Metropolitan Museum, New York, had previously been the primary witness (Maxwell-Hyslop, 1971, 88ff., pl.61-4).

For the Neo-Assyrian period the small collection of gold jewellery from Nimrud is an important complement to the unusually rich documentation of personal ornaments provided in this period by relief sculpture (Curtis and Maxwell-Hyslop, 1971; cf. Hrouda, 1965, pls. 8-9).

(ii) Weapons and tools:

The use of gold for the working parts of weapons and tools is only archaeologically evident in the "royal" tombs of Early Dynastic IIIA at Ur. RT 789 contained sets of spearheads, their shafts decorated with sheet gold (Woolley, 1934, 438, pl.153); PG 755 was furnished with gold axeheads and a gold dagger (ibid., 436, pl.152); RT 580 (ibid., 433-5, pl.151) had a gold spearhead and a dagger, as did RTs 1054 (ibid., 450, pl.15b) and 1618 (ibid., 476). Tools of gold, even here, were less common. RT 800, Pu-abi's tomb, had a gold saw and flat adzes ("chisels"; ibid., 443-4, pl.158B), whilst RT 580 (ibid., 432-3, pl.165) yielded an adze, an awl and a chisel of gold. It is possible that RT 580 was also a woman's grave. Mallowan originally reported a horse-shaped cheekpiece from a Neo-Assyrian horse-bit found at Nimrud as "electrum", but later published it as bronze (Mallowan, 1966, 127, pl.70). It has not been analyzed.

(iii) Cylinder Seals:

In Mesopotamia sheet gold cylinder seals are only known from the "royal" tombs at Ur. RT 1054 (U.11904) contained one made of thin sheet gold, decorated with a banquet scene, over a core; it had sheet caps, one now lost (Woolley, 1934, 127, pl.193:21, fig.23). That in RT 1236 (U.12457) was similarly made, but decorated with a contest scene (ibid., pl.197, 217). A related find was made in RT 800 (U.10444) whence came "gold impressions from cylinder seals; 3 fragments of very thin gold foil which has been pressed into seals so as to take the engraving" (ibid., 81, pl.192:10-11).

(iv) Vessels:

Gold vessels are only known from the "Royal Cemetery" at Ur: a strainer; a goblet (Woolley, 1934, 159, pl.160); fluted bowls (ibid., 159, pl.160, 162); spouted vessels (ibid., pls.41, 161, 164); a fluted tumbler (ibid., pls.157, 162); a cup and cockle shell imitation (ibid., pl. 165); an ostrich shell replica (ibid., pl.170). Gold foil overlays on silver vessels are known from the Neo-Assyrian levels at Nimrud (Mallowan, 1966, figs.356-7, 427-30).

A unique object, whose manufacture in some ways parallels that of hammered sheet gold bowls, is the imitation man's hair style in gold from PG 755 in the "Royal Cemetery" at Ur. Commonly known as "Meskalamdug's helmet", this remarkable object, of Early Dynastic IIIA, is made in repoussé with chased details (Woolley, 1934, pls.21, 150).

(v) Foundation deposits

Ellis (1968, 97ff.; 131ff.) describes both the use of gold for inscribed foundation tablets from the Middle Assyrian period onwards and the appearance of gold among the more varied deposits of valuable materials buried, in the processes of building, for magical or votive purposes. Gold foundation tablets may have a much older history. Nabonidus (555-539 B.C.) claims to have found examples deposited by Naram-Sin (c.2254-18 B.C.) (Kupper, 1971, 93, n.9): a strip of gold inscribed with this king's name was reported from Bismaya (Banks, 1912, 145; cf. Sollberger and Kupper, 1971, 112: II A 5a, also on gold).

(vi) Statuary:

Biblical and Classical references (Daniel iii.1; Herodotus I:183) and local texts have sometimes been read so as to give exaggerated ideas about the currency of gold statuary in ancient Mesopotamia. Undoubtedly gold was lavishly used in the ornamentation of temple buildings and all temple furnishings, including statuary; but the extent to which cast gold statues existed it is impossible to say. Although statuettes, less than a metre high, might have been hollow or even solid cast of gold, it is unlikely that it was used in this way, save quite exceptionally, for larger pieces of sculpture. References to gold statuary of any size are likely to have meant ingenious combinations of hollow cast parts, perhaps sometimes even of gilded base metal, and extensive use of sheet gold on a stone or wooden base, as well as semi-precious stones. This is implied by the description in Diodorus Siculus (ii.9.5-10) of the deity statues set up by Semiramis, with their gold plates hammered into shape. A random selection of quotations from the C.A.D. (s.v. ṣalmu) indicate similar methods of manufacture:

> "I cleverly built a likeness of his great godhead out of the choicest quarried stone and scraps of gold"

> "I had a statue of me as king made out of silver, gold and shining copper"

also C.A.D. ṣariru (a poetic term for fine quality gold):

> "(I decorated the divine images) with fifty (or

thirty) talents of reddish gold, the product of the nether world, (from) mined ore, not melted down (from scraps) for this work"

A revealing sidelight on this general question is offered by a passage in one of the "Amarna Letters" (EA 27:23-33) from which it is clear that Amenophis IV (Akhenaten) had replaced solid gold statues (size not given) ordered by his father for Tushratta of Mitanni, with wooden replicas only overlaid with gold: "And now my brother, you have not sent the gold (?) statues, which your father intended to send, but wooden ones, that are overlaid, you have sent".

No gold statuette has yet been reported from Mesopotamia, though one is known, of the Middle Elamite Period, from Susa (0.75m. high; Amiet, 1966, 420-1, pl.319, on a bronze base). Only very rare, tiny deity amulets of gold have survived (cf. Wiseman, 1960, pl.XXIIIe-h). An Early Dynastic III statuette in translucent green gypsum from Nippur has a sheet gold face (Amiet, 1977, pl.3b (in colour); Spycket, 1981, 108-9, pl.71), perhaps once set on a wooden head. Gold foil overlays were used on bronze statuary in the earlier second millennium B.C. (Amiet, 1977, pls.433-4; Spycket, 1981, 232ff. 288, pl.159, 191) and a detached sheet-gold face overlay, of the Neo-Assyrian period, was found at Assur (Andrae, 1935, 107, pl.48h). Sheet gold pendants or silhouettes were current throughout from the later prehistoric period. An early example from Uruk shows a profile face and torso (Nöldeke, 1936, 20, pl.25a), whilst a Neo-Assyrian example, incised with a female worshipper, occurred at Assur (Andrae, 1935, pl.48L). Mallowan (1966, 121, fig.63) found tiny gold silhouettes at Nimrud.

Already at Uruk, in the later prehistoric period, cast gold was used to make the legs of composite statuary, in this case of animals (Nöldeke, 1936, 14, pl.23q-s) and sheet gold, over a core, was used for animal horns (Heinrich, 1936, 40, pl.30d), as also later in the Early Dynastic III temple at Tell al-Ubaid (Hall and Woolley, 1927, pl.5). The famous composite animal statuettes from Ur already cited (p.49) indicate an aspect of Sumerian production, whose fragility meant they are only likely to be evident elsewhere in stray fragments of stone or metal. Stone statuary was also at times decorated with gold. There is a gold plaque, cut to the shape of a "beard" and prepared for inlay into a statue from Djokha (Umma), inscribed with the name of Gishakidu, a contemporary of Enanatum of Lagash in Early Dynastic IIIB (Thureau-Dangin, 1937). A statue fragment, of the later third millennium B.C., from Telloh, is described as having "les poignets...encore entourés par deux minces bracelets en cuivre plaqué d'or" (de Sarzec and Heuzey, 1884-1912, 345). Another from the same site and period has "un collier dont les grains, de matières et de couleurs differentes, cornaline, turquoise et cuivre doré, sont enchâsses dans la pierre blanche" (ibid., 345-6; illustrated Heuzey, 1900, 10, fig.1).

Woolley (1976, 185) commented of his excavations at Ur in the earlier second millennium B.C.:

"So little gold was found that we have no criterion for judging the work of the period, but its mere absence is perhaps a sign that there was relatively little in us for ornamental purposes...though some of the stone objects found in the Giparku ruins had evidently been enriched with gold".

A sheet gold ear from the Shamash Temple at Mari was probably from a statue (Parrot, 1935, 6). Major deity statues in temples would have had attributes; but very few indeed have been recognized in excavations. The most certain is a lightning fork, presumably of Adad, of the Neo-Assyrian period, from his temple at Assur, made of gold hammered over a wooden core (Andrae, 1909, pl.XXXIV). It was reported to weigh 290 grains, 250 of which represent the gold.

(vii) Monumental decoration and furnishings:

Even allowing for a degree of exaggeration, Sumerian and Akkadian texts make clear the significant role played by goldsmiths in providing temples and palaces with innumerable furnishings, containers and decorative objects of gold, as well as wooden furniture, of all kinds, inlaid or overlaid with the metal (cf. Salonen, 1963, 248-50). Temple foundation documents show that the temples were not only lavishly adorned with gold, but that it was also extensively used for the cult equipment, including the decoration of chariots for deities. One document illustrates this particularly well, for it recounts the return to Babylon of the statues of Marduk and Sarpanitum from exile and the restoration of them and their shrines (Unger, 1931, 276ff.). Although written in the name of the early Kassite king Agum II (or Agum-kakrime), it may have been composed in the first millennium B.C. (Brinkman, 1976, 97). In all this work gold was lavishly employed. Of this nothing has survived that gives any realistic idea of precisely what was involved, though it is known to have been a practice already current in the prehistoric period and is regularly apparent in texts (cf., for example, Durand, 1983).

One of the earliest, and best preserved, examples of the use of sheet gold for temple ornament is provided by the "altar" in the Late Prehistoric "Eye-Temple" at Brak (Mallowan, 1947, 93-4, pl.III-IV, XLVIII.4): "The three sides of the podium were decorated with an empanelled frieze composed of fretted bands of white limestone and grey shale encased by a surround of gold strip" (ibid., 32). Mallowan (ibid., 93) commented in some detaial on this goldwork:

"The gold foil encasing the stone strips is bent over at right angles to the same depth as it is wide, 20mm., giving the total depth of the freize as it was orginally when backed onto wood. The gold foil is solid and heavy, and was hammered down onto the wooden back by 20 nails on the top and 22 nails on the bottom strip. It is curious that although no expense was spared on the nails - the stems are of silver and the heads of gold foil - cf. pl.IV.2 - they were very carelessly hammered in, at irregular intervals, and were perhaps added as an afterthought to give additional grip to the gold".

There may have been somewhat similar altars in this period at Uruk to judge from fragments (Heinrich, 1936, 47, pl.34b, f, h, 35a).

Archaeologically speaking such decoration may only be traced through the historical period by a very thin line of scattered gold fragments found in monumental buildings, such as the Archaic "Ziggurat" Terrace at Ur (Woolley, 1956, 188: U.18225-6, 18236); the

"Mausolea" of the Ur III kings at Ur (Woolley, 1974, 100: U.16216, 16257-8); the Kassite palace at Aqar Quf (Baqir, 1945, 14); and the Assyrian palaces, where, for example, the gilded-bronze column overlays from Khorsabad are amongst the most spectacular (Loud and Altman, 1938, 16). Berthelot (1906, 81) reported that some of the "gold" leaf found by Place was more properly described as "electrum". Layard (1849, II, 417) recorded very thin gold leaf on baked bricks; and on the ziggurat site at Eridu Taylor (1855) found fragments of gold, of gold leaf, and gold-headed nails of uncertain date.

(g) Techniques:

As the known goldwork from Mesopotamia is so concentrated in the few "royal" graves of Early Dynastic III at Ur, it is inevitable that what research has been done on the technology of Mesopotamian goldwork has been almost exclusively concentrated on them. This is not so restricting as it might sound, for, as is well known, there are very few techniques of metalworking known to the ancient world that were not represented in this cemetery. Sumerian metal crafts were so developed that the Ur graves may be taken to represent the whole tradition. Only enamelling, niello, wire-drawing and spinning (in the manufacture of vessels) are techniques absent from the "royal" graves and, so far as present evidence goes, also from subsequent contexts in Mesopotami before the Achaemenid period, though they may in some cases have been practiced earlier elsewhere in the Near East.

No goldsmith's workshop, nor certainly identified tool kit, has yet been reported from Mesopotamia. Nor has there been any systematic study of the texts relevant to the goldsmith's practice of his craft. Though information is scattered, it appears to be rather more instructive in this case than for workers in base metals (cf. Limet, 1960, 140ff.; Weisberg, 1967, 60ff.). At Uruk the "Stone-cone Temple" was believed to have been built in Uruk IV over an area that had previously been used for various workshops, among them a "metal foundry" (Lenzen, 1960, pl.3, 4, 39). None of the installations clearly indicate this. The squat vessels with narrow hole-mouths found there have no obvious metallurgical role. A fragmentary gold ingot, only in the vicinity of the "foundry", provides little more than a possible indication of secondary working hereabouts.

Goldwork generally requires few tools: a wooden block, a hammer, and some kind of anvil, with chisels to cut, awls to pierce, punches, gouges and burnishers. All native gold is very malleable and extremely ductile, easily stretched, beaten or drawn thin. The metal is worked cold, but requires annealing, as it quickly becomes brittle. It may be hammered into sheets of less than 0.0025m. thickness without annealing. The addition of copper to gold makes the alloy more durable and less subject to abrasion, but at the same time less ductile and malleable. Pure gold melts at 1036°C; gold with 10% silver melts at about 1048°C; gold with about 10% copper at about 925°C; gold with 20% copper at about 880°C. Ogden (1982) provides a well illustrated general review of goldsmithing techniques in antiquity, though little of the evidence used is from Mesopotamia.

(i) Hammered sheet gold and gilding:

Before proceeding to any discussion of gilding it is necessary to draw attention to a basic distinction commonly made between "true gilding" and "depletion gilding" (or gold colouring) (cf. Lechtman, 1971, 2ff.; also, in general, Alexander, 1979). In true gilding gold is applied to the surface of another metal; in certain circumstances the two metals may undergo interalloying, with the result that there is complete metallic bonding. In the second case the gold is already alloyed with another metal and it is a question of internal changes. Depletion gilding is based upon chemical reactions that occur only at the surface of alloys: the object is to remove sufficient of the alloyed metals for the surface to be enriched in gold and eventually appear golden. The two processes involve major technical distinctions, with true gilding being the commonest in ancient Mesopotamia from its appearance in the later prehistoric period. With true gilding the gold may be applied in its solid state as sheet or leaf or as an amalgam. Lechtman (1971, 2) "arbitrarily" defined gold sheet or foil as greater than about one micron in thickness, whilst leaf constitutes thickness smaller than this. Since no systematic measurements have been taken of Mesopotamian hammered gold, there is no accurate way of applying this fine yardstick, so terms are used here without any assumptions about thickness.

Gold was beaten into very thin sheets in Mesopotamia from its earliest appearance there in the fourth millennium B.C., probably on flat hard stones with stone hammers; a rounded tool being better suited for this purpose than a flat-faced one. Ogden identified such a tool from Ur: "...like a tiny pestle, slightly bulbous at one end and measuring only 19mm. in length. The material is haematite". (Plenderleith, 1934, 295; Ogden, 1982, pl.4:3, 35); but it seems hardly large enough for the purpose. There is no definite indication of the use of the "goldbeater's skin", although Plenderleith (1934, 295) argued that use of gut for harp and lyre strings at Ur, in the middle of the third millennium B.C., indicated a mastery of this material which might well have been turned to use in making such a "skin". The earliest type of gilding depended, not on a physical or chemical bond between the gold foil and the base metal, but only on a mechanical folding of the foil over the edges of the underlying objects. The joint might be improved by an adhesive or the overlapping edges joined by burnishing them together. A series of silver nails from the Late Prehistoric "Eye Temple" at Brak (see p.120), examined by Oddy et al. (1981), had heads gilded by wrapping gold sheet over the silver. Of "gold" beads from the "Royal Cemetery" at Ur Plenderleith (1934, 295) wrote "beads of bituminous shale, wood, etc., have been found covered with gold which is so thin that when removed and held by one end it hangs down, possessing insufficient rigidity to remain horizontal." This technique was long and widely used, and not only for beads. Early Dynastic III gold bracelets were sometimes made by hammering gold over a core which was then withdrawn (Maxwell-Hyslop, 1971, 66). A gold bracelet in the Ur III grave PG 1847 at Ur is made of thin sheet gold over a copper core (Maxwell-Hyslop, 1971, 69). Two millennia later at Ur a sheet gold pin, with a head modelled as a draped human figure, was made over a wooden core (Woolley, 1962, 106, pl.21, U.456). The sheet gold overlays were themselves sometimes ornamented. The spear shafts from PG 755 at Ur, for example, have gold overlays worked to represent bamboo (Woolley, 1934, pl.154 U.10023).

Sheet gold, too thin to support its own weight, requires a

method of attachment. This may be either adhesive (perhaps one based on egg white) or the leaf may be attached to fairly pure silver or copper by burnishing, provided the metal surface is scrupulously clean when the gold is applied. The adhesion of the gold is enhanced by application of low-temperature heat to promote inter-diffusion with the underlying metal ("diffusion-bonding" or "cladding": Oddy et al., 1981). Such was the prime means of gilding until at least the Achaemenid period.

When fire-gilding first appeared in the Near East is not yet clear, though it was used in Greece by the fourth century B.C. (Craddock, 1977, 109-110; Healy, 1978, 190ff.). It has been suggested that it was an innovation of the Achaemenid period (Cooney, 1964, 75); but hard evidence for this is not yet available. As the new technique was quicker and more effective for larger and more irregularly shaped objects, it superseded earlier methods. In fire-gilding an amalgam of mercury and powdered gold is spread over the surface of the silver, which is then heated to vapourize the mercury, leaving the gold behind in contact with the silver. The earliest analytically proven example of this technique is from Late Bronze Age Ireland (Craddock; pers.comm.). Woolley (1983, 24) found metallic mercury at Al Mina in level III (c.430-375 B.C.) and believed it to be a traded commodity at this time. It was also found in one of the Achaemenid palaces at Susa (Labrousse and Boucharlat, 1972, 86-7). Cinnabar, native mercury sulphide, was available near Kirkuk according to Campbell-Thompson (1936, 29 30), though his translation of the Akkadian kalgukku as "mercury" has been superseded by the rendering "red ochre or similar mineral" (Oppenheim, 1970, n.58). It was also available at various places in Iran (Labrousse and Boucharlat, 1972; Allan, 1979, 1-2) and in Anatolia (Ryan, 1969, 69-72).

Depletion gilding may only be identified through laboratory examination and there has been little of that. It also has to be borne in mind that apparent depletion gilding may be no more than the result of selective corrosion whilst buried in the earth. Plenderleith (1934, 292) reported on the spearhead, U.9122, from RT 580 at Ur (Woolley, 1934, pl.227), as follows: "This specimen had been broken in two by an unfortunate accident...; the fractured surfaces were white, granular and crystalline, whilst the surface of the object was such a rich yellow as to suggest that it was gold of a high carat. The metal proved on analysis to contain only 30 per cent of gold with about twice as much silver the remainder being copper". Smith (1974, 160) has taken this to be the earliest documented example of an attempt to improve the colour of the electrum by the removal of natural alloying constituents, primarily silver, to yield a surface of pure gold. The silver, he points out, might have been removed by cementation (hot) with chlorides or by a (cold) parting reaction. Cementation, done hot, "intrinsically depends upon solid state diffusion of the base metal to the surface of the alloy where it can chemically react and be removed. In most cementation mixtures the base metal diffuses to the surface nearly as rapidly as it can cross it so that a rich gold surface can only be achieved by depletion to a considerable depth". It is usually argued, that parting, a selective corrosion process, could not have been done before the discovery of mineral acids in the Middle Ages. Although there are alternatives (e.g. ferric sulphate), there is no evidence for their use in the ancient Near East. Smith believes there was a probable connection between such early attempts to improve the colour of electrum, as evident in this third millennium spearhead, and the discovery of processes that later became the basis of industrial refining. Certainly, on present evidence, it precedes in date evidence for the refining in bulk of naturally occuring gold to remove the silver content (see p.71 here).

The shaping of sheet gold vessels differs in no significant way from the methods employed for base metal (see p.46). Smith (1972, 124) has given a graphic description of how one of the fine mid. third millennium gold beakers from Ur might have been worked:

"This probably began as a thick cake of gold, which was hammered to a round flat sheet, from which the main cup shape was made by raising, that is, by a sequence of local hammer blows from a slightly convex hammer against the sheet metal held at a slight inclination against the rounded edge of an anvil, so producing innumerable little bends in concentric circles or in a close spiral which eventually integrate into the formation of a progressively deepening cup. Because metal is hardened by cold work, it would have been necessary to soften it repeatedly by annealing to allow the work to proceed. The fluting originated by local punching i.e. a combination of repoussé and chasing done, not against a hard anvil, but against an internal filling of pitch or some other material of strain-rate-dependent properties which provides moderately strong distributed support for the metal under a local blow, followed by slow relaxation to conform to the new shape in preparation for the next stage of deformation".

The use of sheet gold for animal-head fittings, as on the musical instruments from the "royal" graves at Ur, and as parts of composite statuettes, as for the pair of rampant he-goats from RT 1237 at Ur, presents special problems. These were made from thin sheets, in the case of the heads probably hammered first into a bowl form, and then shaped to final form with a hammer and chasing tools. Maryon (1949, 99-100) has described a possible method, implicitly criticizing Plenderleith's (1934, 297) original proposals:

"To do this it is quite possible that the gold was laid upon a boss of wood covered with bituminous compound, which would support the gold mask while it was being chased to shape. I do not feel that the craftsman would have spent his time in carving the complete mask in wood as a preparation and backing for his goldwork. That would not be the way of a metal worker. He would prefer complete freedom for repoussé work rather than the restricted certainty of the carved model... In a number of surviving heads the bitumen backing still remains within the thin gold. I do not know of any example where a wooden form or core comes immediately next the gold". Ears and horns were often fitted separately. This assumes that the bitumen formed part of the forming process and was not just poured in afterwards to support the metal as seems to have been the case with bronze animal heads at Tell al-Ubaid (see p.49).

Plenderleith (1934, 296ff.) found much of the finer decoration on the Ur tombs goldwork to have been done by chasing rather than by engraving. On the diadem (U.8173: Woolley, 1934, pl.139) "all the outlines of the figures are impressed by hammering the flat edge of the chasing tool or die on the gold". Real engraving was noticed on a gold bowl from PG 755 (Woolley, 1934, pl.160: U.10003) "where the fluting has been touched up by the engraver whereas the pattern round rim and foot is chased". Inlaying is also first observed in Early Dynastic III at Ur, when strips of gold were hammered into engraved lines on spearheads. This technique was used two millennia later by Assyrian, or possibly Levantine, craftsmen to inlay gold scarabs on cuboid bronze weights (Layard, 1853, 196, figure).

(ii) Cast gold objects:

In view of the greater quantities of metal required, the role of cast gold in Mesopotamia is likely to have been largely confined to objects in the royal or temple service or to small pieces of jewellery like the copper or bronze pin from PG 755 with a cast gold monkey as its head (Woolley, 1934, 300, pl.115: U.10010). Here again, as with sheet-metal working, methods for the most part would have been those used also for base metals (see pp.40ff.). Maryon (1954, 626) has described the casting of some well known objects from the "royal" tombs at Ur:

"Three- or four-pieced moulds were in regular use. Examination of a little gold monkey surmounting a bronze pin (PG 755), of the electrum onager on Queen Shubad's (RT 800: Pu-abi) rein-ring, and of the silver bull on the king's rein-ring (RT 789) suggests that they were probably first modelled in wax, but then, like the axes, cast in piece-moulds. The monkey and the onager, as their surfaces indicate, were thoroughly "chased" (i.e. cleaned down) after casting, to remove all the casting webs and rough surfaces. The silver bull, however, was not smoothed, and the original cast surface can be seen. Surviving parts of the webs outline a piece of the mould which covered most of the right side of the bull's body. There would be other pieces also. Rough evidence is available to show that the wax or clay in which the bull was modelled could have been removed piecemeal when the side piece was lifted away from the rest of the mould".

(iii) Filigree and cloisonné:

Filigree, the ornamental use of fine wire to make delicate tracery, is used on the famous gold dagger from RT 580 at Ur; but shown at a higher level of technical achievement on a double-conoid bead from the same tomb (U.9779), again combined with granulation (Woolley, 1934, pl.138). Alone, or combined, filigree regularly appears thereafter on the surviving gold jewellery from Mesopotamia, notably on the fine Ur III Abbabashti necklace from Uruk (Maxwell-Hyslop, 1971, pl.45; Nöldeke, 1937, 22ff., pl.39); in the "Larsa Hoard" (Arnaud et al., 1979); on a Kassite bracelet from Aqar Quf (Maxwell-Hyslop, 1971, pl.127); in the remarkable collection of Middle Assyrian jewellery from grave 45 at Assur (Haller, 1954, 140ff.; Maxwell-Hyslop, 1971, 169ff.); and on the so-called "Nimrud jewel" (Curtis and Maxwell-Hyslop, 1971, 102ff., fig., pl.XXXII). From the earliest period gold wire had been used in simpler ways for beads and pendants, as is well illustrated by examples in the "Royal Cemetery" at Ur (Maxwell-Hyslop, 1971, 10-11, pl.B (colour), 9-11).

The manufacture of gold wire throughout the time of this survey was without use of the drawplate, not yet certainly known before the

European Dark Ages (Oddy, 1977; 1984). Carroll (1972, 323) has argued for a neat dichotomy in the early methods of wire manufacture: "Egyptian smiths, and possibly other Near Eastern smiths, drew wire from a straight strip. Greek and Etruscan smiths twisted the strips before drawing them." Oddy (1977 with important illustrations; also 1979; 1984) challenged this simple distinction and offered clear descriptions of the various ways gold wire might have been made before the draw plate was invented. The concomitant, detailed examination of actual examples from Mesopotamia remains to be done. Oddy´s categories were:

1. Hammering: in this, the simplest and most basic (though not necessarily the easiest) method, an ingot is hammered out until a wire with more or less round and even section is produced. The wire, if not polished, has a facetted surface.

2. Block-twisting: this is done by hammering out an ingot to give a rod of the required thickness for the wire, twisted as tightly as possible and rolled between two flat pieces of wood (cf. Plenderleith, 1934, 296).

3. Strip-drawing: this involves taking a strip of metal foil and drawing it through a series of holes of decreasing diameter so that the strip curls in upon itself to form a hollow tube. A steel draw-plate is not needed; the holes could be pierced in a hard wood or bone.

4. Strip-twisting: a strip of metal foil is cut and wrapped round an existing wire which is then removed. This helix is then tightened and gently extended by hand. This type of wire might also have been drawn through dies to finish it off.

A particular application of gold wire is illustrated by third millennium gold chains. Plenderleith (1934, 296, pl.134) examined the brim (male) headdress U.8693 from PG 429 at Ur and commented:

"...consisting of two lengths of gold chain and three large beads of gold and lapis divided by carnelian rings; but whereas the chains seem to have interlocking links giving a square section of traditional form (loop-in-loop) they are not really chains at all, not even flexible, but lengths of cut wire twisted on themselves and soldered together into an imitation which without careful examination would pass muster as the genuine thing".

A piece of fine gold chain was recovered from the area of cemetery "A" at Kish, but without precise context, though probably Early Dynastic III or early Akkadian in date. Mackay described it as follows:

"It is 18.70cm. long and 5mm. thick, and is built up of wire links of such a form as to make the chain square in section. The wire of which the links are made is slightly over 1mm. in diameter and uniform in thickness. On examination under a magnifying glass the wire showed the characteristic longitudinal grooving associated with drawn wire..." (Mackay, 1929, 182, pl.XLIII.3, 5). Mackay suggested drawing "through a carnelian bead". As Oddy has demonstrated, such lines need not be evidence of "drawn" wire in the strict modern sense of the term and the Kish example fits well into his scheme for ancient wire making. A piece of gold chain, of what appears to be comparable type, if slightly later in date, was reported from old French excavations at Asherah (Terqa) in 1923 (Dhorme and Thureau-Dangin, 1924, 288, pl.LX.8-9). A recent attempt to elucidate the technology of third millennium loop-in-loop chains through an art market example, purchased as Anatolian, raises more questions than it solves (Athanassopoulos et al., 1983).

Already in the Ur "royal" tombs *cloisons* were soldered to a base-plate to form cells for inlay on finger rings (Woolley, 1934, pl.138: U.9778) and there is a particularly well-preserved example of this technique of the same mid. third millennium B.C. date from Telloh. This finger-ring is inlaid with lapis lazuli and carnelian (de Sarzec and Heuzey, 1884-1912, 391, pl.44 ter., fig.3a, b; Margueron, 1965, pl.43 (in colour)). Silver hair ornaments from RT 1237 at Ur, terminating in inlaid flowers, are more ornate illustrations of the same technique in Early Dynastic III (Woolley, 1934, pl.136). It became a technique recurrent through the history of Mesopotamian gold working down to magnificent examples produced in the Achaemenid period. Then, and perhaps earlier in the Kassite and Middle Assyrian periods (Maxwell-Hyslop, 1971, 167), it is possible that sporadically closer contacts with Egypt, where this technique was typical, stimulated the use of *cloisonné* work in Mesopotamia and Iran.

(iv) *Joining methods*

Archaeological literature contains many confusions over the terms *soldering*, *brazing* and *welding* that have not been eased by technical experts introducing such words as sweating, fusing and autogenous soldering. An important fundamental distinction has first to be drawn between soldering (whether "hard" or "soft") which involves an agent (the solder), and welding, which does not. As soft soldering is not relevant to gold working, it will be discussed elsewhere (see p.127); but it may be usefully defined here as the employment of lead, or occasionally tin, rarely tin-lead before late Roman times, with a melting point which is below $427^{\circ}C$ ($800^{\circ}F$), that is below the temperature at which the work is done, and also below the melting point of the metals to be joined. "Soldering and brazing are joining techniques employing filler metal at temperatures below solidus of the metals being joined. Conventionally, soldering refers to processes below $450^{\circ}C$; brazing to processes above $450^{\circ}C$ (Fell, 1982).

To promote flow of the braze or solder a liquid flux is required. Pliny was the first to describe specific fluxing agents (oils and fats, wax and honey, and artifically prepared products from dung and urine, etc.); but they are extremely difficult to detect, with the possible exception of borax, in corroded ancient joints. Fell (1982) has tested a whole series of potential fluxes listed in Classical and Medieval literature for low-temperature and for high-temperature work; but his results cannot at present be directly correlated with Mesopotamian evidence.

Brazing (or *hard soldering*) is "the production of a joint by means of an alloy of reasonably high strength having a melting point somewhat lower than that of the metals or alloys to be joined" (Roberts, 1973, 113). Brazers form molecular bonds with the surface they unite and are therefore stronger than soft soldered joints. Whereas soft solders are generally greyish white in colour, brazers may be made to match closely the colours of the metals to be joined. Plenderleith (1934, 296) and Maryon (1949, 108) suggested that the goldsmiths at Ur used a native gold alloy:

"The highly decorated beads like U.9779 (Woolley, 1934, pl.138), are generally sweated, i.e. the filigree and granulations are attached to the background by carefully controlled heat without the extraneous medium of solder; but the flat disk beads made from two thicknesses of metal joined together so as to leave a hole for the thread through the centre were apparently soldered (Type 16, Woolley, 1934, fig.70). Apparently the gold beads of tubular form, either barrel-shaped or in the form of two cones, base to base, all have a longitudinal soldered seam; the solder used would be either a thin strip of the same gold (autogenous soldering) or an alloy richer in silver and consequently of a lower melting-point than gold" (Plenderleith, 1934, 296).

The selection of such a brazing alloy for the manufacture of gold joints would have been a matter of some complexity. The extremely high melting point of native golds make them difficult to use as brazers, whilst native gold alloys containing enough silver to be of use for brazing are distinctly lighter in colour. But the striking feature of some Ur objects is the very uniformity of colour in the area of joins. This may sometimes, if not wholly, be explained by the disappearance of metals other than gold from the surface of the brazing alloy through the action of chemicals in the earth among which it was buried ("surface enrichment"), as Maryon argued (1949, 114). He also cited the view of Littledale that "when a joint is overfired the solder penetrates further and further into the surrounding metal, alloying itself with it, and becoming more like it in quality, till the colour of the solder is indistinguishable from that of the work". Smith (1930, 22) and Plenderleith (1934, 296) explained this lack of colour difference by the use of a piece of metal of the same alloy as the pieces they joined. Even when the ancient craftsmen had solved the problems both of colour and melting point, they still had to control the flow of the brazer, which tends to flow towards the hottest area of the parts being fastened. With a charcoal fire and blow-pipe or bellows, the amount and direction would have been very hard to control successfully.

Roberts (1973, 115) has described early brazing technique, suggesting natron (sodium carbonate) as the most likely flux:

".....small pieces or chips of the chosen brazing alloy would then have been laid in position, or held in position by small pieces of wire, around the joint to be made, a flux of some kind would have been applied to the surfaces to be joined, and the work put into a charcoal fire - or held by tongs in the fire - until the brazing alloy became molten. The flame was then concentrated on the joint, with the reed blowpipe, and the brazing alloy would flow by capillary attraction into the space between the surfaces to be joined. Later the metalworker would clean off residual traces of flux from the work, probably by scraping."

Roberts examined in detail a pedestal bowl (U.10452 (BM 121345): Woolley, 1934, pl.160a) from RT 800 at Ur, which had held green eye-paint. It is electrum (25% silver) with a double-walled upper part, and a brazed joint round the junction of bowl and pedestal foot. There is a mechanical joint at the rim. On a bowl from the same tomb, with a twisted wire handle, the fillet of the brazed joint between the tubular lugs and the body of the vessel is still clear (cf. Lang and Hughes, 1977, 169: U.10851 (BM 121344), Woolley, 1934, pl.161 b).

It is necessary to say that "cold welding", in the modern sense of pressure welding, is not practical under ordinary conditions and is most unlikely to account for any ancient solderless joints. Only the gold beater, who makes gold leaf for gilding, may be accounted an exception, for he repairs a defective leaf by laying a fresh piece of gold over the gap and welds it into position by hammer pressure alone.

(v) Granulation:

Granulation is the most studied (Wolters, 1983), and consequently one of the most controversial, aspects of ancient jewellery, for it may not be assumed that only one method was used throughout antiquity nor that variations are predictable, as has sometimes been done in the past in the large literature on the question (cf. Carroll, 1974). In granulation the craftsman arranges small grains of metal in an ornamental or figurative pattern on a metal surface, with the granules held in place by joining them to the base (Wolters, 1981). Many examples of granulation showing remnants of solder were obviously produced with solder alloys and fluxes, probably in no way different from those used for other contemporary goldwork described above (metal alloy granulation). Modern experiments to replicate such work have used a gold, silver and copper solder for gold, and a mixture of silver and copper for silver (Carroll, 1974, 35). But granulation without evident traces of solder must have been produced in some other way. The extent to which sintered granulation was used in ancient Mesopotamia remains a matter for detailed investigation. In this method a join was made by heating the metal surfaces to be joined to melting point (without reaching the point at which the bulk of the metal would become liquid), so that both surfaces are enriched in copper and silver relative to the insides. Carroll (1974, 36) has demonstrated that such a method is feasible with simple equipment, since temperatures can be estimated by observation of the visual changes at the surface (but cf. Wolters, 1981, 125-6). A grain ring-bead from the "Royal Cemetery" at Ur may be the earliest recognized example of sintered granulation (Plenderleith, 1934, 297; cf. Seeley in Maxwell-Hyslop, 1977, 86); its date, however, is not certain.

Much early research into methods of granulation rested on the false assumption that only one technique had been used in antiquity. This, it was argued, had been "lost" and then "found" again by modern research, notably through the experimental work of H.A.P. Littledale, patented in 1933 (cf. Maryon and Plenderleith, 1954, 657). The key to this method was the use of copper-salt as solder (copper-salt granulation); a method, of greater adaptability than sintering, which might be used both for granulation and filigree, and achieved bonding through surface alloying. Finely powdered copper-salt is laid on the surfaces to be joined, with carbon as the reducing agent. Under heat, the copper-salt is reduced to metallic copper. At sufficiently high temperature it alloys with gold in the surface zones and lowers its melting point. "The carbon monoxide from the charcoal fire, and also from the reduction of the copper oxide, will not only serve as a reducing agent to remove any oxide films on the work but also prevent access of oxygen to the parts being heated. Thus the parent metal surface is free from oxides and the prime requirements for the production of a brazed joint have been satisfied" (Roberts, 1973, 117). Wolters (1981, 127) has argued that, apart from the very obvious use of metallic solders, bonding with non-metallic solders in this way, is the only feasible method with ancient heating resources and produces all the technical characteristics of ancient granulation. As small metal particles tend to assume a spherical shape in melting, their formation would not have presented a problem. They would have been positioned with the use of an organic adhesive before proceeding with the joining process.

As Carroll (1974, 38) pointed out, care is needed in identifying granulation. Among the gold objects in the "royal" tombs at Ur, are the dagger hilts which appear to be decorated in this way, but are not (Woolley, 1934, pls.151, 155, 157, 190). The "grains" in these cases are the heads of tiny nails, possibly made by heating the tip of a wire until it became molten and formed a ball.

(h) Gold Analyses

1. Tepe Gawra: gold bead from tomb 109; Late Prehistoric period (Tobler, 1950, 88, n.30).

   Cu: 0.56%
   Ag: 61.39%
   Au: 38.05%

2. Ur: A: all Early Dynastic III "Royal Cemetery", from table in Woolley, 1934 (Table III); analyses by Plenderleith, Johnson, Matthey and Co.; and E.C. Padgham.

	Gold	Silver	Copper	Approx. carat
1. Spear head (U.9122)	30.30%	59.37%	10.35%	7
2. Ass on rein-ring (U.10439)	65.60%	31.45%	2.65%	16
3. Mes-kalam-dug "helmet" (U.10000)	--	--	--	15
4. Plain beaker (U.10452)	75.62%	23.34%	1.04%	18
5. Fluted beaker (U.10453)	73.48%	24.73%	1.79%	18
6. Round bowl (U.10002)	58.16%	40.04%	1.80%	14
7. Oval bowl (U.10001)	60.63%	37.07%	2.30%	15
8. Fluted bowl (U.10003)	72.56%	25.44%	2.00%	17
9 "Lamp" (U.10004)	60.00%	38.60%	1.40%	14
10.Dagger (U.10020)	91.11%	7.69%	1.20%	22
11.Headdress:leaf (U.number absent)	67.17%	32.38%	0.45%	16

12.Ribbon (U.12388)	77.84%	21.74%	0.42%	19
13.Hair ribbon (U.number absent)	75.11%	24.5%	0.32%	18
14. ibid. (U.number absent)	77.77%	21.16%	1.07%	19
15. ibid. (U.number absent)	77.44%	22.06%	0.50%	18
16. ibid. (U.number absent)	77.77%	21.16%	1.07%	19
17. ibid. (U.number absent)	70.91%	27.52%	1.57%	17
18. ibid. (U.number absent)	77.44%	22.06%	0.50%	18
19. ibid. (U.number absent)	71.36%	27.86%	0.81%	17
20. ibid. (U.number absent)	83.95%	13.95%	2.10%	20
21. ibid. (U.number absent)	76.98%	20.72%	2.30%	18

B: i/6-granule gold ring, said to be from RT 800 (Pu-abi): (Seeley in Maxwell-Hyslop, 1977, 86); this provenance and the early date are not absolutely certain.

Granule 1: 100% Au.

Granules 2-5: 80-90% Au.
9-20% Ag.
0.5-1% Cu.

ii/Fragment of gold strip with attached granules, also said to be RT 800 (Pu-abi):

Granules: 77% Au.
18% Ag.
5% Cu.

Base: 88% Au.
10% Ag.
2% Cu.

VI: IRON

(a) Types and ores

Iron, strictly speaking, is the pure metal (the chemical element Fe) never encountered in antiquity; steel is "iron that has been combined with carbon atoms through a controlled treatment of heating and cooling" (Congdon, 1971, 18). Iron is the only common metal in which carbon can dissolve; but, since charcoal (i.e. carbon) was vital to provide the heat needed for ironworking, the fact that some carbon went into the metal was not realised until A.D. 1786 (Smith, 1981, 33ff.). Until then craftsmen believed that steel was merely a refined form of iron (Wertime, 1980, 5). Carbon atoms find spaces between the layers of atoms in the iron crystals so they move through the hot metal even when it is solid, far below the melting point. Thus carbon may be added or removed at the surface of the metal, whilst longer heating will make the metal more homogeneous. As pure iron has a melting point of 1540°C, and primitive smelters could not reach much above 1200°C, it was not possible to melt this metal in the period considered here. However, when iron is alloyed with 4% carbon, it can be made to melt at 1150°C, though the resolidified metal is very brittle. Inevitably, perhaps often, cast iron would have been made by accident in smelting hearths that were producing wrought iron. Yet it does not appear in the Near East as a material valued for its own sake until relatively modern times.

Wrought iron, steel, and cast iron are in almost every sense three distinct materials, each needing a different treatment, each best adapted to different purposes. But they are all over 95% metallic iron; indeed wrought iron and steel contain over 99%. The difference between them is due mainly to varying amounts of carbon. Wought iron or low carbon steel (less than 0.5% carbon), is a soft, ductile product with small amounts of slag; mild steel begins at about 0.8% carbon; a high carbon steel runs to about 1.5% carbon.

".....unless the speaker is absolutely certain that an object he is talking of (particularly if it is early or pre-Roman) is true steel, he should avoid the use of the term ´steel´, for the sake of accuracy..... It is far more accurate to term the products dating from about 1000-200 B.C. as being carburized iron, or, if one is certain of his ground, as being case-hardened, or semi-steel. As for iron products before 1000 B.C. or better still, 1400 B.C., unless the sample has been analyzed, it is safest to refer to it as iron or wrought iron" (Congdon, 1971, 27). The marked advantages of steel over bronze and wrought iron are its mechanical strength and ability to hold an edge.

Native iron is rare, found chiefly in meteorites; but not all meteorites contain iron in readily available form. Only siderites are made almost wholly of iron; aerolites are largely stone and siderolites are a mixture of iron and stone, with the metal like raisins in a cake. By hammering with stone tools or prying with copper chisels small flakes of iron may be detached for use (Bjorkman 1973). Meteoric iron almost invariably contains the metal nickel in proportions varying from about 5 to 26%, but usually in the range 7 to 8%. Nickel appears only in very small quantities, if at all, in terrestrial iron and its ores, so it has commonly been used to distinguish celestial from terrestrial iron in ancient artefacts (Piaskowski, 1982, challenges this view).

It should be noted that there are some ancient iron objects, notably an axe-blade of the Late Bronze Age from Ras Shamra in Syria (Schaeffer, 1939, 107ff.; Richardson, 1943; Waldbaum, 1980, 76, nn.7,22) that have a significant nickel content claimed to be too low to be meteoric in origin; in the Ras Shamra axe-blade there is 2.5% nickel. It is not yet known either how this was made or from what. It might possibly be a mixture of terrestrial and meteoric iron. However, old analyses of meteoric iron tended to yield too low a nickel content, which might also be affected by weathering. New investigation is required.

In many areas terrestrial iron ores are far more readily available than copper; indeed they are among the most common on earth. Many lie so close to the surface that they are easily recoverable. Nor is a more or less pure mineral required for workable iron, since many rocks of high iron content can be smelted. Red oxide (haematite) and the hydrated form (limonite) both contain a high proportion of iron and, as they are relatively free from sulphur and phospherous, are easily worked to give good iron. The magnetic oxide (magnetite/-lodestone) and the carbonate (siderite/chalybite or spathic ore) are also easily worked. The sulphide ores (pyrites or marcasite and pyrrhotite) are more problematic, since any sulphur left in the metal affects its working properties and use. Most ironstones contain one of the oxides of iron, but with siliceous, argillaceous or even partly calcareous matter, so the yield of metal is very low by comparison with the purer minerals. Iron ores usually contain 30-40% of unwanted materials.

(b) Production:

Many of the techniques developed in the previous three or four millennia to produce and manufacture copper and its alloys do not apply to iron. Iron could not be smelted to a liquid metal in primitive furnaces; iron has to be red hot before it is malleable; iron's impurities have to be squeezed out by hammering at red heat; and it may only be worked effectively with tongs to hold the incandescent metal (above $800^{o}C$).

In ancient Mesopotamia, so far as present evidence goes, iron was not subject to cold working, to raising and repoussé, to punching, tracing, chasing and similar techniques, seen to best advantage in gold and silver. But, as such techniques have now been recognized in level IV (c.1000-800 B.C.) at Hasanlu in northwest Iran, this may just be a fortuitous gap in the evidence for Assyria (Pigott, 1980).

It is still something of a mystery how smiths came to smelt iron ores systematically, for they require a much more precise firing process than copper. The ore had to be completely surrounded with charcoal and the fire fed with a blast of air; but the iron, if subject to air, might re-oxidize; if exposed to too much carbon, it would become hard and brittle. Nor is understanding of the matter eased by the absence of archaeological information on iron-smelting furnaces in the Near East before the Classical periods (cf. Tylecote, 1980, 210ff.; Waldbaum, 1980, 88ff). Petrie's "iron smelting furnaces" at Tell Jemmeh in Israel are an enigma; they might be iron forges (Stech-Wheeler et al., 1981, 260) as there is related evidence for blacksmiths' workshops, probably military ones, at Tell esh-Sharia and Mesad Hashavyahu in Israel in the seventh and earlier sixth centuries B.C. Early iron working methods in the Near East have then still to be reconstructed from a blend of analogy (with primitive furnaces both ancient and modern) and the indications provided by metallographic examination of early iron artefacts from Cyprus (Tholander, 1971; 1982), Syro-Palestine (Stech-Wheeler et al., 1981) and Iran (Pigott, 1980). Actual smelting operations must have taken place where ample supplies of fuel were readily available, as it takes about 8 tons of charcoal to smelt 1 ton of iron ore; though this is considerably less than the 20:1 ratio for copper (Horne, 1982).

When, as in this case, archaeological evidence is not available, valuable evidence on primitive methods is to be found in the publications of nineteenth century travellers to parts of the Near East where technology had changed little in the previous three thousand years. Such an account exists for iron smelting in Turkey:

".....I had brought no interpreter with me, and it was some time before they could be made to understand that I wished to see the mines from whence the iron-ore was extracted. To this they replied that there were no mines, but that the ore was found everywhere about the hills near the surface. This they proved by scraping up the soil near their hut with a mattock, and collecting small nodular masses, which I understood was the form in which it is universally found in this district..... The ore is poor, and the miners.....must lead a hard and laborious life; they are at the same time charcoal-burners, for their own use; removing their huts and forges to a more productive spot, as soon as they have exhausted the ore and consumed the wood in their immediate vicinity.

The mode of extracting the iron is however very slow and laborious; the ore is smelted in a common blacksmith's forge, in which 180 okes* of the rude material produce three batmans or pigs of metal, weighing six okes or thirteen pounds and a half each; consequently the ore only yields ten per cent of metal, and to procure this small quantity 300 okes of charcoal are requisite. The blast of the furnace is kept up for 24 hours, during which time the mass must be constantly stirred, and the scum and scoria raked off, after which the melted iron is found at the bottom, which, from the specimen I saw, appeared of a very good quality.....we passed the remains of several forges in places where the ore had been completely worked out, and where the ground was strewn with ashes...." (Hamilton, 1842, I, 275ff.).

To produce sponge iron from the ore charcoal was layered with chunks of iron ore, commonly haematite or magnetite, in a furnace, at first probably no more than a hole in the ground, or in rock (bowl furnace) (Tylecote, 1980, 209ff.). Air blast from a bellows was used to raise the temperature. Under firing conditions the charcoal oxidized, forming carbon monoxide, which rose through the ore and reduced it. As reduction took place at about 1200°C, well below the melting point of iron, the melt did not liquify, as does copper, but emerged as a spongy mass in which unreduced ore and fuel fragments were caught up (bloom iron). The non-metallic substances arise from the combination of ferrous oxide and silica gangue in the reduction process; the commonest is fayalite. The smith took a mass of spongy iron from the furnace, reheated it at a forge and squeezed out the fayalite by hammering it. The smith then had a tough, malleable bar of wrought iron. From this raw material objects were made by further heating and hammering.

In this state the metal (bloomery iron) is soft and no substitute for well worked tin-bronze, which can be cast and work-hardened, and will only corrode slowly. Iron rusts rapidly, often suffering considerable damage as it does so. The key process in transforming iron from an unacceptable to an acceptable substitute for bronze is the method of hardening known as carburization (see diagram in Congdon 1971): the absorption of carbon into the metal, thereby converting iron into steel. To carburize, or steel, an iron object it was heated in close contact with charcoal in temperatures in excess of 900°C. The amount of carbon absorbed into the iron depended both on the temperature of the fire and the length of time the metal itself was left in it. The ancient smith could not have known

---

* The oke (or okra) is a Turkish measure of weight equivalent to about 2 lbs.

exactly when the appropriate proportion of carbon had been absorbed; but experience had taught him to assess the heat of the fire, from its colour, and the length of time a piece of iron needed to remain in it to provide the required finished product.

Fast cooling (quenching) is a further process for strengthening iron by plunging it into cold water; a process of no significance in copper or bronze manufacture, or for iron without carbon in it. Quenching decomposes the austenite (a solid phase of carbon in steel existing over 710°C) into hard, brittle martensite; if cooled more slowly it produces pearlite, as the austenite decomposes into alternate lamellae of ferrite and cementite, with a "pearly" appearance under an optical microscope. Quenching may be followed by tempering. At one time this term referred to the hardening of steel, but it now describes the relaxing of the hardness of previously quench-hardened material (martensite) by heating the steel to temperatures in excess of 150°C to produce a soft, but tougher metal.

With primitive methods of iron working the penetration of carburizing would have been slight, sufficient for thin, slender objects like knife-blades, but not for heavier objects like axeheads. For such objects the laminated, piled or compacted structure was devised. In this process plates of iron of varying carbon content were bonded together, commonly with a high carbon layer on the surface. This takes advantage of the fact, unusual among metals, that pieces of iron will weld perfectly to each other if simply hammered together when hot. The iron sheets were each carburized separately before being piled and forge-welded together so that there would be a reasonable diffusion of carbon through its mass (see p.107 here). The history of laminated iron in the Near East, from examples dating to the first phase of the Iron Age in Cyprus and Anatolia (Sardis) to those of the Achaemenid period and later, has recently been reviewed by Maddin and his colleagues (Maddin et al., 1983a).

The incidence of carburization and tempering in ancient iron is not yet well understood, since the recognition of intentional, as distinct from accidental, carburization in ancient ironwork is a controversial issue in modern metallographic research. Even into relatively modern times smiths were not sure what they were doing and, although some ancient smiths might have had an empirical knowledge of how to improve the working of iron, it is likely most did not. Accidental carburization might well have arisen from carbon (charcoal) trapped in the pores of the bloom iron and then forged into a finished object. Lamination is a good demonstration of deliberate carburization and its earliest appearance in Mesopotamia is significant. It appears on tripods (Coghlan, 1956, 137, 180, 202, pl.I, fig.2) found at Nimrud in palace stores sacked about 614 B.C.; but much of the associated material is earlier booty or tribute from Syro-Palestine, so the tripods may only be evidence for the skills of western iron smiths at this time.

Deliberate quenching is even more difficult to identify for certain. About the only independent clue to its early development is in Homer's Odyssey (IX: 389-94), indicating knowledge of it in the East Mediterranean, by the eighth century B.C. Water quenching might have been done simply to cool a forged object quickly. The key layer of an object for resolving whether or not it has been quenched is the martensite outer surface, that most subject to rusting and to damage. Although there may be some indication of the deliberate quenching of iron objects from Nimrud (Curtis et al., 1979, 381), the earliest certain evidence for it is on an iron chisel from Al Mina of the fourth century B.C. Its tip had been coated with clay to prevent quenching cracks (Maddin et al., 1977, 130ff.; Wheeler et al., 1976, 110-I).

Direct evidence for the activities of the Assyrian blacksmith is negligible. Pleiner and Bjorkman (1974, 303ff.) have assembled the few documentary references there are. The same workshops were active both in the production of copper and its alloys, and of iron, for at times the two metals had to be combined in the same object when the casting properties of bronze rendered it indispensable (see p.105). Assyrian texts distinguish between coppersmiths and ironsmiths, but texts, as well as objects, attest the overlapping of their functions.

The combination of iron and bronze (bimetallism) is a significant aspect of metalworking for much of the so-called Iron Age in the Near East and attracted the attention of Layard and his colleagues (Layard, 1853, 670). It is particularly well illustrated by finds from level IV at Hasanlu in north-west Iran, destroyed about 800 B.C. (Pigott, 1981, table 10), where considerable numbers of bimetallic swords, spearheads and maceheads were recovered in addition to those entirely of iron. The evidence from Assyria is more scattered and, in the case of maceheads and tripods, rather equivocal, since a number of the published examples have West Semitic inscriptions clearly indicating an origin outside Assyria (Barnett, 1967). But there are others which suggest that, whatever the origin of the technology, it had established itself in Assyrian workshops by at least the ninth century. No technological study of this technique in the Near East is available, though one exists for the European Iron Age (Drescher, 1958) and it is relevant particularly to weapons in Assyria (cf. Hodges, 1962, 132). One dagger from Nimrud (Ashmolean Museum, 1951.60) has a bronze hilt, directly imitating a bone prototype, cast onto an iron blade. This illustrates how, when ornamental features were required, cast bronze allowed for a freedom impossible with primitive iron technology. Such also was the case with the maceheads, which commonly have bronze heads cast-on to iron shafts. The earliest example of a macehead of this standard type, dated by an inscription of Shalmaneser III (c. 858-824 B.C.), has a facetted stone head mounted on an iron shaft capped with bronze (Curtis and Grayson, 1982). By placing a baked clay piece-mould, possibly taken direct from a carved stone macehead, over an iron shaft, smiths were able to cast-on a copper or bronze head decorated in a manner unattainable at the time in iron. Tripods illustrate another technical role for bronze in this stage of iron production. There the bronze was cast round the junctions of iron struts to secure the joint in a rigid "sleeve", presumably made by pouring the liquid metal into an appropriate clay mould encasing the iron parts. Bimetallism continued into at least the Achaemenid period. Two linch-pins have upper parts modelled as human bodies, in bronze cast-on to iron pins (Schmidt, 1978, 19-22, pl.40-2). Comparable linch-pins are shown in place on chariots depicted on the Persepolis reliefs (ibid., pl.22b). An actual example, however, from Neo-Assyrian levels at Nimrud, is entirely of bronze (Mallowan, 1966, pl.142; cf. in general, Calmeyer, 1980).

(c) Sources:

Layard (1849, 225) explored iron mines in Kurdistan, in the Berwari Valley north-east of Amadiya, and also in the Tiyari mountains. Unfortunately there is so far neither documentary nor archaeological evidence to indicate that the Assyrians exploited them, though on the argument of proximity and accessibility it would seem likely. In view of the prevalence of iron ores across the Near East, source location is not so crucial a problem in the study of early iron working in Mesopotamia as it is with certain other metals. Iron ores are to be found in Anatolia, as well as in Caucasia and Transcaucasia (Maxwell-Hyslop, 1974, map on pl.XX), in Iran (Harrison, 1968, 496ff., fig.113) and in Syro-Palestine (Stech-Wheeler et at., 1981, 259), which is referred to in Deuteronomy (8:9) as "a land whose stones are iron-ore". Tylecote (1981a) has investigated the possibility that the iron sands of the Black Sea region in Turkey might have been a useful source of iron from the outset of the local Iron Age (cf. Bryer, 1982).

In a survey of Assyrian iron sources Maxwell-Hyslop (1974) compared the ancient documentary evidence with modern information on sources, under two separate chronological heads: "Middle Assyrian" and "Neo-Assyrian". In the earlier period, Tiglath-Pileser I (c.1114-1076 B.C.) obtained iron ore (unspecified) and haematite from Nairi, west and south of Lake Van. The famous "Kizzuwatna Letter" of the Hittite king Ḫattusilis III (c.1289-1265 B.C.) (Goetze, 1949; Kempinski and Košak, 1977) to an Assyrian king, indicates that Kizzuwatna, which lay about the axis of Taurus-Comana, was already an important centre for iron production, storage and distribution in the thirteenth century, under direct control of the Hittite king. However careful reading of this letter, and of later Biblical reports of the Philistines, does not support popular theories about "metal monopolies" at this time. The metal presumably travelled as ore or bloom iron from the mining areas to the royal ironworking centre, from which the finished objects were distributed. As there is no documentary evidence for Mitannian access to iron, it is not known whether they, or their successors in Hanigalbat (geographically across Assyrian routes to iron sources in Anatolia) affected Middle Assyrian iron supplies.

Data from the tenth to the seventh centuries B.C. is more ample and instructive, as iron then featured conspicuously in Neo-Assyrian lists of booty and tribute. Despite certain known exaggerations, and uneven records, this information is instructive. Iron was usually named as a raw material, sometimes with a quantity given in talents, but rarely as artefacts. Jankovska (1969, 263) summarized the evidence as follows (references are to Luckenbill, 1926-7):

"Though iron was the most important strategic material, it is mentioned much less frequently than other metals. It is recorded in the joint tribute of several areas of Asia Minor, Syria and the Mediterranean coast (I, 722, 801), in Damascus (I, 740: 150 tons), Ḫattina (I, 585: I, 477: 3 tons), Carchemish (I, 601: 3 tons; I, 476: 7.5 tons), Que (I, 588), Bit-Ḫalupe (I, 443), Bit-Zamani (I, 405; I, 406 and I, 501: 9 tons), Šubria (I, 502), the Chaldaean area (I, 625), and in the Laqe of Hammath (I, 412: 30kg.). But to a great extent this lack of data on iron can be explained by the nature of our sources, referring for the most part to treasuries".

Pleiner and Bjorkman (1974, 291ff.) attempted to plot this information as a diachronic distribution graph, but the peculiar nature of the available textual information and its uneven spread, restricts its validity. In the majority of cases iron came from the west and north-west: modern Anatolia and Syria; some also came from the north (ancient Urartu); but regions in Iran do not feature in the tribute lists as yielding iron. Maxwell-Hyslop (1974) has treated in some detail the following potential Neo-Assyrian sources of iron: the Middle Euphrates and the Khabur region (Laqe); Carchemish, Marash and the Amanus; the Elazig region (Iruwa); the Gurun and Malatya areas; the Divrigi deposits, between Sivas and Erzincan, the largest and most important iron producing area in modern Anatolia; Damascus, perhaps only a distribution centre for manufactured iron; and Tabal (anti-Taurus region) and Que (Cilicia), both in modern Anatolia. In all cases what evidence there is remains circumstantial, not direct.

Oppenheim (1967, 240ff.) published two tablets from Uruk, of the mid. sixth century B.C., referring to iron "from Yamana" and from "Lebanon"; the first probably indicating some region on the islands, or the western or southern coast of Anatolia, whilst the latter may refer to native Syrian iron, since the region had previously sent enormous quantities of the metal to Assyria. Other Neo-Babylonian documents from Sippar and Uruk (Oppenheim, 1967, 241) refer to iron from Ḫume, a variant writing of Que (Cilicia). Iron also came from an unlocated area known as Simir, which produced wine and may have been in eastern Anatolia.

No scientific method has yet been applied to identifying individual sources of iron in the Near East.

(d) The Development of iron-working in Mesopotamia:

Snodgrass (1980, 336-7) has most usefully defined three broad stages in the emergence of iron technology, each of which may be observed in Mesopotamia.

"In stage 1, iron may be employed with some frequency, but it is not true working iron..... In the main, its employment is for ornament, as is appropriate for the expensive commodity which we know it to have been in many cases. But this ornamental use must be extended to cover objects that have the outward form of real weapons and tools, yet whose circumstances reveal that they played no practical role. Iron-bladed daggers are worth mentioning specifically because of the time-honoured use of the dagger as a prestige object, a component of 'dress-uniform', and the consequent tendency to choose materials for their costliness rather than their effectiveness.

In stage 2, working iron is present but is used less than bronze for implements of practical use.

In stage 3, iron predominates over bronze as a working metal, although it need not, and usually does not, completely displace bronze even in this role."

One further point made by Snodgrass (1980, 368) is important in any historical survey: "To understand the spread of early iron-working one must distinguish between the essentially technological factors, such as those that brought the initiation of stage 1 and the transition to stage 2, and the essentially economic factors that must lie behind the change from stage 2 to stage 3". This is a significant consideration in Mesopotamia, even though the transitions are not easy to isolate at present. Even in the Near East there are wide variations in the duration of stage 2, considerably shorter in Cyprus and Syro-Palestine, than it is in Egypt, Assyria (perhaps also Urartu), and Iran.

Even though it is known that craftsmen had encountered smelted iron in the Bronze Age (see below), widespread acceptance of the new metal as a utilitarian commodity has still be be explained. Despite the ready availability of iron ores in contrast to those of copper, and the much rarer tin, two key technological factors inhibited the rapid adoption of iron: the appreciable amount of time involved in forging every single iron artefact and the difficulty of achieving consistent results. For most of the Iron Age a lot of heating and forging was needed to produce a workable metal, since craftsmen operated smelting facilities that led to a spongy iron, low in carbon and mixed with slag. Although a single smelting operation, with relatively little subsequent working, would produce a range of copper or bronze objects, it was not so with iron. To be able to produce an iron weapon or tool with a cutting edge of a quality equal, or superior, to that of a fully work-hardened medium tin-bronze, the iron smith had to master a variety of hazards. The homogeneity of his crude iron had to be improved; he had to forge a blank to a thin-sectioned blade; he had to harden the blade to an evenly distributed and appropriate hardness; and, in finishing, any edge had to be ground to a maximum cutting potential. Such tasks demanded many man-hours, much skill, and wide experience of smithing.

Still in the seventh century apparently, Assyrian iron was wrought iron with a heterogeneous carbon distribution. It is difficult to sustain the argument that it was the superior quality of the metal that influenced its adoption in Assyria (cf. Pigott, 1982). Nor, yet, is there any evidence of a tin "famine", or of severe disruptions in the lines of supply for copper, such as have been held to explain iron's widespread acceptance in the Eastern Mediterranean in the final centuries of the second millennium B.C. (Muhly, 1980; Snodgrass, 1980). There seems to be little doubt that an iron industry was established in Syro-Palestine some time before a comparable industry existed in Assyria (Waldbaum, 1980; Stech-Wheeler et al., 1981). It is likely that a combination of factors, probably in the ninth century B.C. (see below), stimulated the industry's growth there. Assyrian campaigns into Syria would have increased her contact with the metal and its production, whilst, at the same time, offering the opportunity to capture trained smiths and raw material. It is possible that the growing commitment of the Assyrian army by then required an investment in weapons and tools on a scale not contemplated before and strained the always uncertain supplies of tin. The Assyrian state also now had the capacity to set up the necessary industrial infra-structure for the extensive exploitation of iron, which, even under the most primitive conditions, requires relatively complex organization (cf. Pigott, 1981, 217ff.); the more so, if efficient and economic production on any scale was required.

(i) Stage 1 (c.5000-1250 B.C.):

The search for pigments may first have familiarized man with the main iron ores. Haematite, for example, is red ochre (Schmandt-Besserat, 1980, 144). Recent research has demonstrated how Bronze Age smiths are likely to have acquired some knowledge of the properties of iron from the accidental production of iron through the use of its ores as fluxing agents in copper smelting or through the use of iron-rich copper ores (see pp.7ff.). Copper ores were regularly smelted with iron fluxes and in the process there was a distinct possibility of iron being reduced in the bottom of the furnace (Tylecote, 1976, 2-3, 40; Cooke and Aschenbrenner, 1975). The problem for the ancient metalworker was how to produce useful metallic iron from the bits of spongy bloom resulting from such accidental smelting. This might be the origin of the four-sided instrument of smelted iron from grave A at Samarra, commonly dated to the Halaf period (Herzfeld, 1930, 5:1, 5, fig.f, pl.47; cf. Perkins, 1949, 7, n.69). In a subsequent publication, however, Herzfeld (1933, 29) argued that all the metal he had originally published with the prehistoric painted pottery from Samarra was intrusive from the Islamic levels immediately above.

Vaiman (1982) has fully documented the textual evidence for iron in Sumerian texts between the later fourth and early second millennium B.C. His evidence correlates with what little artefactual evidence is available (cf. Waldbaum, 1980). Most significantly, in an inventory text from Uruk III, at the end of the prehistoric period, iron objects are listed among weapons and vessels of gold, electrum, silver and just possibly bronze (Vaiman, 1982, 33; cf. Heinrich,1935, pl.32c). A contemporary lexical fragment also lists iron artefacts. A fragment of iron is reported from Jebel Aruda on the Euphrates in Syria, a settlement of Uruk IV type (Van Driel et al., 1979, 19).

Pieces of smelted iron, or traces of it, are reported from third millennium B.C. contexts at Tell Asmar (Frankfort, 1950), and Chagar Bazar (Mallowan, 1936, 11, 26-7, 58, grave 67 in level 5), both perhaps from dagger blades. Unanalysed pieces are also recorded from Chagar Bazar (Mallowan, 1937, 151-2), Tell Asmar (Delougaz, 1967, 198, 244), Khafajah (cited in Waldbaum, 1980, 70) and Kish (Mackay, 1929, 97, 123-4, pl.36:2). Some iron fragments from Mari are probably later than was claimed when they were first excavated (Parrot, 1956, 180-1).

Meteoric iron was also in use at this time in Mesopotamia. It offers much the same problems to the metalworker as native copper. If cold hammered, it tends to crack along well-defined crystal planes; but with care and annealing, or hot forging, it is suitable for the manufacture of small objects (Tylecote, 1976, 2-3). Two occurrences of iron in early contexts have been said, from their high nickel content, to be meteoric: a fragment found between levels D and E of the Anu Ziggurat at Uruk (Protoliterate) (cf. Vaiman, 1982), and rusted fragments of uncertain purpose from RT 580 at Ur (Heinrich in Nöldeke et al., 1937, 53; Woolley, 1934, 49, 542). There is confusion over a supposed piece of meteoric iron from Tell al-Ubaid

(Woolley, 1934, 293; but, perhaps significantly, not listed in Hall and Woolley, 1927).

Iron fragments are rarely reported from the Near East in the first half of the second millennium B.C., though the metal was then more apparent in texts, notably those of the "Cappadocian trade" (Maxwell-Hyslop, 1972; Muhly, 1980, 34ff.; Waldbaum, 1980, 175) and the Old Babylonian period in Mesopotamia (Muhly, 1980, 39). At Ur an "iron peg", in a shapeless mass of copper, was reported from Enunmah with tablets of Rim-Sin (Woolley, 1974, 51). From Nuzi, in the fifteenth century B.C., came a small, spherical bead of iron and a bronze dagger, its flanged hilt inlaid with iron plaques rivetted into place with iron rivets (Starr, 1939, 94, 470, 475, pl.125KK). As neither piece of iron has been analysed, it is not known whether this is celestial or terrestrial metal. A macehead from Aqar Quf may be inlaid with iron (Baqir, 1945, 13). It is not known whether these are local or imported artefacts, as they might well have been, from an area either of Hittite or Mitannian production (Stech-Wheeler et al., 1981, 264-5). An iron ring was recorded in a Middle Assyrian grave at Assur (Haller, 1954, 63, grave 746) and there are iron finger rings in the "Middle Assyrian" cemetery at Mari (Beyer, 1982, 187, n.1.).

By this time texts indicate that iron was a commodity, with a place in luxury trade and exchange more valuable than gold, eagerly sought and not readily available; but nowhere in the Near East is there yet clear archaeological evidence for regular iron smelting, even in Anatolia (Waldbaum, 1980, 76ff.). The status and supply of the metal c.1250 B.C. is well illustrated by the "Kizzuwatna Letter", already cited (see p.97). In it the Hittite king, writing to his Assyrian counterpart, attempts to defer a demand for a consignment of iron with the gift of an iron dagger blade (Goetze, 1949, 30ff.). This letter usefully indicates that production was seasonal, slow and unreliable, and that the individual gift of an iron dagger was of quite exceptional value and status (referring almost certainly just to the blade, as with that in Tutankhamun's tomb).

Thus, although it is clear that iron was current in Mesopotamia for much of the Bronze Age, it was generally a rarity, valued as a curiosity and primarily used for small pieces of jewellery, for ornamentation and for the blades of presentation weapons. A textual reference at Nuzi in the fifteenth century B.C. to its use as armour scales is exceptional (C.A.D. s.v. *binâtu*). It was not until iron could be carburized that it was a challenge to the utility of copper and its alloys. "Figures for the relative hardness of soft wrought iron, and bronze of around 10 per cent tin content, clearly show that a good bronze when fully work-hardened is considerably superior in hardness to the primitive iron" (Coghlan, 1956, 69).

(ii) Stage 2 (c.1250-850 B.C.):

As archaeological evidence for iron working in Babylonia is insignificant, any study of the early Iron Age in Mesopotamia must concern itself exclusively at present with the evidence from Assyria, though even that is by no means without its problems. No smelting or forging installations are known and no textual evidence repairs the deficiency, for none of the documents describe such technical

UNIV DUNELM

processes as the extracting and working of iron. There are no iron objects of any note from Assyrian sites that may be dated before the middle or end of the eighth century B.C., though textual references make clear that iron was available to Assyrian kings from at least the thirteenth century B.C. In the absence of finds from this critical period it is not possible to say confidently when iron production was first extented to cover the majority of tools and weapons, nor may it be established when such fundamental techniques as carburization and quenching first came into general use in Assyria. Even for the later eighth and seventh centuries B.C. the geographical distribution of the best evidence is limited to the Neo-Assyrian royal palaces at Khorsabad, Nineveh and Nimrud, and there the published material is often only a small part of what was recovered.

In a cursory way the documentary evidence alone carries the history of iron in Assyria through stage 2 down to the time when material evidence is available. Shalmaneser I (c.1274-1245 B.C.) lists iron among the materials placed as foundation deposits at Assur (Grayson, 1972, 84) and a contemporary administrative text lists one dagger of iron and a spearhead of ḫabalginnu, perhaps a certain kind of iron (Postgate, 1973a, 13-14). King Ninurta-Tukulti-Assur (mid. twelfth century B.C.) had a blacksmith at his court, but the range and nature of his duties are not specified (Weidner, 1935-6, 19). Tiglath-Pileser I (c.1115-1077 B.C.) used iron arrowheads to hunt wild bulls (Grayson, 1976, 16). By the earlier ninth century iron was becoming more common, for Tukulti-Ninurta II (c.890-884 B.C.) recorded that his army used iron axes (?) to clear a path. He also received one hundred iron daggers from Laqe as tribute and iron from the Nairi-lands (Grayson, 1976, 100, 103-4). It is notable, in view of the huge quantities listed by later rulers, that 30-60kg. of iron was still regarded in the early ninth century as worthy mention in royal annals.

(iii) Stage 3 (c.850-350 B.C.):

It was in the course of the ninth century that iron became common enough in Assyria to be used for those tools and weapons, or parts of them, for which it was particularly appropriate. Under Assurnasirpal II (c.883-859 B.C.) and Shalmaneser III (c.858-824 B.C.) iron as booty and tribute came predominantly from regions adjacent to the Upper Euphrates. Under Shalmaneser III, for the first time, iron was among the metals obtained "without number" and thereafter, until available records cease in the earlier seventh century B.C., very considerable quantities of iron came into Assyrian royal storehouses. What iron came into Assyria by other means, or was exploited locally in mines such as those explored by Layard in Kurdistan (see p.97) is not recorded in published texts, as it may well have passed through private or public hands outside the court.

It has already been noted that the surviving material evidence is predominantly of seventh century date, or slightly earlier. This means that the use of simple proportion to separate stages 2 and 3 in Mesopotamia is not possible, since the available material evidence all falls in so brief a period and that, to judge from the texts, came after the transition. At Nineveh stratigraphic confusions greatly reduce the relevance of the ironwork excavated there; relatively little ironwork, and that not always easy to date closely, was found in graves at Assur; whilst at Nimrud, though the site yielded masses of ironwork, none of it is likely to be much earlier than the seventh century B.C. (Curtis et al., 1979, 370). At Khorsabad the storehouse (room 84) excavated by Place, described as a huge iron wall, held 160 tons (c.5,600 talents) of the metal and that was only part of the iron found at this site (Pleiner and Bjorkman, 1974, 293). Several hundred "ingots" were packed into the storeroom, with pick-axes and hammers, grappling irons and chains, and plough-shares. Scattered finds of iron objects from other sites in Assyria and Babylonia down through into the Achaemenid period do not conspicuously add to the repertory of iron work. Unfortunately, the correct date of an important group of iron tools from Thebes in Egypt, excavated by Petrie and dated by him "to the Assyrian invasion", is an open question. They might be considerably later (Williams and Maxwell-Hyslop, 1976).

It is easier to assess the point at which Assyria entered Stage 3, if the question is set in a wider Near Eastern context. It is now argued that the eastern Mediterranean region in the twelfth century B.C. "holds the key to understanding the transition from the Bronze Age to the Iron Age and the first real use of iron as a utilitarian metal...." (Muhly, 1980, 47). In this area it is argued the change was triggered off by a shortage of copper and tin. Iron technology developed in the Levant, in particular in the region of Cilicia and north-west Syria, and probably passed thence to Cyprus. Although material evidence for the period c.1600-1200 B.C. remains exceedingly rare, documentary indications are varied and widely distributed (Muhly, 1980, 51; Waldbaum, 1980, 76ff.). In the centuries from about 1200-900 B.C., when material evidence from Mesopotamia is absent, most Near Eastern countries, at different rates, passed through stage 2 to 3, with Syro-Palestine to the fore, followed by eastern Anatolia and Assyria.

Comparative study indicates that the change from a predominant use of bronze to a predominant use of iron, in the case of working metals, was fairly abrupt. What evidence there is, albeit mostly indirect, converges to indicate that this change in Assyria fell in the mid. to late ninth century B.C., as Shalmaneser III's western campaigns pierced deeper into regions where iron was abundant, and therefore readily available as tribute or plunder. The ever growing demands of the Assyrian military machine for tools and weapons had increasingly to be met in manufactured iron. Shalmaneser III was the king who commanded "Sharpen the iron swords that subjugate (the foe)" (Lambert, 1961, 150-1).

Curtis (in Curtis et al., 1979, 382ff.) has provided a simple statistical survey of some of the archaeological evidence for iron artefacts. All known plough-shares, sickle-blades, hoes and knives are of iron in the later eighth and seventh centuries B.C. Saws were also then made of iron (Layard, 1853, 195; Curtis et al., 1979, fig.31a-c). Adzes, axe-adzes and axes were commonly, if not exclusively, of iron. Weapons offer equally striking confirmation that Assyria was now into stage 3. Of the spearheads and swords listed by Curtis all are of iron; of 17 daggers, only one is entirely of bronze and another has an elaborate cast bronze hilt on an iron blade. Surviving leaf-shaped arrowheads are overwhelmingly of iron; the flanged, socketed bronze arrowheads of the seventh century B.C.

were an innovation introduced by the Scythians and their confederates. Wrought iron, from which each leaf-shaped arrowhead had to be hand-forged, is not adaptable to methods of mass production, for which the new type of bronze arrowhead was particularly well suited (see p.40). Its currency is a technological phenomenon and in no way indicates a failure in iron supplies. Caution is always necessary in assessing the archaeological record, notoriously inadequate as it is, particularly when some texts indicate a better balance between ferrous and non-ferrous weapons, even into the Neo-Babylonian period (Pleiner and Bjorkman, 1974, 228). Iron was not cheaper than bronze/copper until the Neo-Babylonian period. Similar care is needed in gauging the role of iron in the manufacture of body armour. Helmets both of bronze and of iron survive; the former would have been lighter, the latter more protective. An Assyrian iron helmet weighed about 3,049g against 862g and 700g, respectively, for two Iranian bronze helmets (Curtis et al., 1979, 385). The weight, the rusting, and the heat of iron armour scales in the intense sun, would be a major consideration, though iron scales are better protection against arrowheads and are more easily repaired by army blacksmiths. Consequently, there was a parallel production of the two.

(e) The repertory of iron objects:

(i) "Ingots":

There is a category of iron object from Khorsabad and Nimrud (Pleiner and Bjorkman, 1974, 295; Curtis et al., 1979, 371ff.), rather fish-like in shape when seen in profile, and generally pierced with a small hole towards the end. They are almost rectangular in section at the middle and have tapering ends hammered into broad, flat terminals at 90° to one another. Place reported finding considerable numbers of them in room 84 of Sargon's palace at Khorsabad (1867-70, 84-7; II, 264; III, pl, fr. 5-7; cf. Loud, 1938, 213-7. pl.62). They vary in length between 39.5 and 51.5cm. and weigh between 3.7 and 8.7kg. (Pleiner and Bjorkman, 1974, 295, n.17; Place gives as high as 12kg.). Analysis and metallographic study (Pleiner, 1979; Curtis et al., 1979, 371ff.) have shown them to be of "essentially pure iron with only minor amounts of impurities". Carburization was not consistent in the tested samples, appearing sometimes on the working edge, sometimes on the shaft-hole. These semi-forged iron objects are still an enigma. In some cases the hole contains traces of wood, in others it is not even cut right through. It has been argued that a few appropriately shaped examples might have been miners' pickaxes, with shafts of ash or larch, whose spring would increase the impact, as in more recent tools (Maddin in Curtis et al., 1979, 389ff.). But many of the extant examples could not have been mounted to serve in this way and, for the moment, they are best regarded as iron ingots or bars with the hole to facilitate handling (Pleiner, 1979, 90). The context at Khorsabad is more appropriate to the storage of ingots than mining tools. Although they are assumed to be "Assyrian", if they really are ingots they might well all be imports from a mining region outside Assyrian proper.

(ii) Tools and Implements:

Pleiner and Bjorkman (1974, 296ff.) have succinctly assembled the textual and archaeological evidence for iron tools. Iron axes, adzes and pick-axes were standard equipment by at least the ninth century. Hoes, sickles, plough-shares and probably iron spade blades were current in the farmer´s tool-kit; iron chisels and saws in that of the carpenter. Knives of iron were widely used. If the group of tools from Eqyptian Thebes may indeed be attributed to Assyrian military craftsmen of the seventh century B.C., and that is very doubtful, then it embraces a punch, a carpenter´s chisel, a file, a rasp, an adze, a twist-edged scoop, a cutter, centre bits and crank pieces (Williams and Maxwell-Hyslop, 1976, fig.3).

Various seventh century iron tools from Nimrud have been metallographically studied: a sickle-blade, a knife, a hoe, and two chisels, which might be post-Assyrian in date (Curtis et al., 1979, 377ff.). Carburization was probably deliberate, but not extensive; hot hammering or forging was also evident. "The implication of these observations is that Neo-Assyrian smiths did not know how to manipulate in the most effective ways the operations in which iron attains the greatest strength" (Ibid., 381). The Theban group has been similarly studied, but in view of its uncertain date and place of origin may not be cited as "Assyrian" (Williams and Maxwell-Hyslop, 1976, 303).

Pleiner and Bjorkman (1974, 298ff.) reviewed the textual and archaeological evidence for iron foot-fetters, pokers and tongs, for chains and hooks, for bell clappers, for nails, for some personal ornaments, and for various structural elements such as door fittings. Iron was used for the struts of tripod-stands for cauldrons, jointed and shod with cast bronze (Layard, 1853, 178, 191; 671: report by Percy). Coghlan (1956, 4, 137, 180, 202) examined a fragment of one such tripod strut and found "a piled structure which would indicate that the tripod iron was made from piled wrought material". But many, if not all of these tripods may be imports, as some bear West Semitic inscriptions (Barnett, 1967). From the Nabu Temple at Khorsabad were recovered three iron wheels (c. 22cm in diameter) from a ceremonial wagon. In them iron spokes were inserted into naves of cast bronze and two iron rims and tyres were possibly riveted together (Loud and Altman, 1938, 62, pl.24e). No certainly identifiable fragments of full-scale wheels, or other iron parts of the numerous chariots used in Assyria have yet been reported; bronze linch-pins are attested (Ellis, 1966, pl.VIII, la-b), as are linchpins of iron and bronze (p.96 here).

(iii) Weapons and armour:

The whole range of personal arms, daggers, swords, spearheads and arrowheads, was regularly produced in iron to patterns closely following those of their bronze precursors. Some have received metallographic examination (Curtis et al., 1979, 377ff.); a dagger tip had been carburized and an arrowhead was made of hammered, mildly carburized iron. Maceheads present a rather more varied technology, already considered (see p.96), as has the use of iron for armour and helmets (p.104). Layard vividly described a find of iron scale armour at Nimrud (1849, I, 341; cf. Stronach, 1958); but only two armour scales from Mallowan's Nimrud excavations have been metallurgically examined (Curtis et al., 1979, 378-9). They were of mildly carburized iron, hammered at about 600°C into a stone mould to give a central rib. Such a method would have saved time and insured uniformity for an object needed in some quantity. Shields, when made of sheet bronze, had iron handles rivetted onto the inside (Layard, 1853, 194). Amongst the iron implements he found, Layard noted "a large blunt spearhead, such as we find from the Sculptures were used during sieges to force stones from the walls of besieged cities" (Layard, 1853, 195; cf. 1853(a), pl.66)

(iv) Horse-harness:

To what extent iron was used to manufacture horse-bits it is hard to say, as very few Assyrian horse-bits have been recognized in excavations (cf. Stronach, 1958, 175, pl.XXXV.1). To judge from the very much richer contemporary evidence in western Iran cast bronze was still preferred (Potratz, 1966, 117); iron bits would have been in a minority. Other harness trappings of metal were in sheet or cast bronze (see p.35).

(v) Ornamental work in, or on, iron:

Evidence for the ornamental use of iron in Mesopotamia is still awaited. There is no evidence for swaged and engraved decoration, as on bracelets and swords of iron from Luristan (Moorey, 1971, 315ff.), nor for repoussé decoration as on iron plaques from Hasanlu IV (c.1000-800 B.C.) in northwest Iran (Pigott, 1980, 430). X-ray of an iron helmet from Nineveh in the British Museum (Barnett, 1953, pl.XXXI-II) revealed that it had both a border pattern inlaid with bronze all round the edge and also a similarly inlaid panel at the front. This showed an Assyrian king receiving a cup from his cupbearer, with a third figure behind, all within a frame of linked buds. Layard in reporting a number of iron helmets, which disintegrated during excavation, remarked that "the rings which ornament the lower part, and end in a semicircle in front, were inlaid with copper" (Layard, 1849, II, 339; cf. I, 341).

(vi) Vessels:

Although iron vessels were produced in Urartian workshops, they do not yet seem to have been recognized in Neo-Assyrian contexts (cf. Böhme, 1979).

In the broad spectrum of evidence for iron-working before the Graeco-Roman period now available, the smiths of Assyria, unlike those of the East Mediterranean and the Greek World, do not seem to have been able to control the quality of the metal used. There is no clear evildence for the selection of carbon-poor or carbon-rich pieces for deliberate construction of working edges. Nor is there any certain indication of the regular quenching or tempering of blades.

No metallurgical work has been done on the ironwork of the brief Neo-Babylonian period, though it is represented amongst finds from sites like Babylon and Ur. Only one type of iron weapon, typical of the Achaemenid military repertory, has been subject to such examination. Smith (1971, 43, no.101, fig.2.2d, 2:18) showed that an unprovenanced iron akinakes or short sword, was forged from a single piece of steel save for the pommel, which was a separate piece pierced at the centre to fit a small tenon on top of the grip. Coghlan (1956, 137-9, fig.44) published a technical report on an iron spearhead of the Achaemenid period from a cemetery at Deve Hüyük, near Carchemish (Moorey, 1980, no.170, 64):

"The material of this spear is forged carburized steel made from reduced sponge iron. Average carbon content is 0.18 per cent, but in places as high as 0.6 per cent. The high nickel content of 0.32 per cent is unusual. The spear has been forged from about fifty plates or layers of sponge iron piled into a block.....the majority of the plates have been surface-carburized during heating for forging, the compacted structure resulting in a highly developed water-marked pattern somewhat similar to damascene steel blades produced from cast steel. A noteworthy feature is that the tubular taper socket, which was forged from a "tagged" end, has been brazed and forged together with impure copper. Hardness ranged from 108 Vickers in the low carbon zones to 152 Vickers in the higher carbon bands. There is no indication of hardening and tempering".

VII: NICKEL

As with zinc (p.132), nickel was unknown as a metal in antiquity. It was first discovered and separated in Europe in the middle of the eighteenth century A.D. (Aitchison, 1960, II, 482-4). Nickel occurs as a trace element in objects manufactured in Mesopotamia; what little evidence there is at present indicates higher levels in artefacts from the south than from the north. As it was so frequently detected in the 1920s and 1930s in the third millennium B.C. artefacts analysed by the Sumer Copper Committee, they assumed a copper ore containing nickel as the source of the raw material and devoted considerable research to identifying it. They accumulated a wide range of useful data and eventually suggested that the Jebel Ma´aden deposit, inland from Sohar, in the Wadi Ahin, Oman, had been the primary source of Sumerian copper in the third millennium (Peake, 1928). Broadly speaking renewed research into this problem in the last ten years has endorsed this view (see p.14). As addition of 1% nickel to copper improves the alloy´s basic strength by a factor of two and increases the work-hardening rate of the alloy, it is possible ancient craftsmen had appreciated empirically the value of ores with significant nickel inclusions.

VIII: SILVER

(a) Silver and lead:

SILVER is found both as a metal and in its non-metallic state. It also occurs in practically all gold (see p.69). Native silver is very rare (20% as abundant as gold; 0.2% as abundant as native copper) and is usually found in quantities not worth melting to make larger, workable lumps (Patterson, 1971). It generally appears deep underground as, if appearing at the surface, it is liable to have been converted into chloride by traces of chlorine invariably present in rain water. It is practically pure. The principal ores of silver are the sulphides (silver glance) and the chlorides (horn silver), which yield up their metal by simple smelting. However, the main source of silver in antiquity was not the silver ores. Most silver was then extracted from argentiferous lead ores in a two-step process that included smelting of the lead ores to produce lead metal followed by extraction of the silver from the lead by cupellation (see below). Cupellation made silver "about a thousand times more available" (Patterson, 1971, 312).

As the history of silver in the ancient Near East is closely bound up with that of lead, the two metals are taken together here.

Native LEAD is even rarer than native silver. The principal ore of lead is galena (lead sulphide), which, when it occurs in hydrothermal veins, is frequently associated with silver ore minerals. Cerussite (lead carbonate) is an important, widely distributed secondary ore mineral of lead formed by the action of carbonated waters on galena. Two reasons probably account for the early use of lead ores (see below). They are very shiny, metallic looking, and so attract attention, whilst the metal is very easily obtained from the ore. Tylecote has shown that lead can be smelted from galena in a fire burning charcoal or dry wood at temperatures below 800°C (cited in Gale and Stos-Gale, 1981b, 181).

To obtain silver from argentiferous lead it is necessary first to smelt the ore to obtain metallic lead and then, in a separate operation, to oxidize the lead to litharge (lead monozide), leaving behind the silver, in a process known as cupellation. For this the lead is heated under strongly oxidizing conditions in a cupel. The lead oxide so formed is absorbed in the porous material of bone or ground-up potsherds in the cupel, leaving silver metal behind. This process may be repeated several times to purify the silver; it is very efficient in freeing silver from such common impurities as copper, antimony, arsenic, tin, iron, zinc, (less well from bismuth) in the argentiferous lead. Silver derived from argentiferous galena will be characterized by gold contents from zero to about 0.5%, lead contents between 0.01 and 1%, or rarely a little higher (Gale and Stos-Gale, 1981a, 107). Silver derived from the native metal, with or without admixture of cerargyrite (a "dry" silver ore very easily reduced to silver metal), will generally contain less than 0.01% gold and significant quantities of mercury (ibid.).

Any conjectures about the origins of silver metallurgy in the Near East have to take into account the long interval between the first appearance of lead, sometime in the seventh millennium B.C. in Turkey (de Jesus, 1980, 76), and the earliest manufactured silver, some three millennia later, when it appears relatively suddenly over a wide area (Prag, 1978). Lead, which could only have been obtained by smelting, had long been experimented with before the appearance of silver. The view, argued over a long period by a number of scholars (cf. Gowland, 1920, 132; Hoover, 1950, 390; Wertime, 1973, 883), that silver was discovered in the course of the accidental cupellation of lead, remains a strong possibility in the Near East. But, even after the initial discovery, the recovery of worthwhile quantities of silver required solution of a number of technical difficulties, for a ton of smelted lead will only contain a few ounces of silver.

The view (Smith, 1967, 35) that the earliest silver artefacts were made of native crystals or of cerargyrite (silver chloride) cannot be ruled out; but hard evidence for it in the Near East has yet to be provided. Wertime (1973, 883), following his expeditions to Turkey, argued that in the Anatolian-Iranian highlands the lead miners of the early periods concentrated on the rare cerussites rather than mining for galena. It was below the water table and its smelting was far less rewarding than the working of cerussites. Meyers (in Waldbaum, 1983, 188) is of the same opinion:

"Based on comparisons of elemental compositions of silver coins, silver objects and ore samples and using information from geological and archaeological surveys and from literary evidence, it appears that most silver produced in Anatolia, including that excavated at Sardis, was extracted from the ore cerussite (lead carbonate), the major component in the oxidized portions of lead deposits".

Research on Sasanian silver from Iran has shown that it was also extracted from cerussite, not galena. At least one mining site, with associated smelting operations, has already been recorded. At this time, c.250-650 A.D., it was customary in Iran to add about 4-10% copper to improve the silver's mechanical properties. The copper was probably produced from copper ores at or near the site of silver mining and smelting (Meyers, 1981, 243).

It is possible that silver was recovered sometimes from the cementation process (see p.71) through which electrum was purified into gold. But there is, as yet, no hard evidence for this either from literary, archaeological or analytical sources, in the area and time range considered here.

The role of silver from an early date is illustrated by a Sumerian literary text recounting a dispute between copper and silver (Kramer, 1963, 265). Here silver is judged by copper to be a worthless metal, only good for hoarding away, whereas copper is a useful metal, as it forms the tools which man uses in his daily tasks. The use of silver as a means of exchange is first evident in the "Fara Texts" of Early Dynastic IIIA, in the middle of the third millennium B.C., but it is not until the Akkadian period that its primacy in this role is clear (Limet, 1960, 46-7; 1972, 30). From texts of Early Dynastic III onwards degrees of purity in silver are regularly apparent (Limet, 1960, 46-7); earliest among such references is the inscription on the silver vase made for King Entemena of Lagash (c.2450 B.C.), whose metal is explicitly said to be of "purified" or "fine" silver (Sollberger and Kupper, 1971, 69: IC7e). The point is made clearly in an Old Assyrian text of the earlier second millennium B.C. apparently referring to cupellation (C.A.D. s.v. kaspu): "I refined the silver and from five minas (only) 3 minas came out (of the kiln". Akkadian has a variety of terms to describe silver alloys (Eilers, 1954-9), none yet matched up with the analyses of surviving objects.

Oppenheim (1966) has published and commented on a text from the library of Ashurbanipal (668-27 B.C.), for making a silver-like alloy from base-metal ingredients. How much older such deceptions were has yet to be established. The later economic texts from Mesopotamia are much concerned with the quality of silver and carefully stress the percentage of permitted additions; but far too few analyses have been done to offer any information on the ways in which, for specific purposes, silver might be debased. Nor, owing to the way in which buried silver corrodes, is it likely that archaeology will ever provide any check on textual indications that silver surfaces were variously treated with heat (Limet, 1960, 49-50). Moreover, under field conditions corroded silver may easily be confused with copper or bronze, so the number of silver objects from excavations may at present be underrated.

(b) Sources:

(i) The evidence of ancient texts:

As Limet has pointed out (1960, 94), textual indications for the ultimate sources of the silver used in Mesopotamia are singularly rare and meagre. Pettinato (1972, 80-1), in his review of the Sumerian literary evidence, listed such relatively well known regions in Iran, the Gulf and the Indus region, as *Aratta*, *Dilmun*, Elam and *Marḫaši* and *Meluḫḫa*. "The Silver Mountains" limited the campaigns of Sargon of Akkad to the northwest (Hirsch, 1963, 38, lines 22-8) and are usually identified with silver mines at Keban on the Upper Euphrates, just south of its junction with the Murat river. Manishtushu recorded a campaign in which his army was divided into two, one part invaded "Anshan and Sherikhum" in Iran, whilst the other waged war "up to the silver mines" (Hirsch, 1963, 69; Gadd, 1971, 438ff. considers the geographical problems). Gudea wrote just of silver from its "mountain", taken by Limet (1960, 94-5) to be reference to a source east of the Tigris in Iran. In his summary table of the evidence provided by the *lipšur* litanies and the HAR-ra series 22, Snell (1982, 212) lists as sources *Zar-šu*, *Hašbar*, *la-an-na-ki-ta* and *Kù-sù*; but of these places only one may be tentatively identified: *Kù-sù*:? Kush (Nubia).

It is to the earlier second millennium B.C. that some of the best evidence for the use of Anatolian silver belongs. Larsen (1967, 4) has succinctly described the famous trade between Assur and eastern Anatolia at this time:

"The pattern of the trade as revealed by the texts is clear; tin and textiles were imported into Anatolia and in return silver and gold were sent back to Assyria. The trade in copper was vigorous but seems to have been mainly an internal Anatolian affair."

The silver was either in ingots passing by weight, or sometimes in rings or packages containing bars of metal. Different types of silver are distinguished, probably by quality; various towns in the vicinity of Kültepe (Karum Kanesh) are listed as the sources of silver, but they are probably just the principal centres for its distribution (Garelli, 1963, 265ff.). Leemans (1960, 130ff.) preferred not to speculate on the origin of the silver current in Babylonia during the Old Babylonian period, as it was then widely used as currency. But there is some indication of silver coming up the Gulf (Oppenheim, 1954). Hittite inventory texts of the thriteenth century B.C. attribute silver to *Saqqamaha* and *Arpa* in the north-central region of Turkey (Košak, 1982, 195).

For the Neo-Assyrian period, on the evidence of the tribute lists analysed by Jankowska (1969, 261), in the ranking of the regions with the most silver, Carchemish (Khatti) and Damascus come first, with the centres of Tabal (east Anatolian plateau), Karduniaš (Babylon) and Judah next, Muṣaṣir ranked third. A mention of "silver (capital) of the Habur river" refers only to the direction of trade (Oppenheim, 1970, 10, n.22). However, Shalmaneser III wrote of going to "Mount Tunni (Taurus), the silver mountain, (and) Mount Muli, the marble mountain" (Meissner, 1912; Luckenbill, 1926-7, I, 246). According to the "Foundation Charter" of Darius I, the silver used in his time at Susa was from Egypt (Vallat, 1971).

Three circular silver ingots ("cake-ingots") bearing the name of Bar-Rakib were found in an eighth century context at Sinjirli in southeast Anatolia (Andrae, 1943, 119-21, pl.58); the only complete example weighs 497.37 grammes. They were accompanied by a number of small cast silver blocks of irregular weight. Silver ingots are otherwise rarely found in Near Eastern excavations (cf. Bivar, 1971; Curtis, 1984 (Nushi-i Jan, Iran); Karageorghis et al., 1983 (Cyprus and Egypt)).

(ii) Mining Areas:

In any review of potential sources for Mesopotamian silver in antiquity it is still only possible to follow Forbes (1971, 209) in mapping sources of argentiferous lead, for that reduces as far as possible the numerous hazards in such an inquiry; even one conducted at the very superficial level followed here. This is not to say that these were the only sources of silver; for Turkey de Jesus (1980, 64ff., map 13) concluded that "the lead-silver deposits are widespread in Turkey, but they do not seem to have a dense concentration like those of copper or iron. However, there is a tendency for the deposits, to limit themselves to the northern Anatolian fold and the southern Anatolian fold". According to Marco Polo (i.4) Armenia was rich in silver. The richest silver deposit in Turkey, that of Bulgar Maden, has produced a sample containing 1.8% silver. Although this is commonly taken to be one of the major Anatolian sources of silver in antiquity (Meissner, 1912), the case, as so often, must be declared as yet unproven (de Jesus, 1980, 64). In Iran argentiferous lead ores are again widely distributed. Ladame (1945, 276-89) listed occurrences in Azerbaijan, in the Miyana-Zanjan and the Sava-Qazvin-Teheran regions, in Khurasan, in the Isphahan-Kashan and Anarak-Yazd areas, and in Kerman. In Medieval times two Iranian provinces particularly, Transoxiana and Khurasan, were renowned for silver (Allan, 1979, 13). Afghanistan also has such ores and again Marco Polo (i.24) documents productive mines in Badakhshan (cf. Chmyriov et al., 1973). India is also a possible source (Strabo, 15.1.30, cap.700; Ktesias, Indika, cap.11), though Ratnagar (1981, 140ff.) argues that silver's first appearance there, in the Harappan period, came through trade with Mesopotamia. The Arabian peninsula is poor in silver and lead ores, save for those east of the Gulf of Akaba exploited by the Nabataeans (Strabo, 16.4.26, cap.784), and reports of lead in Oman (Tosi, 1975, 202). The origin of the silver used in ancient Egypt is a long standing problem (Lucas, 1962, 254ff.). Neither native silver, nor the "dry" silver ores, nor jarositic, nor other gossan-associated silver ores, have been reported in Egypt. Analyses also indicate (Gale and Stos-Gale, 1981) that the silver content of local galena is too low for it to have been exploited in ancient times. It is increasingly thought, as a result of more analyses (Gale and Stos-Gale, 1981, 109ff.), that the primary local source of silver was aurian silver (with more than 10% gold) extracted from gold mines, perhaps in the eastern desert; but how early this was the case is still an open question, since the silver found in Predynastic Egypt may have been imported (Prag, 1978).

(iii) Scientific determinations of source:

Ordinary chemical analyses of objects made from copper, or alloyed copper, may reveal something about how the metal was extracted from its ore (see pp.13ff.). But with lead and silver such analyses are of restricted use for demonstrating the kind of ore used, or the location of the mines, as the processes of smelting and cupellation considerably disrupt the pattern of major, minor and trace elements characteristic of the parent ore. An alternative is offered in the case of lead (and thus by extension in certain cases to objects in which it occurs) by isotope analysis. The isotopic compositions of lead samples remain essentially constant no matter what is done, short of mixing together leads from different sources. By determining the isotopic composition of lead extracted from ancient objects it becomes possible in some instances to match these compositions with those of ores from known mining regions. But, as ratios may not be unique to a deposit, conclusions are predominantly of a negative kind, and even then must be used with caution. Lead isotope provenancing of silver sources only works for deposits of argentiferous galena. For other non-lead sources of silver they are not relevant or potentially misleading, as the lead had to be brought from elsewhere to recover the silver. Even when direct connections cannot be made with specific regions (as is very much the case at present for the Near East), such data may serve to group ancient artefacts which contain the same kinds of lead and which, therefore, might have had a common origin.

The basis of the technique is relatively simple. Lead has four staple isotopes: $Pb^{204,206,207,208}$ which represent atoms weighing, respectively, 204, 206, 207 and 208, atomic mass units. In the case of the great majority of chemical elements the relative proportions of their isotopes are the same regardless of where the elements are found. But lead is unusual in that the relative proportions of its isotopes vary somewhat among ores occurring in different geological regions. The variations arise from the different geological ages of the ore deposits and stem from the fact that $Pb^{204,207 \text{ and } 208}$ are found as end-products of the radioactive decay series of uranium and thorium.

The pioneering work in this field was done by Brill and Wampler (1967), beginning with galena ores and samples from lead artefacts, and proceeding to leads in lead-containing ancient glasses (Brill, 1978, with bibliography on 30, n.17). The following pre-Achaemenid Mesopotamian lead artefacts were sampled for Brill and Wampler (1967, 75-6):

Sample 53: fragment of heavily corroded lead from Nippur; Protoliterate period; 0.04% Ag; Group "X".

Sample 121: (illustrated Brill and Wampler, 1967, pl.5): a folded sheet lead pot cover from Ur (U.20073); late Prehistoric to Early Dynastic I; 0.05% Ag; Group "S".

Sample 182: band of lead from Nimrud; Neo-Assyrian; 0.03% Ag; matches only sample 90 (Nimrud: red glass).

Sample 183: folded lead sinker from Nimrud; Neo-Assyrian 0.005% Ag; ?Group "L".

Sample 203: small piece of pyramidal-shaped lead; Tell al-Rimah; site A (?date); 0.02% Ag; Group "L".

The results on this occasion were too general and the sample too small to be of any significance. A number of lead deposits were clearly involved. Subsequent work on metallic lead (three pieces from Nimrud) and lead-containing glasses did isolate a distinctive type of lead used over a long period of time at a well-defined group of sites for a particular purpose:

"Leads of exactly the same isotopic composition (i.e. as at Nimrud) have been found in some of the other ancient glasses we have studied namely the remains of yellow-opaque glass threads in cored vessels from Tell al-Rimah and Nuzi (c.1500 and 1450-1375 B.C.) and also a yellow opaque glass bead and a yellow-glazed faience bead from Ḥasanlu (both dating from 1000-800 B.C.)." (Brill, 1978, 33).

More recently Japanese scholars have begun to work on leads in bronze objects from Tell Gubba (level VIIa) in the Hamrin Region of Iraq. Preliminary reports indicate that the lead isotope ratios at Ur and Gubba are closely comparable (Curtis; pers. comm.). Work on Anatolian lead sources, which will be of considerable importance for the study of leads in Mesopotamia, has only just begun (Karageorghis et al., 1983).

Another approach is to be seen in Meyers' work (1981) on Sasanian silver from Iran. This demonstrated that "the elemental compositions of the silver alloys, especially the concentrations of gold and iridium and to a lesser extent those of arsenic, antimony, zinc, tin and selenium - allow a number of conclusions on the use of ore sources and the distribution of silver to mints and workshops." Again, although it may not be possible to demonstrate the location of ore sources, it is possible to define compositionally homogeneous groups of artefacts probably made with metal from a single ore source and, when the evidence of silver coinage offers a chronological control, to chart such changes across time.

Dayton's (1981, 36; 1982) categorical statements about the European sources of the silver (and lead) used in Mesopotamia in the third and second millennia B.C., based on his interpretation of a few lead isotope analyses, do not correspond with any other evidence at present available.

(c) History and repertory of objects:

In the second half of the fourth millennium B.C. there was a "surprisingly widespread use (and preservation) of silver objects, which have been found in relatively large numbers, relatively that is in metallurgical terms, in Egypt, Levant and in Anatolia" (Prag, 1978, 36). Although the indications remain meagre, this silver was, at least in part, cupelled silver (Prag, 1978, 40; Gale and Stos-Gale, 1981). This marked increase in the presence of silver objects co-incides broadly in Mesopotamia with an unusual occurrence of lead vessels, not so well documented before or after in the archaeological record (see p.122 here). The silver objects of the later prehistoric period from Mesopotamia are not numerous: vessels (Heinrich, 1936, 45, p.35); parts of animal statuettes (Nöldeke, 1936, pl.23q-s); a stamen in an artificial flower (Heinrich, 1936, 41, pl.30c); cast recumbent animal handles on large cylinder seals (Heinrich, 1936, pl.17a,b; cf. Hamilton, 1967); a recumbent stone bull "support" with a silver spike projecting upwards from its back (Heinrich, 1936, pl.6); and a silver harpoon head (Lenzen, 1959, 10, pl.18b). There are isolated earrings from the so-called "Jamdat Nasr" graves at Ur (Woolley, 1956, 116, gr.219; 118, gr.239). A child's necklace from Susa, dated to the Ca(?) phase, in the middle of the fourth millennium B.C., has seven silver pendants with cloisons, inlaid with black rutile (an ore of titanium), partially gilded (Amiet, 1966, pl.46).

As Prag (1978) argued, the pattern of silver finds in the Levant and in Egypt towards the end of the fourth millennium would be consonant with an Anatolian source for the metal (de Jesus, 1980, 75ff.; the Beycesultan silver ring should be dated c.3000 not 4000 B.C., cf. Prag, 1978, 44, n.60). In northern Mesopotamia, apart from silver nails in the "Eye-Temple" at Brak (Mallowan, 1947, 93), the isolated early silver finds are not closely dated. They are probably of the third rather than the later fourth millennium B.C.; a silver pin and finger ring from a tomb in stratum 6 at Tell Billa (Speiser, 1931, 12); another pin attributed to level 5 in Mallowan's deep sounding at Nineveh (Campbell Thompson and Mallowan, 1933, 145, pl.LXVIII:7); and a disk from Brak (Mallowan, 1947, 171, pl.XXXII.7).

By the time of the "royal tombs" at Ur, in the middle of the third millennium B.C., the silver repertory was more extensive: belts (Woolley, 1934, pl.12-13); a "comb" (ibid., pl.20); bowls (ibid., pl.33, 42, 167, 171-2); an ewer (ibid.,, pl.172); a fluted bowl (ibid., pl.172); tumblers (ibid., pl.172); lyre fittings (ibid., pl.75 6, 111-2); a cow's head (ibid., pl.120); the head of a lioness (ibid., pl.127-130); hair ornaments (ibid., pl.136-7); imitation cockle shells for cosmetics (ibid., pl.137); sheet silver bindings for spear shafts (ibid., pl.149); pins (ibid., pl.159, 189); a boat model (ibid., pl.169); a lamp (?) (ibid., pl.170); a libation vessel (ibid., pl.171); vessels with electrum handles and electrum ribbons "sweated on" (ibid., pl.173); a tub with holes at regular intervals; double pipe (ibid., p.258); nose rings for oxen (ibid., 64, pl.35); and a rein-ring (ibid., 78, 301, pl.166-7). Four silver spearheads were found in RT 789 (U.19472: Woolley, 1934, pl.153) and silver axeheads in PGs 250 (U.8428), 560 (U.9247) and 1422 (Neo-Sumerian: U.12478: Woolley, 1934, pl.223). RT 580 contained "two long spindles of silver with lapis lazuli spindle whorls" (Woolley, 1934, 53: U.9777).

Millichamp and Levey (cited by Organ, 1967) published a series of analyses of silver found in some of the more massive pieces of jewellery from Ur, now in the University Museum, Philadelphia. The range of concentration for all objects registered silver as the major constituent, but with copper in a range of 1 to 5%. The presence of this copper might have been an inevitable impurity in repeated remelting, or an economy, or an attempt to harden the silver.

Away from Ur there is a wide scatter of evidence. Coiled silver wire was found at Tell al-ʻUbaid (Hall and Woolley, 1927, 103, fig.) in an Early Dynastic IIIB context. Some particularly fine filigree work is encountered on silver medallions during Early Dynastic III (Maxwell-Hyslop, 1971, 15, pl.14 (Uruk)). A grave of a child at Abu Salabikh contained a remarkable collection of silver objects: a roundel; an "eye-patch"; a pair of sandals and a long bead (Postgate, 1980, 94, pl.X). The so-called Early Dynastic III "Treasure of Ur" from Mari contained silver, gold and lapis lazuli pendants, silver bracelets and pins, silver "rosettes" and, on a necklace, a silver "wheel" and bead separator (Parrot, 1968, pls.XI, XIII, XV, XVII).

The outstanding silver object to have survived from Early Dynastic Mesopotamia is the display vase of King Entemena of Lagash found at Telloh and now in the Louvre (Maryon and Plenderleith, 1954, 653). It was hammered up from a single sheet of silver except for the inside of the neck, which is lined with a plain deep silver collar, silver-soldered into position. The vase is set on a copper foot, now much corroded. It is engraved (?) with an inscription on the tall neck, then with a frieze of calves couchant on the shoulder. On the body of the vessel, framed by herringbone bands, are four repeats of the lion-headed eagle (Imdugud). First he grasps a pair of lions in his talons, then two goats, again a couple of lions, then a pair of oxen (de Sarzec and Heuzey, 1884-1912, 261-4, pl.43, 43bis; Parrot, 1948, 108-9, pl.VIIIa). A fragmentary silver bowl from PG 1130 at Ur (Woolley, 1934, pl.167, 217: U.11794) is decorated round the side with a frieze of mountain goats moving through hilly country. It is all in repoussé, finished by chasing.

So far unique is a bronze bull statuette, said to have been bought in Mosul, inlaid with silver, now in the Louvre. It is commonly identified as the ornament of a rein-ring and dated to the middle of the third millennium B.C., though it might be later (Heuzey, 1900; Calmeyer, 1964, 74, fig.16). The animal is "piebald"; inlaid across the body with variously shaped patches of silver. The base-plate on which the beast stands was analysed by Berthelot (1906, 81), who reported Cu: 82.4%; Sn: 11.9%; Fe: 4.1%; no lead, antimony, zinc nor arsenic was traced; so far as I am aware the "silver" identification is only by eye.

After the middle of the third millennium B.C., as with gold, the range of evidence for silverwork is only sustained archaeologically through personal ornaments (Maxwell-Hyslop, 1971, passim). Limet (1960) concluded that the Sumerian metalworkers primarily used gold and copper and it is not, apparently, until the Akkadian period that silver comes more commonly into use for jewellery (Maxwell-Hyslop, 1971, 19, 59, 77-9). There are isolated surviving silver vessels of the Old Babylonian period (cf. Uruk, Lenzen, 1962, pl.27f, 29a-e) and of the Neo-Assyrian period (Nimrud, Mallowan, 1966, pl.356-7). Even in the Achaemenid period they are not so commonly encountered in Mesopotamian excavations as in other parts of the western Empire, perhaps suggesting that they were less commonly placed in graves there (cf. Nippur, McCown, 1967, pl.108: 8-9; Ur, Woolley, 1962, 194, pl.21; fragments: Robinson, 1950). Though well documented in texts, and known in isolated cases in neighbouring countries, examples of silver statuary have yet to be recovered from Mesopotamia (cf. Spycket, 1981), though they have been found at Susa, as have silver inlay for statuettes made in other materials (Amiet, 1966, pl.318, 355, 404). Salonen (1963, 250-1) has listed some of the textual evidence for the use of silver to decorate furniture.

One of the more important functional distinctions between silver and gold in Mesopotamia (for in much else they were interchangeable) was the general role of silver, passing by weight, as a primary means of exchange and payment. As in French so in Akkadian the word for silver came to mean "money". The earliest coins of local origin from Babylonia, are silver staters issued in the governorship of Mazaeus, 331-28 B.C. They have very erratic weights (Newell, 1978) suggesting that they were still passing by weight rather than by a standard face value. Powell (1978, 217) notes payments given by weight but to be paid in staters (cf. C.A.D.: istatirru) in the Seleucid period. It has surprised many scholars that coinage, devised in Asia Minor as a convenient way of storing and dispensing wealth in the earlier seventh century B.C., had not appeared even earlier in Mesopotamia, where many textual references were assumed to imply its existence. Passages in cuneiform texts formerly translated as "Heads of Ishtar", or of other deities, were taken to indicate the existence of standard currency units stamped with such representations (Smith, 5, 1922; cf. Lipinski, 1979). In the continuing absence of material evidence to this effect, and in view of more recent renderings of these phrases as references to particular stores of wealth, or treasuries in temples of the deities referred to, such an hypothesis may be seen to be redundant.

However, if true coinage was absent from Mesopotamia until the introduction there of Achaemenid Persian coinage (from the later sixth century B.C.), this does not mean that there were not standard media of exchange in silver in addition to miscellaneous collections of scrap metal passing by weight. Silver was certainly cast as standard ingots, or cut from ingots in pieces sufficiently regular, in terms of an established system of weights, to pass as currency ("Hacksilber") (cf. Herodotus III. 96). Such money took distinctive forms from an early date, most commonly rings (hullu; šeweru) (Powell, 1978; Michalowski, 1978) or miniature axeheads (haṣṣinu) (cf. Bottero, 1958, 332). This range of silver is particularly well illustrated in a hoard of the seventh century B.C. from Tepe Nush-i Jan in western Iran (Bivar, 1971; Curtis, 1984, 1ff.).

Silver used for payment was often of poor quality and needed close controls. Even though it is not until the Neo-Babylonian period that common terms for assaying and refining regularly appear in texts, it is clear that checks and balances had long existed in the system to counteract fraud. Even small quantities of silver are mentioned as being marked "normal" (GIN) and there may well have been sheets of silver stamped all over with such a mark, then cut up and weighed out as payment (Oppenheim, 1978, 665, n.223). When payments were received in temple or palace in uncoined silver currency (rings, wires, cut pieces, objects fragments etc.) they were dropped into a box, the contents of which were then smelted in the official foundry. When refired, the metal was cast as ingots of standard size and fineness and deposited in the treasury (Oppenheim, 1947; cf. II Kings XII. 9-10 for temple collections of this type).

There has not yet been a systematic chronological study of the metal media of exchange in Mesopotamia as evident in texts. Copper predominates before the Akkadian period, when silver first emerges clearly in this role (Lambert, 1963; Limet, 1972). By the earlier second millennium, as at Mari, silver both in scrap and in manufactured form as bowls, axeheads and ingots is commonest for payments; it passes by weights on a given standard. Copper, lead and tin do not then appear to have been used for currency; gold only occasionally (Bottero, 1957). In the Old Assyrian colony at Kültepe (Karum Kanesh) in Anatolia large payments were made in silver; but smaller ones in "tokens" of tin (or lead). In Kassite texts, and at Nuzi, silver was the primary medium, though for international purposes at this time there was a gold standard (see p.75). In Middle Assyrian texts silver did not occur so regularly for payments, which were more often of lead when paid internally, of bronze and tin when paid externally (Freydank, 1982). In the Neo-Assyrian period, the situation is not so clear; before the eighth century payments, at least internally, were in copper/bronze; thereafter silver returned as the primary currency through into the Neo-Babylonian and Achaemenid periods (cf. Curtis, 1984, 17).

Some surviving hoards of scrap silver may have been personal "bank deposits" (bullion), buried at times of peril, or jewellers' stock-in-trade for recycling; but they all illustrate the role of silver as bullion and are not paralleled in gold or base metals, though pieces of gold may sometimes appear mixed in with the silver. Relevant hoards are known from the later Early Dynastic and Akkadian periods at Tells Agrab and Asmar, and from Khafajah (Delougaz, 1967, 45; Powell, 1978, pl.IIIA,B) and in the north at Tells Brak, Chuera and Taya (Mallowan, 1947, 166, 176-9, pls.XXXIII-XXXVI; Moortgat, 1960, 7ff., figs.11-12; Reade, 1968, 248; 1973, 165). For the Old Babylonian period there are hoards from Larsa (Parrot, 1933; Arnaud et al., 1979). A hoard from the Neo-Assyrian level V in area TA at Nippur was "in a small pot, which contained numerous chunks of silver as well as beads" (McCown, 1967, 98, pl.147:1); and another of cut silver, ingots and rings found at Assur in a jar under a private house floor (Andrae 1908, 22; Preusser 1954, 35). A sixth or fifth century hoard of silver fragments (bracelets, complete rings and earrings) concealed in a pot was found in the ruins of a Kassite period building at Qalat al-Bahrein (Bibby, 1964, 102-3, fig.1; Krauss et al., 1983). Scrap coined money appears for the first time in Mesopotamia with silver vessel fragments and jewellery in hoards of the Achaemenid period. One, now partly in the British Museum, is reported to be from Rassam's excavations at Babylon (Robinson, 1950). Analysis of some items: a vessel handle and a sheet fragment, revealed between 98.1 and 99% silver, with up to 1.1% copper (Curtis, 1984, 18-19, Appendix I). Another fourth century hoard, discovered by chance on the line of the Tigris, early in the nineteenth century, included ingots of silver with coins (Jenkins, 1964).

A hoard of jewellery, both gold and silver, from under the Persian period level and above the Nebuchadnezzar pavement in room 5 of the E-nun-mah at Ur, might be considered a votive deposit, rather than a capital or currency hoard, were it not for the presence of a pair of silver balance pans, which seem unlikely in a votive deposit (U.456-500: Woolley, 1962, 106-8, pl.21-3).

(d) Manufacture:

The principal uses of silver in Mesopotamia were for bowls and other vessels, for personal ornaments, for plating wood, bricks and other metals, for small statuary and for currency, either specially made "rings" or scrap-metal passing by weight. Silver is not so malleable as gold and requires more annealing in manufacture. It may be beaten out to form sheets of foil, but not to such thinness as the finest gold leaf. Berthelot (1906) reported a piece of Egyptian silver sheet as varying in thickness from as little as 0.001 to 0.0025. Pure silver is generally too soft for use in itself. As the few analyses of objects from the "Royal Cemetery" at Ur (Woolley, 1934, 294) make clear, it was alloyed with copper (comparable to modern sterling silver) for purposes of manufacture from at least the middle of the third millennium B.C.

The methods for working gold described above (see pp.81ff.) also apply to silver, whether it be hammered or cast, decorated with granulation or filigree; but less laboratory research has been directed to silver jewellery. It has not, for instance, yet been shown whether diffusion bonding to produce silver plate was a technique used in Mesopotamia, as it was to the west in the second millennium B.C. (cf. Charles, 1968; Khalil, 1980, 50, pls.58-9, no.34); nor has niello yet been identified on an object manufactured in Mesopotamia. Hammering was the main technique for vessel production. Relief decoration was in repoussé; the high relief may most often have been achieved with insets: separately manufactured pieces placed in hollowed-out slots and secured by hammering the lip of the slot over the inset to crimp it. Such methods may have been used in the fine Neo-Assyrian silver bowl with lion-head protomes all round the side found at Nimrud (Mallowan, 1966, 430, pl.357). Further decoration was by chasing or engraving.

The silversmith's tool-kit is unlikely to have differed in any significant way from that of the goldsmith. One of the later third millennium hoards of silver from Tell Taya was associated with a "chisel" and three other tools (Reade, 1968, 248), perhaps constituting the stock-in-trade and equipment of a silversmith.

Graham, who restored some of the silver objects from the "Royal Tombs" at Ur, allocated to the University Museum in Philadelphia, commented briefly on the metallurgy of the silver vessels. "The structure is that of annealed metal with numerous cases of twinning indicating previous working. One may conclude, therefore, that the silver alloy was first cast and then alternately annealed and hammered until the desired object was obtained" (Graham, 1929, fig.on p.257). In forming a bowl "at least three to five annealing operations would be required depending on the skill of the workman, in order to keep the metal soft enough to work into shape" (Graham, 1929, 253). Oddy (1984) has reported on the silver wire jewellery from Tepe Nush-i Jan in western Iran, found in a hoard deposited in the seventh century B.C., though the contents may be older (Curtis, 1984). The wire was hammered and then smoothed, perhaps with sand.

To judge by the evidence of a text from level VII, seventeenth century B.C., at Alalakh (Atchana) in Syria, even divine statues (once they had become obsolete) were recycled to provide silver for mundane objects. In this case the statue provided 685 shekels (about 5.5kg.) of silver for the manufacture of vessels and for plating various wooden and ivory objects (Na´aman, 1981). Similar recycling of precious metals, including silver, is evident in the Mari texts (cf. Durand, 1983).

(e) Silver Analyses

(1) Unprovenanced: Proto-Elamite silver figurine of a kneeling bull holding a spouted vessel: Metropolitan Museum, New York, No.66.173 (Hansen, 1970). Analyses by Sayre and Meyers (Hansen, 1970, 23: Thermal Neutron Activation Analysis):

		Silver	Copper	Gold
A. Solders:				
U.1 :	solder (neck overlap)	95.3	4.6	0.1
U.2 :	solder (near base)	86.7	13.3	0.004
B. Figurine:				
SB1 :	left hoof	99.4	0.6	0.008
SB2 :	left haunch	98.3	1.7	0.01
U.2 :	back of head	99.1	0.9	0.006
U.6 :	left haunch	99.3	0.7	0.03
U.8 :	across edge of sternum	98.6	1.4	0.04
C. Vessel:				
U.9 :	vessel held by figurine	96.8	3.2	0.01

Meyers also undertook spectographic analysis in certain areas; the silver sample contained the following elements:

Ag(XO.O), Cu(X.O), Pb(.X), Sn(.OX), and Au, Ni, Fe, Al, Mn, Sb, Ca, Cr and Mg (trace amounts); not detected: Zn, As, Co, Bi, Ti, V, and Ba.

Analysis of a solder sample: Ag(XO.O), Cu(10-20%), Pb(X.O-X), Sn (.X-.OX), Mg(grater than trace), and Au, Ni, Fe, Al, Mn, Sb, As, Bi, Ca, Cr, V, Ba(trace amounts); not detected: Zn, Co, Ti.

(2) Ur: silver rein-ring (Woolley, 1934, pl.166) Early Dynastic IIIA: Analysis by J.R. Ogden (Woolley, 1934, 293):

Silver	93.5%
Copper	6.10%
Gold	0.08%
Zinc	0.15%

Although no further analyses are published here, Plenderleith commented that "Traces of gold are commonly found in the silver from Ur."

(3) Ur: silver pin with lapis head (Metropolitan Museum, New York: 33.35.44) Early Dynstic III: Analysis by Pieter Meyers (Thermal neutron activation analysis) (Hanson, 1970, 23).

Silver	95.5%
Copper	4.4%
Gold	0.07%

A series of analyses of silver found in some of the more massive pieces of jewellery from Ur, now in the University Museum, Philadelphia, have been published by Millichamp and Levey (1963). They may be summarized as follows:

	Range of concentration (%) for all objects:
Ag	Major
Cu	1 to 5
Pb	0.01 - 0.5
Au	0.02 - 0.1
Sn	0.1) present in one object only
Zn	1) present in one object only
As	ND to 0.2
Bis	0.02 to 0.2
Fe	0.01 to 0.05

Analyses done by Dayton (1978, 85, 448ff.) have not been fully published.

(4) Telloh: Analysis published by Berthelot (1906, 78); the objects are not dated, but came from Heuzey´s excavations: "Lingot et rognures (anciennes) d´un métal blanc, trouvés avec des objets chaldéens dans un vase de grosse poterie.

La limaille du lingot renfermait, pour 100 parties:

Argent	96.3
Cuivre	3.5
	99.8

pas de plomb."

(5) Kish (as cited in Aitchison, 1960, I, p.180): "Sumerian" finger ring: Ag: 94.86%; Au: 0.29%; Cu: 4.85%. "Later" fragments: Ag: 92.98%; Au: 1.57%; Cu: 4.23%; Pb: 1.22%.

(6) Brak: Akkadian period; Mallowan, 1947, 258, no.4: pin (F.447): Cu present in "analytical quantity"; traces of Pb and Bi.

(7) Khorsabad: Neo-Assyrian period (Berthelot, 1906, 81): piece of sheet overlay from Sargon´s palace (Place): "Cet or ne referme ni cuivre, ni plomb, ni fer en proportion sensible. Il contient une dose considerable d´argent..." i.e. "electrum".

(8) "Babylon": zoomorphic handle and fragments of sheet silver; Achaemenid (cf. Robinson, 1950, pl.XXIII, XXIV, 27-8); see Hughes in Curtis, 1984, 59; all in the British Museum:

		Ag. %	Cu. %	Zn. %	Au. %	Pb. %	Sn. %
1. BM 120405: handle	(AA)	98.3	1.1	0.02	n.a.	0.49	0.1
	(XRF)	98.1	1.1	trace	0.3	0.6	n.a.
2. BM 82-12-20, 27: Sheet	(XRF)	99.0	0.7	0.1	0.3	0.05	n.a.
3. BM 82-12-20, 28: Sheet	(XRF)	95.9	3.4	0.1	0.4	0.3	n.a.

AA = Atomic absorption; XRF = X-ray fluorescence.

IX: LEAD

(a) Manufacture:

Lead is easily recovered through roasting galena (see p.108) to remove the sulphides or through heating cerussite with charcoal to remove the oxygen so that it passes off as gas. It is also easy to work and, because of its weight, useful for stabilizing objects, for making weights and anchors. It was generally used in the condition it came from the furnace. Archaeological and documentary evidence (C.A.D s.v. abāru) reveal that it was employed in Mesopotamia for vessels, for pipes, large and small (in medical contexts), and as covering in sheets, both for artefacts and parts of buildings. Cast lead formed the cores of statues, particularly of their bases, and of metal weights. It stabilized door sockets and was used to plug casting faults in copper or bronze statuary. Rivets and clamps of lead provided re-inforcement in buildings and in repairing stone vessels. Cast in stone moulds it was made into trinkets and statuettes. It was added to copper alloys for three possible reasons: to ease the filling of complex moulds; to lower the temperature to assist with casting in lengthy or slender moulds, or to assist the casting of several objects from one melt; and to increase the weight of metal available. Lead was also used in the production of glass and glazes, though not so commonly as is often assumed (see p.166), in cosmetics, and in pigments. It may be alloyed with tin as pewter; but this is not yet evident from any analytical data at present available for Mesopotamia and is rare elsewhere before late Roman times. For a comment on leaded bronzes see p.33 here and the possible ambiguity in translating annaku (tin or lead) see p.124 here.

(b) Sources:

After the consideration given to argentiferous lead ores on pp.111ff. here, there is little more that may usefully be said about lead sources and their distribution in the Near East. de Jesus (1980, maps 12-13) comprehensively summarized the evidence for Anatolia and there may be little doubt, for ancient texts from time to time confirm it, that this was the primary source of the lead used in Mesopotamia. After Egyptian conquests in Western Asia in the Late Bronze Age Egyptian historical texts reveal that lead was imported from Djahi, Retenu and Isy; regions in Syria adjacent to Anatolian source zones (Lucas, 1962, 244). Lead was commonly collected from peoples of southern Anatolia by the Assyrian kings from Tukulti-Ninurta I (c.1244-1208 B.C.) onwards (Jankowska, 1969). Emphasis has commonly been put on sources accessible from Diyarbakir and what is said to be a considerable outcrop in the Tiyari mountains, north of Nineveh (Layard, 1849, 223). Lead ores also occur widely in Iran and Afghanistan (Harrison, 1968, 505ff.; Chmyriov et al., 1973; Allan, 1979, 28ff.), and in Oman (Tosi, 1975).

(c) Historical survey of uses:

Lead was one of the first metals to be exploited by man and might well have been the first metal obtained by smelting an ore, since it is so easily recovered by the simple heating of galena (see p.121). Lead ores were being smelted sporadically from the seventh millennium B.C., when it appears in level IX at Catal Hüyük in Anatolia (Mellaart, 1967, 217-8, pl.104 rt.). The evidence of a single lead "bead", of the seventh millennium B.C., at Jarmo remains equivocal, as there was some doubt about its context (Braidwood and Braidwood, 1983, 542, fig.136:19). It occurs in manufactured form in Mesopotamia as a bracelet in Yarim Tepe I, in the sixth millennium B.C., (Merpert, 1977, 84, pl.XII.2). In TT6 at Arpachiyah, of the Halaf period, a small conical piece of lead was found in the "burnt house" (Mallowan, 1935, 104). That lead facilitated copper casting was apparently recognized by the later prehistoric period to judge from a free-standing copper lion amulet (4.7x3.4x1.1cm.) from Uruk containing 9% lead (Heinrich, 1936, 25, 47, pl.13a). Lead was used for vessels ("bowls, tumblers, trays") in the later phases of the prehistoric period at Ur and into Early Dynastic I (Woolley, 1956, 104ff.), and in the Diyala area (Delougaz, 1967, 58ff.), to an extent unparalleled in the archaeological record of any other period. In the so-called "Jamdat Nasr" cemetery at Ur, extending in time well into the Early Dynastic period, out of a total of 84 metal objects catalogued four were of silver, (two pairs of earrings), 42 of copper and 38 of lead, a remarkably high percentage of the total (Woolley, 1956, 104ff.).

There is some scattered evidence from analyses to suggest that at a time when lead was unusually common for vessels it was also being experimented with in binary copper alloys (Berthoud et al., 1982, 45ff.). In the Uruk amulet, cited above (cf. p.24), its purpose seems clear, but its presence, up to 15% in a pin from Susa or 20% in cylinder seals, is difficult to explain save as an aberration at a time of general experimentation with copper alloys (cf. p.24). What little analytical evidence is at present available later, for the third millennium, does not indicate recognition of the special properties of lead-copper alloys; indeed lead is markedly absent from the few statues so far analysed. Nor do isolated references in Ur III texts indicate that it was then deliberately alloyed with copper (Limet, 1960, 54-5). Metalworkers in Iran worked with copper and lead in the same workshops on various sites; but, it would seem, for distinct purposes (Berthoud et al., 1982, 45).

A fragment of sheet lead from level 5 (?) of the deep sounding at Nineveh is of the earlier third millennium B.C. (Campbell Thompson and Mallowan, 1933, 146). At Ur it was used as a cover for pottery vessels (Woolley, 1956, 220: U.20073; Brill and Wampler, 1967a, pl.5). In Early Dynastic I it was also used, very skilfully, at Kish to repair stone vessels (Moorey, 1966, pl.Va,b) as also, somewhat later, in House D at Khafajah, room XVI (room 1 in final report) (Frankfort, 1932, 100), and, no doubt, elsewhere. At this period it also appears concealing casting faults in base metal statuary. The copper support in the shape of a nude male figure from the Temple Oval at Khafajah is almost pure copper (Cu: 99%; Sn: 00.63%; Pb and Fe: traces; Ni: nil) with "faults such as show in the back of the right calf of no.181 (pl.101A) were filled out with lead" (Frankfort, 1939, 42).

By Early Dynastic III mortuary evidence indicates a marked decline in the popularity of lead for vessels; they do not appear at Khafajah, ʿUbaid or Ur in graves of this period and only two example were reported from cemetery "A" at Kish (Mackay, 1925, pl.XX.2: graves 9 and 12), though Mackay did report that "melted fragments (of lead) were found in several parts of Kish in buildings of the earlier type of plano-convex brick" (ibid. 17). Only one occurrence is recorded in the "Royal Cemetery" at Ur: PG 55 yielded "some sheet lead... shapeless and pierced with holes due to decay" (Woolley, 1934, 148). In the metal statuary of Early Dynastic IIIB at Tell al-ʿUbaid "the antlers (of remarkable size and number of tines) were wrought and hammered, and soldered into their sockets with lead. This lead had so expanded as to burst the heads..." (Hall and Woolley, 1927, 29). The use of lead in casting is not as apparent as might be expected; it is not, for example, used in the cast copper statue inscribed for Naram-Sin from Bassetki (Al-Fouadi, 1976, 65-6); but Berthelot (1906, 79) reported that a copper foundation figurine of Amar-Sin (c.2046-38 B.C.) contained between 17 — 18% lead. In the later third millennium levels at Nuzi lead was only occasionally reported, as needles and pins (Starr, 1939, 384).

One of the distinctive aspects of the material culture of the Anatolian sites of Alishar and Kültepe in the "Colony Period", in the first quarter of the second millennium B.C., was the production of figurines and trinkets of lead in special stone moulds, a number of which have survived (Opitz, 1933; Canby, 1965; Emre, 1971). They were also used at this time across north Syria and Mesopotamia, perhaps as an aspect of the commercial contact at that time between Assur and eastern Anatolia (Canby, 1965; cf. Atshana: Woolley, 1956, 273, fig.73 rt.; Chagar Bazar: Mallowan, 1937, pl.XVII; Brak: Maxwell-Hyslop, 1971, pl.28; cf. Mallowan, 1947, 170, pl.XXXII.1 (lead trinket); Assur: Andrae, 1922, pl.29p; Rimah: Oates, 1967, 289). One such mould, now in the British Museum, is said to be from as far south as Sippar (Abu Habba) (Opitz, 1933, 189, pl.VII:7). Texts refer to such lead trinkets and statuettes as an "open hand of lead" and "figurine of the dead (made) of lead" (C.A.D s.v. abāru). Small lead figurines occurred in graves of the Ur III to Old Assyrian periods at Assur (Haller, 1954, 9ff.). A lump of lead formed part of a large hoard of Old Babylonian copper and bronze agricultural implements at Tell Sifr (Loftus, 1857, 268-9; Moorey, 1971a, 62). Small lead plaques appeared in two graves of the Isin-Larsa period at Ur (Woolley, 1976, graves 125, 134). Lead wire was used in the construction of sheet silver bracelets from tombs of the Old Babylonian period at Tell ed-Der (Lerberghe, forthcoming).

No site so far excavated in Mesopotamia has illustrated the varied uses of lead so well as Assur, where large circular plates of it formed the bases of stone pivots (Andrae, 1913, 26, fig.17, weighing 90kg: Shalmaneser III); large inscribed plaques were placed in foundation deposits by Tukulti-Ninurta I (c.1243 - 1207 B.C.) in the Ishtar Temple (Andrae, 1935, 42ff., pl.21-22), and numerous small lead objects were also encountered, primarily in the Ishtar Temple, in the so-called "New Palace" of Tukulti-Ninurta I, and at a point on the northwest fortifications (Andrae, 1923a). In recent years a peasant found a hoard of lead objects, model tools and "tokens" stamped with rosettes, at Kar-Tukulti-Ninurta (now in Assur Museum; M. Roaf, pers.comm.). The Ishtar Temple roundels have been published in detail (Andrae, 1935, 102ff.). Whether they were just personal ornaments, tokens or some form of "money" was once much discussed (Smith, 1922; Andrae, 1923a). In addition to the roundels, Smith also called particular attention to many "lumps of lead, bearing stamped impressions, now invisible", varying considerably in size, and to small pieces of folded lead that could have been threaded (Smith, 1922, figure, nos.1-2). The latter need be no more than weights for fishing nets. The known use of AN.NA (annaku) as money in Middle Assyrian texts, and its unanimous translation as "lead" rather than "tin" in 1923, inspired the view that here were actual examples of the manner in which lead was cast for use as "coinage". This is now widely discounted (on this see p.116, under silver).

Despite Landsberger's magisterial discussion (1965), ambiguity still surrounds the terminology for "lead" and "tin" in Akkadian, notably in the second half of the second millennium B.C. (see also p.130 here). In a Middle Assyrian text (VAT 18062: Freydank, 1982; Müller, 1982, 272) a distinction is made between AN.NA BABBAR (pesû) and AN.NA abāru. The first is fifteen times more valuable than the latter. AN.NA without qualification is also used and has a low value comparable to AN.NA abāru. In this context the expensive metal is taken to be tin, the cheap one lead. This also seems to be the case in Middle Assyrian legal texts, where fines are given in terms of AN.NA (annaku), and its silver value ratio runs as low as from 180:1 to 240:1, when in the Old Assyrian period it had run as high as between 10:1 and 14:1 (Muhly, 1980, 38ff.; Freydank, 1982, 74, n.27; Müller, 1982, 272). There are also Middle Assyrian cases where bronze is twice as expensive as annaku. Yet still, in the Middle Assyrian period, as before and after, in texts describing the manufacture of bronze AN.NA (annaku) is most unlikely to be "lead" rather than "tin" (Müller, 1982, 277, n.40). Then again, when quantities of annaku are as large as they are in some Middle Assyran texts from Tell Rimah, as much as 50 mana on some occasions (Wiseman, 1968, 175-6), it is not easy to understand why a merchant should deal in such large quantities of "lead" (rather than "tin") in a relatively small rural town. AN.NA. BABBAR has been translated by some scholars as "good quality" tin (Landsberger, 1965, 295; Muhly, 1980, 32-4). As "white tin" it has been suggested it might be metallic tin; but then annaku would always represent the oxide (cassiterite), which seems unlikely (cf. Muhly, 1980).

At Nuzi in the fifteenth century B.C. "lead is relatively rare in comparison with bronze and copper, and is found only in needles and long headless pins" (Starr, 1939, 470). A T-shaped ingot or clamp of lead was found in a level of the temple courtyard at Tell Rimah dated to the earlier thirteenth century B.C. (Oates, 1965, 75). What is meant at Assur by textual references to the "lead courtyard" in the temple of the god Assur is not clear, though excavation revealed nothing to indicate that it was paved with the metal (C.A.D s.v. abāru; Landsberger, 1965, 287, n.12).

In the Neo-Assyrian period lead is only sporadically reported in excavations. At Nimrud Layard (1853, 357) found "a mass of lead melted by the fire, for embedded in it was the iron head of a hatchet". At Khorsabad the American excavators refer only to "plain fragments of lead, impossible of identification" (Loud and Altman, 1938, 16); but one of the inscribed foundation plaques found by Place was lead (Ellis, 1968, 194). The British Museum has a bronze gate-socket set in solid lead, from Birs Nimrud, probably of the Neo-Babylonian period (BM 104419: Guide, 1922, 73). Lead was sometimes used in adjusting the weight of bronze lion-shaped weights, bearing the names of Assyrian kings. One example (BM 91223) has a rectangular cavity in the base filled with lead, whilst others do not (BM 91222, 91224). A Neo-Assyrian dedication, either by or for a king, now in the British Museum, is cut on four faces of a broken strip of lead (18mm high; 4.5mm; 36mm long as extant) differing in shape from the usual foundation plaques (Lambert, 1968-9).

The role of lead in buildings where baked brick was the primary medium, and in a region where bitumen was readily available, was naturally restricted. Even in Assyria, where advanced stone construction was not uncommon, architectural uses of lead are not yet as evident as might be expected, though it was used in fixing sculptured slabs together (Reade, 1983). Nylander (1970, 63ff.) has examined the history of clamps in Mesopotamian stonework. The simple dovetail clamp, working horizontally, was mostly not of molten lead, but rather of wood or else prefabricated lead, bronze or stone inserted into a cutting, as at Babylon (Buddensieg, 1912, fig.3: cf. also Nylander, 1970, 64, n.160 with references to Nimrud, Assur and Khorsabad). A clamp combining vertical with horizontal effect is only evident in a developed form, of molten lead, at Khorsabad, where it may have been one of the architectural features introduced into Mesopotamia from Syro-Anatolia in the ninth to eighth centuries B.C. (Nylander, 1970, 65 n.162; Place, 1867-70, pl.70). A new type, the double clamp, reached the Near East from western Anatolia in the sixth century B.C. and became part of Achaemenid clamping techniques. Nylander (1970, 66, n.165) has commented critically on certain, regularly quoted, classical references:

"Herodotus (I:186) and Diodorus (II:8) mention the use of clamps, embedded in lead, in the Euphrates bridge in Babylon of the latter half of the seventh century B.C. (R. Koldewey, Das Wiedererstehende Babylon, 193ff.). Nothing has been found of the stone materials of this bridge. It is thus hard to know whether this was true or whether the Greek writers made anachronistic references to a construction practice common in their own day (see Thucydides I:93, 5). Nebuchadnezzar used clamps, though wooden, in the north wall of the Kasr in Babylon (MDOG 48 (1912), 10ff., fig.3)."

A most unusual use of lead, to create a composite statuette, is illustrated by finds at Toprak Kale near Van in Urartu. For this lead was cast in sheet with a network of tiny square cavities to take inlays of glass or faience and ivory to serve as the garments, whilst other pieces are partitioned into wavy strands and circular curls

(Barnett, 1950, 16, pl.XIII:3; Mitchell, 1983).

Lead as a medium for writing tablets was used in the Neo-Hittite states of Syria (Bossert, 1951); whether this was also the case in Mesopotamia is a more open question (cf. Lambert, 1968-9, 64).

(d) Select Lead Analyses

The extent to which ancient metallurgists were able to extract silver from lead is still uncertain. By the Hellenistic and Roman Periods at least they were capable of extracting silver from lead down to a level of about 0.01 per cent residual silver. As there were galena deposits worked in antiquity that had natural silver contents of less than 0.01 per cent, it is not valid to argue that all ancient lead objects containing little silver (of the order of 0.00x per cent) must necessarily have been desilvered (Brill, 1967, 73, n.22). Aitchison (1960, I, 185) cites the following approximate standards for desilvering: Roman Period: 0.01%; later Medieval Period: 0.008%; modern practice: 0.002%.

(1) Ur: Late Prehistoric/Early Dynastic I:
Padgham (Woolley, 1934, 295) undertook "two assays of a leaden tumbler from a grave of Jamdat Nasr date, which each gave a quantity of silver amounting to 0.07 per cent". Reported in Woolley, 1956, 165, as "100 grams lead gave 0.0255 grams silver, approximately 9oz. silver to the ton".

(2) Ur: Early Dynastic III:
"some sheet lead...shapeless and pierced with holes due to decay" (Woolley, 1934, 148).
Desch analysed it as: Pb: 98.29%; Cu: nil; Sn: 1.30%; Fe: 0.41%; Ag: trace; Au: nil; (Woolley, 1934, 293). Padgham (Woolley, 1934, 295) reported that a "50-gramme sample of lead of the Cemetery Period, from PG 55, contained an unweighable amount of silver".

(3) Tell al-´Ubaid: Temple: Early Dynastic III:
Lead from stag´s head: Scott reported (Hall and Woolley, 1927, 38): "the lead...from the setting of the base of the antlers was pure lead, analysis indicating at least 99.8%... No silver or copper could be detected"; but Friend and Thorneycroft (1929) reported it as containing 0.0131% of silver.

(4) Assur: foundation plaque (Ass.22804) of Tukulti-Ninurta I (c.1244-08 B.C.) from the Ishtar Temple at Assur (Andrae, 1935, 45):
Pb: 99.60%; Cu: 0.14%; Sn: 0.15%; 0.06% iron; traces of arsenic and antimony.

(5) Friend and Thorneycroft, 1929, 105: reported Neo-Assyrian lead as having 0.11% silver.

X: TIN

(a) The metal and its uses:

Tin (Charles, 1975) does not occur in nature in metallic form, but in the combined state as a mineral, either the oxide (cassiterite or tin stone) or the much less common sulphide, combined with sulphides of copper and iron (stannite or tin pyrites). Tin minerals are only formed in situ where the mineralization is genetically related to granite or to the granite family of rocks, though not all such deposits contain tin. The oxide occurs both in veins and lodes, occasionally in association with copper, and alluvially as debris from disintegrated rocks bearing vein ore. Stream tin was always a particularly rich source of the ore when veins of tin running through granite may well have been inaccessible to miners and prospectors. In his review of the ancient Near Eastern terminology for tin Muhly (1973, 178) noted "that tin has no connection with mines, mining, or with any verb that refers to a mining operation". Alluvial tin normally appears as clean grains or pebbles free from any gangue. Only its weight, not its form or colour, suggests that it is a metallic compound. It may have been initially discovered as part of gold panning operations (Lucas, 1962, 255), though, in Mesopotamia at least, gold had been used for centuries before there is any evidence for the local use of tin. Even if a stream once bearing tin has been exhausted of the mineral, the pegmatites of the source should still be there for modern geological identification. They are the main key in modern searches for potential ancient tin sources in such likely regions as Turkey, Iran and Afghanistan.

The complex problem of the meaning of the Akkadian word annaku was considered in a paper of fundamental importance by Landsberger (1965) and more recently, in response to misleading metallurgical arguments, by Muhly and Wertime (1973). Although "tin" was its primary meaning, the possibility that it was from time to time applied to "lead", is further considered here on p.124. In so far as this section is concerned any reference to textual evidence for tin is to the word annaku (C.A.D s.v.).

In Mesopotamia, as elsewhere in antiquity, tin was used primarily for alloying with copper to make bronze (see p.17 here). At times it was used alone or, passing by weight, as currency, whilst texts refer to tin "utensils", tin foundation tablets, tin beads and bracelets. As a pure metal tin has drawbacks, since objects cast with it have a tendency to break and it has to be worked slowly to avoid cracking. Tin's earliest regularly recorded appearance in the form of the alloy bronze falls in Mesopotamia sometime in the first half of the third millennium B.C. at the time of the "royal" graves at Ur (E.D.IIIA) (see Bronze, p.17). By the middle of the third millennium copper bowls, found at Tell Asmar, were being dipped in tin, presumably for decorative effect (Muhly, et al., 1980a, 254) and tin was being used as a solder for silver lugs on a copper bowl at Ur (Collon, 1982; Craddock, 1984). It is quite a good solder for silver, copper and its alloys; Craddock has suggested this might have been its earliest use, before alloying with copper.

(b) *Production:*

Nothing is known of tin production methods in the ancient Near East, since no certain workings have yet been identified and no ingots are known from workshops, only from shipwrecks in the east Mediterranean (Muhly, 1978, cf. Artzy, 1983). The mineral cassiterite (see p.127) is easily smelted in the same way as the oxide ores of copper (see pp.5ff.). Once that technique had been mastered, cassiterite could have been smelted either in the same furnace charge as the copper ores, or separately. It is not known whether cassiterite itself was transported and traded much outside the source zone. Mesopotamian textual references to "bricks", "sheets", or "sticks" of tin indicate that the metal most often travelled already processed in the form of ingots (C.A.D. s.v. *annaku*, *ittû* and *zakû*; Veenhof, 1972, 32ff.; Wiseman, 1967), as finds from the sea off Israel illustrate. Two fragments of tin discovered in a Late Bronze Age context at Timna, a copper mining area in Israel, also indicate that it was added in metallic form to the copper (Craddock, 1980, 171). Analysis of metal from the Timna site indicated that very small percentages of tin, commonly attributed to copper ores, can be deliberate additions, though why, unless as a deoxidant, is uncertain.

(c) *The pattern of trade and the identification of sources:*

(i) *Ancient textual references:*

Few questions in the economic history of the Near East have been so much debated as the ultimate origin of the tin used there in antiquity. It remains a notorious enigma. Muhly (1973; 1976) has thoroughly reviewed the primary textual evidence. As the "archaic" texts, of Early Dynastic I from Ur, distinguish copper from bronze (Sumerian *URUDU* and *ZABAR* (Akkadian *erû* and *siparru*), its presence has been assumed a few centuries before its earliest material occurrence in the "royal" graves at Ur in Early Dynastic IIIA (see p.26). Gudea identified *Meluḫḫa* as the source of his tin (Falkenstein, 1966, I, 48, Cylinder B XIV); but the mention of tin in "Enki and the World Order", cited by Muhly (1973, 309) does not in fact refer to *Meluḫḫa* as a tin source. A Sumerian hymn to the god Ninurta refers to the copper and tin of *Magan* (Cohen, 1975, 28, line 144). Whether a poetic description of *Aratta*, in the Sumerian Lugalbanda Epic, stating that "its soil is 'tin-stone" (*im-an-na*)" may be taken to identify another source for alluvial tin or cassiterite, falters on uncertainties of translation. A text of Shu-Sin (c.2037-29 B.C.) refers to tin captured as booty in a campaign against *Zabšali*, a neighbour of Elam somewhere deep in Iran (Sollberger and Kupper, 1971, 152:IIIA4e). References to "tin-mountains" in Sumerian literature, generally to the east of Mesopotamia, are too vague to offer any reliable information on sources (Muhly, 1973, 288ff.). Yet, if taken with second millennium B.C. evidence from the Mari texts, they do not entirely lack credence. It is recorded that tin was sent to Mari from "Sheplarpak", usually taken to refer to Siwe-palar-ḫuppak, king of Anshan, in Iranian Fars (Dossin, 1970, 97). This re-inforces the view that some at least of the tin reaching Sumer came from a considerable distance to the east, either by land through southern Iran or by sea up the Gulf. In view of recurrent hostilities between Elam and Sumer in the third millennium B.C. the land route may well not have been used regularly. This tin largely by-passed the local

Iranian metal industries of the third millennium B.C., where tin bronze is rare; only at Susa is its presence better attested at this time (Berthoud and Francaix, 1980). Its occurrence in the Gulf metal industries has yet to be fully elucidated though it is present (Berthoud et al., 1982).

The situation in northern Mesopotamia in the third millennium B.C. remains obscure, not least on account of unresolved questions about the possibility of tin sources in Anatolia, where, on present chronologies, the earliest tin-bronze artefacts appear. "In spite of the fact that the meagre geological evidence we have does not point to an Anatolian tin source, archaeological logic still maintains that Anatolia should have produced - if only locally - some of its own tin in the Early Bronze Age (de Jesus, 1978, 101; 1980, 51ff.; see p.130).

The pattern of the tin trade west of the Zagros, in the earlier second millennium B.C., is reasonably clear from documentary sources. Tin, reaching Assur from somewhere to the east, was traded into Anatolia in the Old Assyrian period:

"What we can say with certainty is the following: the texts available (i.e. from Kültepe) refer to exports from Assur to Kanesh of ca. 17,500 pieces of textile and about 13.5 tons of tin. This documentation accordingly covers part of the total exports over a period of some 40-50 years. A very conservative estimate of the total would be at least 100,000 textiles and 100 tons of tin" (Larsen, 1982, 40).

In the Old Babylonian period Sippar was re-exporting tin (price ratio: 16(tin):1(silver)), received from Susa, either by way of Der or Eshnunna (Tell Asmar), or Larsa, to Mari, whence it travelled to the cities of Syro-Palestine (Muhly, 1973, 292ff.). Tin is also known to have been passing through Tell Shemshara (ancient Šušarra), which controlled the Sungasur gorge leading into the Rania Plain from northwest Iran (Laessøe, 1959). Thence it is likely to have passed through Erbil and Kirkuk into the Diyala Basin. The price of tin was 11-12 times higher than that asked for copper at Mari and in the Assur trade, and more or less 11-12 times cheaper than silver (Helzer, 1978, 109; assuming that weights at Kültepe, Mari, Eshnunna, Sippar, etc., were uniform and that is not certain).

For the Middle Assyrian period there is textual evidence from various sites in the north (Assur, Tell Billa, Tell Rimah) and the east (Nuzi), all indicating a relative abundance of tin at this time (Muhly, 1973, 290, n.357ff.; Saggs, 1968). This is confirmed by the prevailing prices (but cf. comments here under lead on p.124). In Ugarit tin was a little cheaper than the average price for copper and at Nuzi, where copper relative to silver was lower in price than at Ugarit, the average price of tin was 20% lower than that of copper (Helzer, 1978, 109). Indeed, at Nuzi the price of tin was at its lowest. But here again the equivocal identity of the metal (tin or lead) must be kept in mind. Only the texts from Tell Rimah give any indication of the tin's source "...brought from Nairi" (TR 3019: Wiseman, 1968, 183). This name embraces the region between the Upper Euphrates and Lake Van (perhaps extending further east towards Lake Urmia) known in the first millennium as Urartu. Hittite inventory texts of the thirteenth century B.C. record tin from Kizzuwatna (Cilicia), presumably only a transit zone (Košak, 1982, 195).

Of the Neo-Assyrian period Jankowska (1969, 265), in her review of booty and tribute lists, concluded:

"Although any definite conclusions may be hazardous owing to the scarcity of specific figures in the annals, we may still mark out Bit-Zamani as having the greatest mentioned reserves of tin (annaku), all other districts being incomparably poorer in this respect... Next to Bit-Zamani, the Laqe of Hammath are mentioned as being the most considerable reserves of tin...the locality ranking third among those richest in tin, namely Bit-Ḫalupe, lay midway between Laqe and Bit-Zamani". Bit-Zamani significantly occupied much of the western part of the earlier Nairi-lands.

In summary, the Sumerian and Akkadian sources all seem to focus on Iran, or lands to the east of it, whilst from the middle of the second millennium B.C. the region between the Upper Euphrates and northwest Iran comes to the fore. In the second half of the second millennium B.C., if annaku invariably refers to tin (see p.124), the metal is unusually plentiful and correspondingly cheap.

A review of this kind may do no more than leave a whole range of questions open, since the available evidence is all too inconclusive. Before moving on to the geological possibilities, three outstanding questions must be spelt out:

1. Was there a local tin source(s) in Anatolia exploited in the early third millennium B.C. (if not subsequently) that would explain the apparently precocious appearances of bronze in that region?

2. Was a source(s) in Afghaniṣtan (or eastern Iran) serving Mesopotamia predominantly by sea up the Gulf, rather than overland through southern Iran, in the middle and later third millennium B.C.? The reported absence of tin in a series of objects from Danish Excavations in Bahrein (levels of City 11), c.2500-1900 B.C., is not an absolute objection to this view (cf. McKerrell, 1977, 167: no details), since it is now reported in artefacts from Oman (Berthoud and Cleuziou, 1983).

3. Was a fresh source(s) to the north of Mesopotamia, with very considerable yields of tin, opened up in the second half of the second millennium B.C. and then exploited through into the first millennium B.C.? This might account for the fact that the price of tin (annaku) declined more markedly even than that of copper between the earlier and later second millennium B.C. (cf. Helzer, 1978, 109).

(ii) Evidence of modern geology

In the close study given to potential tin sources in recent years the old literature has been very thoroughly scrutinized and tested against the most reliable modern geological evidence. This examination has left much less room for manoeuvre than was once the case, as the possible sources for ancient Near Eastern tin have been very considerably narrowed down (Franklin et al., 1978).

Tin has now been detected in Egypt (Muhly et al., 1980). At several locations in the eastern desert the presence of alluvial tin has been confirmed; but there is no clear evidence for their exploitation in antiquity. Bronze was not regularly manufactured in Egypt until the earlier second millennium B.C. Evidence for potential tin sources in Anatolia is still equivocal though ideas of its existence have been encouraged by the granites of the Black Sea mountains (de Jesus, 1978, 101; 1980, 51ff.). A tin mine with signs of old workings has now been reported near Bursa (Yakar; pers.comm.). Long standing advocacy of tin sources in the Lebanon or Syria have not survived modern scrutiny (Muhly, 1973, 258). The question of tin sources in Transcaucasia and Caucasia, where again there seems to be no firm geological evidence for its presence (Dayton, 1972, 51-2; 1978, 66; Selimkhanov, 1978, 57), is complicated by reports of ancient metallurgical operations at Metsamor. This is a site on a large, low outcrop of rock in the Araxes Plain upstream from Erevan in Soviet Armenia (Mkrtiachan, 1967). Lack of any primary publication by an archaeometallurgist of this potentially very significant site precludes conclusions about its real purpose (cf. Burney and Lang, 1971, 68; Pigott, 1981, 74ff.). The original report suggested that the cassiterite found at Metsamor might have been recovered from alluvial deposits near Paleoaraks, at the foot of Mount Aragats (Alagöz), to the north, where a recent geological report was said to have confirmed the presence of tin, gold and silver. If this really was a source of tin in antiquity, local artefactual evidence, including actual objects of tin (Selimkhanov, 1978), as well as bronze, do not support exploitation before sometime in the local Middle Bronze Age at the earliest (cf. Mellaart, 1968, 200ff.). This would, however, correlate well with the trend in tin prices just reviewed (p.124).

Although geological circumstances in isolated parts of modern Iran are suitable for the occurrence of tin, possibly once yielding alluvial tin as they may have yielded alluvial gold (Muhly, 1973, 260-1; Muhly and Wertime, 1973, 119), hard evidence for tin there is still elusive. Tin is not mined in Iran today and there is no Medieval evidence for its exploitation locally (Allan, 1979, 27ff.). Muhly (1976, 98) notes a tin source in the area to the south of Birjand, in the Dasht-e Lut, reported by the Geological Survey of Iran. It is hardly surprising then that attention has increasingly concentrated on possible ancient tin sources east of Iran. Worked tin deposits certainly exist in the region southwest of Samarkand and along the Kok-Su river in Turkmenia (U.S.S.R.), with associated artefacts suggesting exploitation by the third and fourth centuries A.D., if not earlier (Ryzanov, 1979; Crawford, 1974). However, analytical evidence (Terekhova, 1981, 319) indicates that tin-bronze was not used in this area until the late third or earlier second millennium B.C.

Strabo (XV.2.10) referred specifically to Drangiana, the modern region of Seistan in southwest Iran, into Afghanistan, as a source of tin and Muhly (1973, 260) associated this directly with Gudea's report of receiving tin from Meluḫḫa. It is now known that

Afghanistan has two zones of tin mineralization. One embraces much of eastern Afghanistan from south of Kandahar to Badakhshan in the northeast corner of the country (Shareq et al., 1977); the other lies to the west and extends from Seistan north towards Herat (Cleuziou and Berthoud, 1982), the valley of the Sarkar river, where the hills are granitic. Here tin appears commonly as cassiterite, frequently associated with copper, gold and lead, and in quantities sufficient to attract attention in antiquity. Bronzes at Mundigak, and the controversial Snake Cave artefacts, indicate local use of bronze by at least the third millennium B.C. (Shaffer, 1978, 89, 115, 144). There is, as yet, no evidence of the early exploitation of tin sources in India or Pakistan, such as those in Bihar, whilst evidence for tin in Rajasthan is uncertain (Hegde, 1978; Chakrabarti, 1979; challenged by Craddock, pers.comm.). But there is every likelihood that tin was reaching Mesopotamia, along with semi-precious stones like lapis-lazuli and carnelian, from the region of modern Afghanistan (Ratnagar, 1981, 39ff., 70ff.; cf. Agrawal, 1982). Sources substantially to the east, in Thailand, now seem much less likely, particularly if the local Bronze Age eventually proves to have started later than the earliest use of tin further west; but tin from Etruria in Italy, from Spain, even from Cornwall, through Phoenician enterprise, in the first millennium B.C., may not be ruled out. Opinions differ radically about the possibility of Bohemian tin entering the Near East. Placer tin deposits in Bohemia appear to account for the precocious appearance of high tin bronzes there in the Early Bronze Age, so Mellaart (1968) may be correct in advocating it as an early source for Turkey, though Muhly (1973; 1976) is opposed to the idea, largely, it appears, from misconceptions about the nature of Bohemian tin deposits (cf. Taylor, 1983).

(iii) Objects:

The only objects of tin so far identified from Mesopotamia are a small group of bracelets from children's graves (nos.148, 158, 227), of the Old Babylonian period, at Tell ed-Der (Lerberghe, forthcoming).

XI. ZINC

There is no evidence, artefactual or textual, that zinc as a metal was known in Mesopotamia before Seleucid times. Its role, alloyed with copper, is treated here under BRASS (p.20). This metal was not separated in Europe until the eighteenth century A.D. (Aitchison, 1960, II, 480-2). Its role, as a trace element, in the copper used in Mesopotamia, is discussed elsewhere (see p.12; cf. Berthoud et al., 1982, 50-51). There is a possible reference in Strabo (13.1.56) to the production of zinc in antiquity in Anatolia by a kind of distillation process (cf. Waldbaum, 1983, 26-7).

Chapter 2. GLAZED AND RELATED MATERIALS

Whereas there is already a considerable literature on the development of metalworking in Mesopotamia, and other parts of the ancient Near East, there are very few complementary studies of the faience and glass industries. It is a subject even more fraught with problems than metallurgy, since so many of its products were tiny and fragile, particularly subject to decay and discolouration in the soils of the region. Moreover, as Stone and Thomas (1956, 38) observed, "far too frequently excavation reports, especially those dealing with the Middle East, lack precision, and accurate descriptions of the substances found are mainly wanting or may be wholly misleading". This potential hazard is not eased by classifications that also differ according to the native language of the excavator. Any remedy for these deficiencies will only be made gradually, as more and more objects receive proper scientific study. In the meanwhile it is possible to provide no more than a broad picture of the main trends, in which the details remain very much open to correction. A preliminary step has to be taken with terminology even if (as is all too often inevitably the case) good intentions are thwarted by the variety of the older literature and a necessary popular terminology.

I: Terminology and the Problem of Origins:

(a) Faience:

The ancient material known as faience has no relationship to objects of earthenware decorated with glaze and enamels, such as the faience of Faenza in Italy from which the term was originally derived. It is a long-standing misnomer for a composite material consisting of a sintered quartz body and a glaze. However inappropriate the term may now be, it is too deeply entrenched in the literature to be discarded. Synonyms used by various authors are: sintered quartz, glazed frit, composition, "Egyptian faience", paste, etc. The body of ancient faience is of "finely powdered quartz grains cemented together by fusion with small amounts of alkali or lime or both, and is usually colourless though impurities have occasionally tinted it brown or grey" (Stone and Thomas, 1956, 38) (see p.159). Although glass is treated separately here (pp.194ff.), there is an important distinction which has to be emphasized in any full definition of ancient faience. In a true glass the various ingredients must become completely fused together in a liquid melt, which is then cooled and acts as glass. The synthetic ancient material known as faience has only been sintered: "the ingredients were heated sufficiently so that some of them melted or reacted to form a very small proportion of liquid, which upon cooling served to cement together the unfused grains" (Brill, 1970, 115). This state is often, if rather confusingly, termed "glassy"; but it needs to be sharply distinguished from glass, particularly when the relevant ancient documentary sources, not in themselves explicit, are under consideration.

Faience can be modelled by hand or pressed into a mould. When dried and fired it becomes harder, though the hardness varies. The material ranges from soft and crumbly, through very compact, to a very close approximation both in hardness and in appearance to glass ("glassy faience").

In practical terms in Mesopotamia, where the surviving body is commonly in a friable condition and the glaze often only evident from tiny surviving specks, faience objects are just "sintered quartz", explaining why such misnomers as "white frit" or "white paste" are regularly encountered in the literature to describe them. In the nineteenth century accounts they sometimes appear as "porcelain" (Layard, 1853). Paste, like frit (see below), is a particularly unfortunate designation, as "in modern usage it is loosely used to describe a material formed by fusing powdered quartz and alkaline salts, and coloured with metallic oxides, for making imitation gems. It has no scientific meaning" (Stone and Thomas, 1956, 39).

(b) Glaze (and glass):

Glaze is a discrete layer of glass on the surface of crystalline (faience) or ceramic (baked clay) bodies.

Glass is a material made by cooling a silicate melt in such a way that it does not crystallize and retains some of the characteristics of a liquid. Ancient glass was made by melting together silica and impure soda. However, ancient glazes on beads, which date from a much earlier period than glass, were much more probably formed by the high-temperature reaction of an alkali on silica. It will be seen that the raw materials of faience and glass are the same but employed in very different proportions and with a different technology. Glazes are not found on baked clay (terracotta) until the sixteenth century B.C. in the Near East. The glazes were coloured (most commonly blue or green) with metallic compounds. True vitreous glazes must be clearly distinguished from slipped, painted or burnished terracotta surfaces. A glaze differs from a paint in that it has to be fused at high temperatures whereas the medium of paint is organic and will burn on heating. But, since mixtures of frit, silica and lime, may be "painted" onto a baked clay surface, and then fired to form a glaze, the term "paint" inevitably appears as a verb in the following pages.

(c) Frit:

This is the most problematic of the terms relevant to the subject, since it is often used as if it were synonymous with faience (cf. Perkins, 1949), from which it should in fact be clearly distinguished. Faience has a glazed surface and only when this is totally worn off might an object be mistaken by eye for frit. A frit has a sintered polycrystalline body and no glaze (cf. Kuschke, 1970). Frit has a very specific meaning in glass technology: "a calcined mixture of sand and fluxes ready to be melted in a crucible to form glass" (Oxford English Dictionary). It is the first stage in production; the raw materials are "fritted" at a relatively low temperature before regrinding and subsequent melting. In the present context the term "frit" is used to describe all sintered polycrystalline bodies which were not glazed. (Fritting is usually considered a process in which the soluble salts are made insoluble by breaking down the carbonates etc. and forming a complex mass of sintered silicates). It has been suggested that "unglazed faience" might be an appropriate synonym to avoid confusion with glass technology.

In the case of a coloured frit the colour runs throughout; there is no distinction, as there is in faience, between an inner body of one colour, usually white, and a distinct surface colour. Such cakes of differently coloured frits, and the pans for producing them, were found at Tell el-Amarna in Egypt, dating to the first half of the fourteenth century B.C. This is the oldest glass factory yet excavated in the Near East or Egypt (Petrie, 1894, 25ff.; Lucas, 1962, 19ff.). Once produced, cakes of frit may be ground into a fine powder for use as a pigment, or for further combination and treatment to produce objects of moulded frit or glass. In the period covered by this survey, blue was by far the most popular colour for all frit artefacts; other colours occur relatively rarely, usually only for beads or similar small items of personal ornament.

The blue objects are usually sintered mixtures of quartz and the mineral "Egyptian Blue", which is a silicate of copper and calcium ($CuO.\ CaO.\ 4SiO_2$) identical to the natural mineral cuprorivarite (see p.188). The term "Egyptian Blue" is frequently used to describe both the frit itself as well as its mineral constituent. Indeed, it is not yet established if there were blue frits in the ancient Near East from which the mineral "Egyptian Blue" was absent. The term "glass paste" (pâte de verre) is often applied rather indiscriminately to frit (blue or any other colour) or any opaque glass, most significantly, in this context, by Woolley in his reports on the excavations at Ur.

The problem of ancient terminology for the materials discussed here is considered on pp.224ff.

(d) Glazed stones:

Both in Egypt and Mesopotamia the stones most commonly glazed in prehistoric times were either magnesium silicates, like steatite or serpentine, or silicates, such as quartz or flint, which frequently contain iron as an impurity.*

If one thing emerges clearly from this preliminary survey, it is the prevailing confusion in the popular terminology for ancient glazed and related materials. But, as this is progressively refined by proper scientific investigation, it will be no easier to apply the emerging terminology acceptable to the scientists, since much of it inevitably turns on features not evident to the naked eye, even a trained one. In such circumstances there is much still to be said for unambiguous, non-committal phraseology for common use, such as: glazed stone; "glazed composition" for faience (giving colour of glaze, and of body if possible); "unglazed composition" for the

---

* I propose to discuss stones in a future volume in this series.

frits (giving the colour of the body and evidence that traces of glaze have been sought in vain); and glass, which may usually be recognized by eye and is familiar to everyone. Even the current suggestion that "sintered quartz" or "soda-lime-silicate" be used for "composition" prejudges the vital questions.

Although in Mesopotamia it is broadly possible to separate "glazed stones" from "faience" and "glazed terracotta" for coherent treatment, this is not yet the case with "frits". In the present state of knowledge, and then only with reservations, for it is not yet known whether there were any other blue frits, may "Egyptian Blue" be treated independently. "Frits" of other colours, though known at sites like Nuzi, have not yet been systematically identified elsewhere, so any separate treatment of them is impossible at present. This is anomalous, but unavoidable.

(e) The emergence of glazed materials:

The organization of the following sections, for convenience rather more than from any conviction, reflects a prevalent view about the way in which faience was first developed in Western Asia. It is usually argued that the initial impulse came from a desire to reproduce rare stones of bluish-green colour by glazing commoner dark or dull coloured stones a blue-green colour. It is assumed that this originally happened in a situation where siliceous stones, such as steatite, the ashes of alkaline-rich salt water plants, and some copper, were in fortuitous combination under heat, presumably in a fire-pit or hearth. The use of malachite as a cosmetic from an early date, and the palettes on which it was ground for this purpose, notably those of sandstone, would have been a possible combination for accidental glazing (cf. Lucas, 1962, 172; Peltenburg, 1971): Lucas, in fact, successfully experimented with a two-stage method for reproducing faience in which he first glazed solid quartz pebbles using a mixture of natron and malachite. This glaze was then chipped off, ground to a fine powder and applied to the surface of a ground-quartz body, which was subsequently fired.

How the glazing of stones, or artificial bodies, might have related to the development of contemporary copper working, if at all, has yet to be clarified; nor is it clear whether the marked absence, in both Egypt and Mesopotamia, of attempts to glaze baked clay at this initial stage is indicative of a particular context for the development of faience. Attention has been called in this connection to Perrot's observation that the clay walls of furnaces at the Chalcolithic site of Abu Matar, near Beersheva, in Israel "had become glazed by the products of the combination of metal, silica and bases" (Perrot, 1955, 79); but it is not an obvious analogy for the early glazing of soda-lime-silicates.

It is, at present, impossible to say with any real confidence whether the production of faience appeared in Western Asia before it did in Egypt; comparative chronologies are still too imprecise for the question to be resolved. In Egypt beads of green or greenish-blue glazed steatite (perhaps also of other stones) appear first in the Badarian Period (Early Predynastic), currently dated to the last quarter of the fifth millennium B.C. or soon after. Glazed stones, predominantly steatite, continued as one of the commoner materials for beads in Predynastic Egypt. Green-glazed faience beads appear first in the Amratian Period (Naqada I; Middle Predynastic), c.4000-3500 B.C., though not as commonly as was once assumed (cf. Vandiver, 1983a, A-64ff.). Caution is also required with larger objects, like the hawk pendant from tomb 1774 at Naqada, since it may be made of glazed quartzite or glazed sandstone rather than of faience (Vandiver, 1983a, A-71ff., pl.29c). The calibrated Carbon-14 chronology on which the absolute dates given here are based is still open to modification, as there are isolated indications that the Egyptian Predynastic cultures may have begun earlier and changed more slowly than current hypotheses assume. With the exception of an isolated find in Iran (see p.142) no faience has been reported from anywhere in Western Asia that may be certainly dated before the Ubaid Period.* Within that long period, currently dated on a calibrated Carbon-14 chronology from about 4500-3800 B.C., it is hard to place surviving faience beads with any precision (see p.142) though they are certainly in production before its final phase (Ubaid III). This being so, if the absolute chronologies for the respective regions given here are correct, the present weight of evidence would be in favour of first production of faience somewhere in Western Asia, probably in Iran or Northern Mesopotamia, with diffusion into Egypt to the west, the Indus region to the east (cf. Stone and Thomas, 1956).

Whatever the force of traditional speculations about origins, it is likely that the heat treatment of stones, from the Upper Palaeolithic period, to facilitate flaking had familiarized men from an early date with the virtues of applying heat to stone. In areas where early faience industries may at present be noted, in Egypt, in northern Mesopotamia and in the Indus Valley, there is invariably a previous tradition of heat treating steatite, in some cases actually of glazing it (in the precise sense of the word). This craft continued to flourish for some time side-by-side with production of faience and frit. A "burnt" steatite bead (see p.138), painted with powdered malachite, a common prehistoric pigment, when fired would produce a greenish-blue glaze. The steatite would provide silica, the whitening process alkali and the malachite the colourant. From this it is a small step to faience.

In those regions where innovatory faience production is evident by the later fourth and third millennium B.C. contemporary potting and metalwork exhibits a considerable mastery of pyrotechnology and a willingness to experiment both with ores and pigments. Craftsmen were also skilled in working on a small scale with such materials as stone, bone, shell and ivory. It is difficult to gauge whether a general restriction on the supply of the real stones stimulated imitation or whether it was in places where the semi-precious stones themselves were readily available to the rich and privileged that craftsmen were moved to produce imitations for those among the less affluent wishing to emulate the privileged. The history of faience in the later second millennium B.C. (cf. Foster, 1978, 173) might be taken to suggest that innovatory faience and glass manufacture was more likely where the naturally coloured stones were readily available and widely used; widening social demand, rather than deprivation in the richer sections of society, was the stimulus.

---

* The evidence from Arpachiyah is equivocal (see pp.142, 193).

II Glazed stones:

(1) Steatite:

As steatite is soft, it may be cut easily and its fine-grained structure makes it a very suitable medium for small objects like beads and amulets. It is particularly well suited to glazing, as it can be heated without distortion. The heating dehydrates it and dramatically increases its hardness from 1 to 6 or 7 on the Moh's scale. This seems to have been appreciated very early; presumably small objects were first carved in the soft stone then heated to harden them.

Beck (1934a), whose study of glazed stones is still fundamental, defined three different ways through which the surface of steatite was chemically altered in Egypt and the Near East in antiquity:

(i) "Glazed" this involves either the application to the surface of the stone of a vitreous glaze, which is then fired on, or use of the cementation process as in faience manufacture (see p.161). Even when the glaze has flaked away, a very hard surface remains on the steatite. Beck (1934a, 69; cf. Lucas, 1962, 155-6) regarded this procedure as "almost entirely Egyptian", an observation which still stands. Even from Egypt objects of considerable size in glazed steatite are rare, mostly dating between the XIIth and XVIIIth Dynasties, c.2000-1200 B.C. (Bannister and Plenderleith, 1936), and none has been published from Mesopotamia. Noble (1969, 438) has described the Egyptian method of glazing steatite and pointed out its hardening effect on the mineral. Vandiver (1983a, A-64ff.) has published a careful examination of Egyptian glazed steatite beads and seals.

(ii) "Burnt": Beck (1934a, 69; cf. also in Vats, 1940, 393-4) explained this as the application of an alkali which was then fired, or of a glaze which was of such a nature that it almost always flaked entirely away and left a very soft surface on the steatite. This is the category to which the commonly termed "glazed steatite" of Mesopotamia belongs, though it is best known from Indus Valley seals and beads of the third millennium B.C. (Hegde, 1983).

(iii) "Etched": for this the surface of the steatite was whitened, probably only with an alkali, and heated, after which it was decorated with a pattern in ferruginous clay or some salt of iron. This is an Indus Valley technique not yet independently reported from Mesopotamia.

The history of burnt steatite in Mesopotamia is obscured by the absence of any systematic study of beads from the region. In only one case is it certainly known for any other range of objects from the area: a well-defined group of cylinder seals current in Late Protoliterate to Early Dynastic I times. "Burnt Steatite" beads are listed in the Ubaid period graves at Arpachiyah (Mallowan, 1935, 38), when it was used for "minute glazed steatite ring beads". In reporting on the beads from the deep sounding at Nineveh Beck (in Campbell Thompson and Mallowan, 1933, 179) noted "a third bead from the layer Nin.2 at a depth of -63' is made of chemically treated steatite. This bead is exactly similar to the white steatite beads...from Nin.3 and Nin.4, and it is of interest to know that this process is so early". Nin.2 would fall somewhere in the early Ubaid horizon (?Arpachiyah TT 8-6); 3 and 4 cover the Uruk IV - III horizons in the south. Some of the treated steatite beads of Nin.3 were identical in shape and material to Nin.2, others were long cylinders or barrels (ibid., pl.LXXIX, fig.20-23). In a child's grave, as at Arpachiyah, treated white steatite beads were combined with untreated black; two faience beads were associated. In Nin.4 there was more variety, but the great majority were tiny white steatite beads (cf. Beck, 1934a, 76-7).

A cylinder seal attributed to "Nineveh 4" in the later fourth millennium B.C. is of considerable interest (Campbell Thompson and Mallowan, 1933, 140, pl.LXV, 16). Beck (ibid., 181-2) commented:

"The stone appears to have been treated by the application of an alkali combined with great heat, but no silica has been applied; in this way the treatment differs from that of the Egyptian specimens". Frankfort (1939a, 34) isolated the group of seals to which this example belongs, noting that shape, material and designs were at this time, in the late Protoliterate to Early Dynastic I periods, without precedent. Seals of the same general type were also made in faience. The group is represented among sealings at Nineveh (Campbell Thompson and Hamilton, 1932, pl.LXIII.II,14; Campbell Thompson and Mallowan, 1933, pl.LXV.21-35, LXVI. 3,4,7,18-26), and at Jamdat Nasr (Buchanan, 1966, no.72), and by actual seals in the Diyala (Frankfort, 1955, 17ff.), at Kish (Buchanan, 1966, 16ff.), at Fara (Heinrich and Andrae, 1931, pls.691, m; 72 c, d, e) at Telloh (Frankfort, 1939a, 34, n2), and in considerable numbers at Susa (Amiet, 1972, 143ff.). The predominance of these seals in northern and eastern Mesopotamia has been commented on. Le Breton (1957, 108) challenged Frankfort's view that there was no reason to consider them other than of local Mesopotamian manufacture, suggesting that "their chief centres of fabrication and dispersion should be sought somewhere along the Eastern piedmont" (Buchanan's "Piedmont Jamdat Nasr" style). Even if they may not be considered specifically "Susian", Amiet (1972, 143) has emphasized their close affinity with "Proto-Elamite" glyptic. In the Susa sequence they first appear in level 17 (Amiet, 1971, 222); significantly some take a different form, the height and diameter being approximately equal, and have a very close affinity with a contemporary series of such seals in faience (cf. Amiet, 1972, 111ff.).

The production of burnt steatite objects in Mesopotamia after the beginning of the Early Dynastic period is not well documented and may have ceased. Beck (1934a, 76; in Vats, 1940, 393, pl.CXXXV.4) cites minute burnt steatite beads from PG 55, perhaps a robbed "royal" grave of Early Dynastic IIIA at Ur; but they are not specifically mentioned in Woolley's final report and seem to be those listed there as "shell", presumably on the basis of field records (Woolley, 1934, 414, 527). They were said to have formed part of a girdle with lapis and carnelian beads. "The great number of these small beads (i.e. of Steatite) in this grave is very extraordinary, since the use of burnt or treated steatite is very rare in Mesopotamia at this time" (Beck in Vats, 1940, 410). What is meant by Beck's remark that "a number of beads of a very much rougher nature, also found at Ur, are dated to the Persian period" (Beck, 1934a, 76) is equally uncertain, since Woolley does not list them specifically in the appropriate excavation report (Woolley, 1962). But, as etched carnelian beads re-appeared at Ur in the Achaemenid period, following the revival of contacts with the Indus region after a lapse of more than a millennium, comparable treatment might well have been given once again at that time to other stones (cf. Reade, 1979).

(2) Quartz:

Beck (1935) is so far alone in examining the evidence from Mesopotamia for the surface modification of quartz beads, pendants and inlay fragments. In each case the surface had been coated with a substance and then heated until it melted. Some of the specimens he tested from Mesopotamia (and the majority from Egypt) "appear to have had an already made glaze or else the ingredients to make a glaze powdered up and applied to the surface, and then to have been fused". Another method had also been current in Mesopotamia, and in Iran: "to these an alkali only was applied instead of a glaze; they were then heated until the alkali fluxed the surface by combining with the silica of the bead and flowed over the surface. These beads have an extremely high polish which was at first attributed to sand polish, but a careful examination soon proved that this was not the case" (Beck, 1935, 21). Some of the Mesopotamian specimens have the surface completely covered with small conchoidal fractures and generally only slight signs of having been glazed (frosted). Beck (1935, 25) explained the difference between high polish and frosted as being "only that, whereas the latter had a completed glaze powdered and applied to them, the former had an alkali only added which formed the glaze by combining with the quartz of the bead. The object of frosting the surface seems to have been to give a greater surface for the alkali or glaze to work on. In the case where an alkali only had been added, this seems to have attached itself so firmly that it never flakes off, whereas in the cases where a glaze had been added, it seems to have flowed at a temperature too low to attach itself properly and had generally almost entirely disappeared".

The specimens tested from Ur were "supposed to date from before 3500 B.C., whilst others which are practically identical are dated 900 B.C." This is presumably information supplied by Woolley. As he dated the "royal graves" at Ur, in 1934, to between 3500-3200 B.C., it may only be assumed that the beads in question date before E.D.IIIA; but their significance is impossible to establish in retrospect. An illustrated bead (Beck, 1935, pl.II:9) would be acceptable as of Late Prehistoric or Early Dynastic date; a small rougher pendant ("shield") is dated to the later third millennium (Ibid., pl.II.6). Disc beads of glazed milky quartz were reported from Nineveh, but without exact context (Campbell Thompson and Mallowan, 1933, 182, pl.LXXIX: 30-2).

The later third millennium "bead layer", excavated at Nineveh in 1929-30, yielded evidence for a distinctive combination of glazed quartz and faience tooth-pendants: "These are made by inserting a large piece of quartz in a piece of faience somewhat resembling a fang, in which the perforation for suspension is made. The quartz portion was covered with soda and then the whole was heated to a great temperature so as to harden the faience and produce a sort of fire polish on the quartz" (Beck, 1931, 428; cf. 1935, 28). In a variant of this combination of materials, spherical in shape, a considerable number of tiny pieces of quartz were inserted into faience and then glazed overall (Beck, 1935, 28).

Mallowan (1947, 160, pl.XXVIII) reported that some of the rock crystals beads in Late Prehistoric or Early Dynastic contexts at Brak were glazed.

(3) Carnelian and other stones:

Although not in a strict sense "glazed", this section may appropriately end with the best known of all surface decoration on stones found in ancient Mesopotamia: the "etching" or "bleaching" of carnelian. This method may also be applied to agate, chalcedony and felspar. Beck (1933) distinguished between decoration in white lines on the background of the natural colour of the stone and a design in black on the artificially whitened surface of the stone. The latter technique is relatively rare. There are no grounds for believing the stones were, strictly speaking, etched, with white material inlaid in a groove. A pattern was drawn on the stone with a solution of alkali (generally soda) and the stone was then heated through a 300° to $400^{o}C$ temperature until the alkali entered into it, making a permanent white design. Mackay (1933) described the actual method as he saw it done by an old man in 1930 at Sehwan in middle Sind. As the soda treatment has a very different effect on different stones, and on different levels of the same stone, this method was sometimes used to heighten colour contrasts in beads (cf. Beck, 1935, 34, pl.V:1). The layers most resembling crystalline quartz are those least affected.

Reade (1979) has assembled and discussed the etched carnelian beads reported from Mesopotamia. They first appear there at some point in Early Dynastic IIIA, in the middle of the third millennium B.C., and are commonest between then and the Akkadian period. Thereafter they decrease and by about 1900 B.C. have virtually disappeared until they re-appear during the Achaemenid period. The technique of decoration was intrusive, from the Indus Valley (Tosi, 1976-80), where supplies of carnelian were locally available, and the beads themselves found in Mesopotamia are predominantly of foreign, again probably Indus, manufacture. Variants are occasionally identified, usually by anomalous patterns or shapes; but whether these were manufactured in Mesopotamia or somewhere along the way from the Indus is an open question.

(4) Summary:

If comparisons with predynastic Egypt are made, it is likely that faience all but replaced "glazed" stones in Mesopotamian workshops after the early third millennium B.C. In the first instance the bead-makers had been concerned with reproducing rare blue-green stones, notably turquoise, by glazing common dark stones. Once they realised that the same effect might be achieved more easily and flexibly with faience they abandoned the glazing of dull-coloured stones.

III: Faience (glazed sintered quartz):

(A) An historical survey:

(i) The Prehistoric period (to 3000 B.C.):

The first appearance* of faience objects in Mesopotamia is a matter of some uncertainty, since Mallowan's claim for "frit" beads and vessels in levels of the Halaf Period at Arpachiyah remains isolated and, in itself, unsatisfactory. As Stone and Thomas (1956, 41) commented "Mallowan recorded a frit (faience) seal or discoid pendant, unfortunately without stating its provenance, but by implication from Halaf levels" (cf. Mallowan, 1935, 91, 98). Both Amiet (1961, 15) and Buchanan (1967, 266) accepted the early dating, though Buchanan's phrase "found at such a depth in Halaf levels", is not substantiated by Mallowan's publication. Stone and Thomas (ibid.) took this seal to be a stray from Ubaid levels at Arpachiyah. As there is a comparable seal of faience, close in shape and design, from Gawra XII (Tobler, 1950, pl.CLIX.27), of the early Uruk period, the case for caution is re-inforced. Even more problematic is Mallowan's identification of two "frit" vessels from Halaf levels at Arpachiyah, since these would be an even more strikingly precocious innovation. In the most important instance Mallowan cites the find spot, in Halaf levels (TT 6-7), and illustrates the vessel (Mallowan, 1935, 175, pl.XIX.9), which is certainly of a shape known in the Halaf ceramic repertory. It is said to be "now bleached white". So long as this early date is unsubstantiated by finds elsewhere, there is always the possibility that both these vessels were of weathered gypsum, or a related stone, not of faience (cf. Fukai et al., 1974, 58).

However, such an early date may not be ruled out on these grounds alone, if only on account of one presently isolated find in Iran. There is a single, well-preserved, blue faience bead from level 2 of the Japanese excavations at Tall-i Muskhi in Fars. This is part of a Hassuna-related chronological horizon (c.5500-5000 B.C.) (Fukai et al., 1973, 65; colour plate; pl.XL.1.1). Even allowing for intrusion, and stratigraphically this was not evident (Fukai, 1973, 65), this is a "one period" site, with a homogeneous material culture in the main occupation levels. There were isolated intrusive graves.

It is from sometime in the Ubaid period, in the fifth millennium B.C., that beads of artificial materials are first clearly evident on a few sites both in northern and southern Mesopotamia and may already have been widely traded. The terminology used in reports is too various in the absence of scientific examinations to allow for rigorous comparisons. Faience ring-beads were reported from graves of the Ubaid period at Arpachiyah in the north (Mallowan, 1935, 38, 92, 97). At Tepe Gawra Tobler (1950, 192) reported that "white paste" was commonly used for beads in strata XIX-XVI (later Ubaid) and more so in the strata succeeding XIII, that is into the Uruk period. It was used for a ring pendant in stratum XII (Tobler, 1950, 195, pl. CLXXIII:41); from strata XIII-XI came a series of faience stamp-

---

* See note on p.193.

seals (Tobler, 1950, 178). In the south in the "Ubaid Cemetery" at Ur only one grave (PFG/L; Woolley, 1956, 88), of the 45 published, contained "small paste ring beads". In these graves however beads of any kind were rare; one grave held steatite beads, three had shell beads. At Eridu the earliest reported "frit" bead is in temple XI (Ubaid 3); later Ubaid period contexts in temple VII yielded a "frit" stamp seal and in temple VI three beads "in bluish frit" (Safar, 1981, fig. 117:28, 32-3). At Eridu a late Ubaid period cemetery of 193 graves was cleared, some with more than one skeleton; of the 175 graves reported 6 contained beads of "frit", sometimes long strings of them, for women and children (Safar, 1981, graves 90, 110, 114, 158, 174, 186). At Qabr Sheykheyn in Khuzistan, a Susa "A" settlement, of the later fifth millennium B.C., Weiss (1972, 172) excavated a child's grave "with bracelets of minute paste beads on its ankles and wrists"; old excavations at Susa itself indicate the appearance of faience personal ornaments in graves of the later Ubaid period (Stone and Thomas, 1956, 43).

It is not until the fourth millennium B.C. that a more varied production of faience objects is evident, primarily in the north. The "bead stratum" at Nineveh, cited by Stone and Thomas (1956, 42; also Foster, 1979, 23) in this connection, is in fact a millennium later in date (see p.145), so cannot be drawn upon. But Gawra and Brak offer sufficient information to establish the general status of the industry at this time. Faience beads were common in burials at Gawra of the fourth millennium (Tobler, 1950, 87-8). The "Eye-Temple" at Brak was set on a platform incorporating a succession of earlier structures. These had been plundered in antiquity through a whole series of tunnels and pits: "we entered these plunderers' tunnels which led to a stratum of grey mud-bricks from which we removed many thousands of objects.....beads of which there were many hundreds of thousands" (Mallowan, 1947, 33). The beads had not only been scattered liberally in the foundations of the "grey-brick" temple (Stratum C), but actually moulded into the bricks. They were predominantly of faience and of glazed steatite (ibid., 254-6, pl.LXXXIV). Mallowan published Stone's reports on three faience beads: a segmented one of "soft faience core with traces of light greenish or bluish glaze remaining.....it exhibits characters which are normally associated with beads of a very much later date, in particular the highly developed technique of manufacture"; a mulberry-shaped bead "of white faience, and traces of blue glaze remain in the crevices"; and a fluted, biconical bead "of white faience which was originally coated with a black glaze". In addition to the faience beads, this general context yielded a remarkable series of faience amulets or seals, kidney or rosette-shaped, or with backs moulded as bears, ducks, hedgehogs, hares and monkeys (Mallowan, 1947, 102, pl.X:9; 108-9, pl.XIV:9-13; 117-8, pl.XVI:1: 119, pl.XVII:5-8; Buchanan and Moorey, 1984, nos. 146-151, 178-181). There is no reason to assume they are of southern origin as Perkins did (1949, 188).

A late prehistoric house at Khafajah yielded a "paste mouse amulet" (Delougaz, 1967, 27: House 11). At Jamdat Nasr itself faience "does not seem to have been in very common use" (Mackay, 1931, 273); in two cases it was used to imitate forms otherwise made in shell or mother-of-pearl (ibid., 273-4, pl.LXXII.6, 27-9). In view of this scattered evidence any categorical statements about comparative degrees of use and production are impossible; but, at the moment, it does not seem that faience was much used in the south for beads and amulets. Faience beads are only very rarely mentioned by Woolley in his report of the prehistoric levels at Ur (cf. Woolley, 1956, 66, 178:U.14469; 67, 179:U.14943). Faience is not listed as one of the materials used for beads in the "Kleinfunde" found at Uruk (Heinrich, 1936; unless it is meant by "kalksandstein", 26, 41, pl.13h), nor does it seem to have been employed for the many animal amulets and seals favoured in the south at this time. In Behm-Blancke´s (1979, 66-9, no.129) catalogue of the excavated prehistoric animal amulets only 1 out of 138 is queried as being of faience, and it is from Khafajah.

At some time towards the end of the fourth millennium B.C. the use of faience was significantly extended to the production of small vessels and for this there is evidence from southern sites, though there is every reason to believe, as later evidence confirms, that they were luxury items available only in small numbers. A faience jar, originally blue-glazed, with impressed chevron ornament on the shoulder, was found at Ur in stratum ´H´ of Pit F: "so far as the decayed state of the frit allows of judgement, the vase was moulded round a core which was subsequently removed" (Woolley, 1956, 30, 63, 177:U.14422, fig.9d). Slightly lower in the same sounding were "fragments of a small vase [in fact a small open bowl] of white frit covered with pale turquoise glaze" (ibid., pl.26g:U.14908). Contemporary vessels were found at Khafajah (Delougaz, 1942, 137, 142). Scattered information from Telloh, where the shapes are those used in contemporary pottery and the glazes when preserved are simple blues and greens, extends the evidence for vessel production at this time (de Genouillac, 1934, 36-7, 61). The occurrence of a "rectangular faience vessel, about 9cm. high" in one of the rooms at Jebel Aruda on the Euphrates in Syria, a site characterized by Uruk IV material culture, is of equivocal significance. It is not clear whether it is of northern or southern origin (van Driel, 1983, 7).

Although Proto-Elamite Susa yielded evidence of inlay plaques in faience, as well as bear-shaped amulets, there does not seem to be any evidence of vessels (de Morgan and de Mecquenem, 1912, figs.22-3; Le Brun, 1971, fig.70:23). Complementary to the "burnt steatite" cylinder seals of the later prehistoric period in Susiana and inner Iran was a series in faience (Frankfort, 1939, 35; 1955, no.51, 53, 55, 63, 129, 143, 162; Buchanan, 1966, 15-16), produced somewhere on Mesopotamia´s eastern periphery.

(ii) Early Dynastic to Ur III Periods (c.3000-2000 B.C.):

The production of simple faience beads and amulets, and the use of faience for globular pinheads and "handles" (cf. Mackay, 1929, pl.XLIII.2), continued through the Early Dynastic period, though the evidence for Early Dynastic I-II is sparse (cf. Khafajah: Delougaz, 1967, 27-9: graves 14, 25; Ur: "Jamdat Nasr" graves: Woolley, 1956, 104ff., nos.16, 82, 103, 149, 159b, 161, 239, 295, 325, U.19249, 19316; Kish: Y graves: Moorey, 1978, Fiches E09 to G03). By early Dynastic III they were more common in graves on certain sites. In virtually every grave in cemetery A at Kish, from which beads were recorded, some are of faience (Mackay, 1925, 1929; Moorey, 1978, Fishes C08 to G01), whereas in the Early Dynastic III graves of the

"Royal Cemetery" at Ur it is not so common as might be expected. Of the 650 or so private graves at Ur attributed to Early Dynastic III, less than ten are reported as containing faience beads or faience pinheads; there was faience in two "Royal Tombs" (Woolley, 1934, 524ff.; Nissen's dating and tabulation, 1966, 82, pl.22). Cylinder seals were occasionally still made of faience; but as Mackay (1929, 193) remarked of those from Cemetery A at Kish, they are commonly in very poor condition. Faience vessels are still extremely rare from excavations of this period. None was reported from the "Royal Cemetery" at Ur; but there was a miniature bowl from an Early Dynastic III level in Houses 6 at Khafajah (Delougaz, 1967, 28) and others are reported from Tell Agrab (Delougaz, 1942, 268, 281). A macehead of green glazed faience from level 4 of the Kl.Anten-Tempel at Tell Chuera has been dated to the second quarter of the third millennium B.C. (Moortgat, 1967, 32-5, fig.23). It may have been a macehead carried by a deity statue; such maceheads in glass are known later (p.223). A fine green-glazed spouted bowl, decorated with a bearded bull couchant in low relief, was found in grave 317 on Tell Ingharra at Kish; this dates late in the Early Dynastic or early in the Akkadian period (Watelin, 1934, pl.XXXI.7; Moorey, 1970, 127-8; 1978, 74). A "glazed tile fragment" was reported from the surface at Khafajah and might belong to the mid.third millennium B.C. (Delougaz, 1942, 163).

Such meagre evidence does no more than indicate that the repertory of faience makers by the third quarter of the third millennium B.C. encompassed not only seals, amulets and beads, but also moulded vessels, votive weapons and other small objects. The use of moulds recognized the potential of faience for large-scale, standardized production and began to emancipate the material from a lithic-based technology. The suspicion persists, though it may not be put more strongly, that this industry was more active in Akkad and the north than in heartland Sumer. However, in the first half of the third millennium B.C. faience beads were already distributed far down the Gulf to Oman (de Cardi, 1979, 80ff.) and an unusually vigorous and varied faience production was established in the "Indus Valley" civilization (Allchin and Allchin, 1982, 192) by the third quarter of the millennium.

In the North chronologically sound archaeological evidence for the study of faience in the Akkadian period is provided by finds in levels VIII-IX at Tell Taya (cf. also Chagar Bazar: Mallowan, 1936, 56-7; Wilayah: Madhlum, 1960, 91-2). These levels yielded beads, "buttons" and pendants, simple bowls, some with lids and the fragments of an inlaid beard (Reade, 1968, 244-5, 249; 1971, 98, pl.XXVd; 1973, 165, pl.LXXVd, LXVIIIb, 167, pl.LXVIIb; Bimson, 1973). Vessels similar to those from Tell Taya are reported from Mari (Parrot, 1956, 121, pl.LII.1047).

Parallels from Tell Taya also indicate that the majority of faience beads in the so-called "bead-layer" excavated at Nineveh in 1929-30 are likely to be of the Akkadian period rather than a millennium earlier (Campbell Thompson and Hutchinson, 1931, 82; Beck, 1931; 1934a; 1935). This washed-out level included "a number of pieces of slag". For the first time there is clear evidence of a wider range of colours than the traditional blues and greens, though black (manganese) had already appeared on "handles" in Early Dynastic

III (at Kish: Moorey, 1978, Fiche F04). Beck (1931, 427) reported of beads that "the core was usually black but in some cases it is a rather dirty white. In one case the core nearest to the glaze is white, whilst that nearest the perforation is black. The glaze which is very opaque, was originally white, green, blue, red and black. The white glaze was sometimes on a white core and sometimes on a black one". Two unusual types of bead are the bicolour inverted drop pendants and the "crumb beads". The former "have the upper and lower parts of different colours.....not merely glazed with a different colour but the cores of the two parts are made of quite different materials..... The upper part, which is perforated, is almost always black, whilst the lower part has a white or colourless core. This part was glazed in colours.....whilst some were red others were yellow and blue". Crumb-beads, oblate or spherical and with "crumbs" of faience over the surface, mostly had a dark core, and were originally glazed blue or red. Some white faience beads had black stripes and vice versa. Beetle and vase-shaped amulets appear, as well as tiny faience ducks, known on other sites in the north and at Ur (cf. Foster, 1979, 27-8). Only 19 of the thousand or more private graves at Ur dated to the Akkadian or Neo-Sumerian periods are reported as containing faience beads; 10 held "glass paste" beads (Nissen, 1966, 82, pl.22).

Small moulded statuettes of animals couchant in faience reported from Assur (Andrae, 1922, pl.49a 60, lower; Haller, 1954, 6ff., pl. 7f), Kish (Watelin, 1934, pl.XXXI.4-5) and Susa (see below) are less easy to attribute precisely to the Akkadian period. They probably appeared then, but may still have been in production towards the end of the third millennium B.C., the period to which the Assur graves probably belong. The most interesting single tomb-group for the history of faience at this period is one at Susa, since it included both a jar and a small bowl of faience with one of these animal statuettes (Amiet, 1966, 234-5, fig.172). This grave also contained a copper or bronze macehead of a group which includes an example inscribed for King Naram-Sin (c.2254-18 B.C.) (Calmeyer, 1969, group 10, 26-7). Another aspect of the faience repertory in the Akkadian period is illustrated in House K (room 291) of level XI at Nippur, where the American excavators found "part of a mosaic with composition tesserae adhering to or set into backing preserved as white film in the ground". An ochre red border and tiny squares formed a pattern with pale blue squares and crescents. It was too damaged to be lifted (McCown, 1967, pl.153:26). Many of the so-called "Guti Seals" were made of faience (Boehmer, 1957, 71) in the period c.2250-2100 B.C.

(iii) Ur III to Old Babylonian periods(c.2100-1550 B.C.):

For the first time, at the very end of the third millennium B.C., documentary evidence may begin to throw light upon the production of faience and related materials. A well-known text from Ur (Legrain, 1937, no.1498) of this date describes the activities of a set of craftsmen for one year, listing raw materials, finished products and scrap items for reprocessing. In the section on lapidaries (Loding, 1981, IIff) the first stone mentioned is the unidentified dušû stone (?agate). It was used predominantly in jewellery, among other things to represent sun-disks; but of particular interest in this context, are later references to the manufacture of artificial dušû, of glazes with the colour of this stone, and of leather dyed this colour (cf. C.A.D. s.v.dušû). In the Ur text it is followed by carnelian and lapis lazuli, and then by a section indicating that the stone cutters may have been involved in the manufacture of objects from artificial materials. First listed is bu-uz-hi-li (stone) in two blocks; this term is very rare in Ur III texts. The first element in the name (bu-uz) has been related to the material later known as būṣu, a "primary glass" (see pp.224ff.) in Oppenheim's terminology (1970) or "faience" in Foster's (1979, 17), whilst hi-li ("abundant") is just a descriptive term. Ur III texts also refer very occasionally to anzaḫḫu, at least once as the material from which a bowl was made. It is grouped in an inventory list, not among the stone vessels, but with the more precious items. In Oppenheim's view (1970, 19ff.) this material was also a "primary" glass (see p.226 here); but Foster (1979, 17) believes it was a natural stone (?quartzite). The artificial materials in the early texts were predominantly used for bead manufacture. What exactly they were is still an open question when the textual evidence is ambiguous and the material remains so meagre.

Archaeological evidence for the relatively short Ur III period is inevitably restricted. The graves at Assur included more animal figures of faience and at least one small bowl (Haller, 1954, 8-9, graves 5-6, 8, 11, 17). Of the animals Andrae (1925, 61) noted that "glaze (was) now mostly white, but occasionally traces of light blue and yellow can be found". At Nippur there is no publication of glazed vessels in the relevant levels (McCown, 1967, 78); but there was a concentration of "frit" beads in room 197 of House I (ibid., 97: table III). At Ur stratigraphical problems confound attempts at accurate dating, as in the case of the faience beads "found against the outer SW wall of Dungi's (Shulgi's) mausoleum" (Woolley, 1974, 102:U.17891). Only one instance is reported where faience inlays, "presumably from wall decoration" may certainly be associated with structures of the Ur III period: in room 1 of Shulgi's "Mausoleum" (ibid., 9:U.16298); comparable pieces from Enunmah may be slightly later (Woolley, 1974, 47, 50:U.225). Although it is difficult to date them closely, it is likely that a distinctive series of faience frog amulets first appeared at this time. This animal had long been a favourite with Sumerian jewellers (van Buren, 1939, 101-2); de Genouillac reported a yellow-glazed example from Telloh (1936, 107, fig.46; cf. Woolley, 1976, pl.93j:U.1274) probably of Ur III date.

It has been argued that north Syria, not Egypt, was the source of inspiration for the development of faience production in Crete sometime in the later third millennium B.C., when Minoan metalworking was also influenced from this region (Foster, 1979, 56ff.). Foster's view, however, that "....north Syria and north Mesopotamia were dynamic and innovative centers of faience production, actively involved in the spread of the art throughout the Near East at the end of the third millennium" (ibid., 59) might be true; but the material evidence for it is, at present, very restricted.

In south Mesopotamia Ur is a key archaeological site for the earlier second millennium B.C., though once again caution is necessary in studying the artefacts, since their stratigraphical contexts are often equivocal. Faience beads and amulets were very rare; out of almost two hundred published graves, only one was reported to have contained faience beads (Woolley, 1976, LG/27:U.16319). Animal amulets in faience, probably of this period, were more commonly picked up in an area, to the east and north-east of the city, known as Diqdiqqah, where Woolley believed many of them had been made (ibid., 217:U.1269-86; 218:U.1580, 1715; 226:U.6863; 236:U.16128; 243:U. 16735-6; 248:U.17002) "the frog is the most common subject, the duck and the squatting monkey are common, the fish and the tortoise occur, but rarely; the flower-petalled rosette....model hair-combs" (Woolley, 1974, 86). "Humbaba" masks were also made in faience. Comparative evidence suggests that a minature faience bucket from a disturbed grave at Ur should not be dated before the fourteenth or thirteenth centuries B.C. (Woolley, 1974, 227 U.7057, pl.96d; see p. 152 below). A fluted blue-green faience bottle (Woolley, 1974, 58, 253, pl.96A) was reported from the <u>Giparu</u>, room C.25; but there is no certain evidence that the context is earlier than the middle of the second millennium B.C. The same applies to small fragments of blue glazed faience vessels, reported to be like pottery type 63, found in room A.9 of the same building, but in the upper part of a drain shaft (Woolley, 1974, 46, 183, pl.105:U.6871). A "very delicate" faience bowl from a grave in room B.12 of the <u>Giparu</u> (Woolley, 1974, 51, pl.96:U.6829) was associated with a faience face-pendant of a type known to date to the thirteenth century B.C. (Woolley, 1974, U.6820, cf. also U.3254; for date, Peltenburg, 1977, 177ff.).

A child's grave of the early second millennium B.C. at Susa contained a faience doll's head (0.027 m. high), set on a silver shank, with cavities for inlay (Amiet, 1966, fig.215). In the palace at Mari isolated finds indicated the use of faience for mosaic inlays, probably in pieces of wooden furniture, as in room 49 (Parrot, 1958a, 37). The best recorded was an example in court 70 (Parrot, 1958a, 234-5; 1959, 105-7, fig.76, pl.XXXIV), where faience mosaic seemed to be reproducing the patterns of a carpet. In the north, in the Old Babylonian period at Chagar Bazar, amulets, animal figurines, beads and face pendants of faience continued in production as before (Mallowan, 1947, 174).

The sparse Mesopotamian evidence indicated no fundamental changes in the faience repertory from the Late Protoliterate to the end of the Old Babylonian period, with a concentration throughout on personal ornaments, an intermittent use for inlays, and a restricted production of vessels. In view of changes to be seen in the second half of the second millennium B.C. a wider perspective on vessel production is instructive. Faience vessel manufacture was introduced into Palestine from Egypt sometime in the first quarter of the second millennium B.C. Although it developed some characteristics of its own, it remained very much an offspring of the Egyptian tradition (Sagona, 1980) as did the faience industry of Byblos (Dunand, 1958, pp.741-80) and perhaps also that of Syria.

The contemporary situation in Anatolia is less easily summarized. Foster (1979, 46) denies the existence at this time of faience manufacture there. In view of the exiguous published data it is better to leave the question open, particularly when sites like Acemhöyük and Kültepe have provided some indication of the very varied luxury goods, including faience, available in court circles at this time in Anatolia.

Some of the currently most relevant finds from Anatolia were excavated from level II at Alishar. It has long been known that the excavator's designation of this level at Alishar as "the period of the Hittite Empire" was incorrect; but its precise time span is still debated. This stratum was between three and four metres thick, embracing two major building phases, with subdivisions within each of them. The oldest phase was 11T (terrace), the most recent 10T (von der Osten, 1937, 3). Most of the cuneiform tablets were found "in the refuse of level 10T"; "three tablet hoards were on actual floor levels, all belonging to level 10" (von der Osten, 1937, 108). All the datable tablets from Alishar belong to the period of Karum Kanesh Ib, contemporary with Shamshi-Adad I of Assyria (c.1813-1781 B.C.), whose reign provides the *terminus post quem* for the abandonment of level 10. That it survived very considerably longer, if only in restricted use, is indicated by various seals and seal impressions; an impression from 10T bears a hieroglyphic Hittite inscription (von der Osten, 1937, 227, fig.254:c.1456, pl.XXV; Boehmer, 1967, 61, n. 11) of the "Old Hittite Period"; an agate cylinder seal ("found in Q28 on the floor level of the large mansion of Level 10T") is in the Kassite style (von der Osten, 1937, 205, fig.246(d.2235)) and two "Mitannian" cylinder seals (ibid., fig.246:c.200, d.2365) were found in Alishar Stratum II. A faience cup found in the same level (Schmidt, 1932, 111, colour pl.III, pl.X:b 1868) was compared by Schaeffer (1948, 322) to similar unpublished vessels found at Ras Shamra in *niveau* II, either of *Ugarit Moyen* 2 (c.1900-1750 B.C.) or early 3 (c.1750-1600 B.C.). Von der Osten (1937, 286) also reported that "in 1929 and 1931-2 one fragmentary frit cup and several sherds of frit vessels were found. They show on the outside a greenish to greenish blue glaze, whereas the inside is sometimes dark brown, almost black. The rim piece (ibid., pl.I:4) was found in M 33-4 in close association with cuneiform tablet d.2200". Of the excavations in near contemporary levels at Alaca Hüyük Arik (1937, 40) wrote: "L'abondance des morceaux de faïence de couleur jaunâtre, blanche ou bleu claire, aux parois très minces, supposés avoir été importés de l'Egypte ou de la Mesopotamie et dont semblables ont été recueillis en quantité à Alisar; les débris ornés des dessins en relief en forme de grappe de raisins (Al.633), également abondants à Alisar; les récipients à fonds pointus, munis, de passoire et ornés de petits reliefs en forme de mamelons, confirmait notre attribution". Alishar II also yielded faience figurines of nude females (von der Osten, 1937, 193, fig.230:d.2966, 2971) and the feet of animal figurines (ibid., 205, fig,243: d.2210, e.1631); Dayton (1978, fig.94) has published sketches of comparable objects from Kültepe.

At Tell Mardikh (Ebla) in Syria a rich grave dating to Middle Bronze IIA (about 1900-1800 B.C.) contained a faience vessel, now without its original pair of rim handles, in shape and decoration very like a rock crystal vessel from Acemhöyük in Anatolia (Matthiae, 1980, 13-4, fig.14). Faience vessels, in a variety of shapes, were found in the slightly later "Palace of Yarim-Lim" (level VII) at Alalakh (Woolley, 1955, 297). Thus in Syria, if not also in Anatolia, there was an industry producing faience vessels and figurines, as well as the more common beads and amulets, through the Middle Bronze Age (c.2100-1550 B.C.). There is no certain evidence, by contrast, that this was also the case in Mesopotamia. Compared with the faience obvious at Tell Taya in the later third millennium B.C. and its popularity at neighbouring Tell Rimah after about 1400

B.C., its present absence thereabouts in the intervening period is striking.

(iv) Kassite, Mitannian, and Middle Assyrian Periods (c.1550-1150 B.C.):

In the history of faience (and frit), as in that of glass (see p.201) and glazed terracotta (see p.167), there is a profound change in the range and intensity of production in the Near East in the middle of the second millennium B.C. The earliest phase for which there is good archaeological evidence of this revived and expanded industry in Mesopotamia is represented by finds from Stratum II at Nuzi, which may not have been destroyed much before about 1350 B.C. (Wilhelm 1976), though its contents are commonly dated to the fifteenth century (cf. also p.202). Here frit and faience were most commonly encountered in the form of beads. In her technical study Vandiver (1982a, 74) has noted that "in the Harvard Semitic Museum collection as well as in the field accession records, the relative numbers of beads occur in the following descending order: glass, frit, stone, shell, glazed quartz body, and no glazed clay beads. An estimate of the ratio of glass to frit beads is that the glass exceeded by a factor of ten the number of frit beads". Starr (1939, 94, cf. 446) found enormous quantities in Temples C-A at Nuzi, notably in Temple A. The frit beads, predominantly blue, "seldom occurred in more elaborate form than representations of frogs and flies.....invariably small".

Frit was produced in five colours: blue, brick red, white, black and yellow, each with a different composition and structure. Blue is the commonest; Starr (1939, 447) noted the red as very rare. Blue frits were sintered mixtures of the minerals quartz, calcite and "Egyptian Blue". The ratio of Fe:Si was higher for the red than for the blue; the ratio of Ca:Si for the blue. The black frit was high in a volume fraction of dark blue glass. Yellow frit was considered to be iron-stained white frit. Lime is present in the blue and white frits, but not in the red and black, where zinc oxide is greater. Copper only occurs in the blue and red beads. "Unground joining seams along the hole axis of cylinder beads are evidence that the material was plastically formed around a polycrystalline brown core. The exterior is harder than the interior indicating that prior to firing, the dried surface of the frit beads often was ground differentially.....several red beads were formed by stacking several small frit spheres around a core" (Vandiver, 1982a, 82ff.).

Vandiver (1982a) found that "beads made of a glaze coating on a white, friable polycrystalline body rich in quartz (often known as Egyptian faience...) are relatively frequent in the Nuzi collection". The tubular beads fall in this category. The melon-shaped beads were, however, "more similar to the frit beads (in that there is a friable brown core and the surface is ground) than to the above category". The glaze did not adhere as well to the melon beads. The melon bead glaze was higher in Mg, more nearly approximating to the glaze composition and maybe the local product, whilst the cylinders could be imports.

In his final report Starr carefully distinguished between "frit" and "faience". In Temple A he found a simple blue "frit" cup of delicate proportions and the fragments of another "frit" vessel (Starr, 1939, 91, cf.pl.56, I, L). Only two fragmentary vessels of "faience" were published (Starr, 1939, 461, pl.119 I, J). Both were found in the palace rather than in the main temple complex, where so many of the faience/frit objects were concentrated; Starr considered them to be imports. Beck and Jackson reported on them as "faience", but Barag (1970, 137, 139-40) included them as mosaic glass in his corpus. Starr (1939, 461) quoted Beck and Jackson's spectroscopic examination as follows:

".....A spectroscopic examination shows that the Nuzi examples most certainly have a lime flux.

On examination, it is evident that this bowl contains red, yellow, and white patches in it, and possibly black, although the last may be accidental. The colouring matter of the red is iron in the form of ferric oxide in great numbers of comparatively large particles which are totally opaque to transmitted light. The yellow colouring matter is the same as that used in the yellow beads of Mesopotamia and Egypt. It is probably due to ferric oxide, but the exact way in which it is introduced is not yet known.

The second vessel was similarly of "kneaded" white, red and yellow, but with the addition of "occasional spots of deep blue like that of the well preserved Nuzi glass".

Subsequent research has shown the yellow colourant to be lead antimonate in Egypt and in Mesopotamia.

The only other major group of faience objects presently published from Mesopotamia, likely to be as early as those from Nuzi II, was found in the temple at Tell Rimah (phase II, room 5). The destruction level from which they were recovered is not independently dated. The excavator, on the strength of close parallels with the material culture of Nuzi II, has dated it to "about 1450 B.C." (Oates, 1982, 97). In view of the previous comments about the <u>terminus</u> of Nuzi II, it might well be later by half a century or more. The repertory at Tell Rimah, in faience and frit, is in some ways more varied than at Nuzi, embracing cylinder seals of the "Mitannian Common Style" (a widely distributed category); tiny cylindrical cosmetic containers with lids, their sides decorated in sunk relief with facsimiles of the seal designs, and with human faces; a bird-shaped vessel decorated on the outside with brown glaze, on the interior with yellow (Dayton, 1978, pl.11); and a range of beads and zoomorphic amulets (Carter, 1965; this paper does not always distinguish finds from phase II clearly from phase I).

A small group of comparable frit and faience objects was found at Tell Chuera (Moortgat and Moortgat-Correns, 1976, 38ff., figs.14-15); as at Chagar Bazar (Mallowan, 1937, 122, 137, pl.14A).

The first phase of expanded faience production (cf. glass, pp.201ff.) in Mesopotamia and Syria has sometimes been termed "Mitannian" (c.1550-1350 B.C.), not merely in chronological terms, but because it has been assumed that their empire stimulated and facilitated the production and diffusion of such luxury objects. At this stage the manufacture of vessels and of other more elaborate objects was largely, if not exclusively, a "prestige" technology, situated in the workshops of the "Great Organizations" (temples and palaces), as at sites like Nuzi and Tell Rimah. Their output found its way largely into shrines and palace apartments, or into the graves of wealthy servants of palace or temple. At this time local production seems largely to have served local purposes. In the following or "Middle Assyrian" phase (c.1350-1200 B.C.) this situation persisted and expanded to more prestige production centres. However, the wide circulation of a standard repertory of certain vessels and personal ornaments, now found more commonly in graves, may indicate a more broadly based hierarchy of production centres, some outside the "Great Organizations".

Assur and its neighbourhood has provided one of the most important groups of faience for this later phase. Apart from finds in the Assur Temple at Kar-Tukulti-Ninurta, hardly used after the reign of its founder Tukulti-Ninurta I (c.1243-1207 B.C.), these objects are concentrated in the Ishtar Temple at Assur, itself founded by the same king, but used until the reign of Assur-resh-ishi I (c.1132-1115 B.C.). The variety of the repertory is remarkable and likely to represent a local industry, though some individual objects may be imports. There is a range of beads, amulets, rosettes and gaming pieces (Andrae, 1935, 96-100, pls.39-40); various "nail-heads" or pommels (ibid., pl.41); fragments from furniture (ibid., pl.38); human statuettes and model footware (ibid., pls.34-5); human eyes and genitals (male and female) (ibid., pl.36); animal statuettes and protomes (ibid., pl.37), and a small repertory of vessels (ibid., pl.42). Tell al-Fakhar (Al-Khalesi, 1970, pl.27) illustrates the availability of faience at this time in a less important northern city.

Phase I at Tell Rimah, dated textually to the thirteenth century B.C. (tablets of the reign of Tukulti-Ninurta I), produced another collection of faience/frit objects, notably face-pendants (cf. Peltenburg, 1977) and small "rosettes" said to be "characteristic of phase I" (Oates, 1965, 74; Carter, 1965, 53, fig.7). Such "rosettes" have a wide distribution, appearing for instance, in contemporary levels at Nimrud (Mallowan, 1950, 174) and at Assur (Andrae, 1935, 97, pl.40) in Iraq and at Tell Fakhariyah (Kantor, 1957, 43-4) in Syria. As they commonly have button-backs, it has been suggested that they were ornaments sewn onto the robes made for deity statues (Andrae, 1935, 97); they have now been found in ordinary graves at Tell Mohammed Arab (M. Roaf; pers.comm.).

It is the vessels at this period which particularly well illustrate a commerce in standard shapes. Any attempt to establish whence individual types, or specific examples of them, originally came is hazardous. Long ago Hall (1928) isolated and discussed the striking similarity between the faience vessels found as far apart as Enkomi in Cyprus and Assur, though his concept of Mycenaean-Minoan production centres for them has not survived. Peltenburg (1972; 1974) has more recently considered the problem in detail. In a summary of his examination of faience vessels from Cyprus he classified them into two major groups: Egyptian or Egyptianizing and Western Asiatic, sub-divided into North Levantine Style and International Western Asiatic Style: "there are other vases from Cyprus which show no Mycenaen influence and which, because they are also made in

Mesopotamia, and because there is no good evidence as to where they first occur, are best considered as belonging to an international Western Asiatic Style" (Peltenburg, 1972, 136). He cited in particular the "blossom bowls", with oblong pentagons arranged like petals around the exterior (cf. Foster, 1979, 51), and a type of small bucket found in Mesopotamia at Babylon (Reuther, 1926, 15, fig.10A, C, pls.47:12b, 48:17.4), Kish (de Genouillac, 1924, pl.50:143, 65:143, VIII.4), Ur (Woolley, 1962, pl.58:233) and Mari (Parrot, 1937, pl.XV.1). By way of contrast a distinctive group of face goblets, represented at Assur (Andrae, 1935, pl.33), belong "almost wholly to the north Levant where they were also made of wood" (Peltenburg, 1972, 134, pl.XXIV; cf. Culican, 1971); though even they show marked stylistic differences. A characteristic of Peltenburg's "Western Asiatic" group is the use of yellow glaze either to pick out details or to line the inside of a vessel. Channel spouted bowls, as Hall appreciated, are another category of vessel found at Assur (Haller, 1954, 42, 64, pl.16e,h; cf. also Layard, 1853a, pl.85), and at Tell Fakhariyah in Syria (Moortgat, 1959, 36, fig.4), as well as in Cyprus (Foster, 1979, 51). Even simple hemispherical bowls of standard types are widely dispersed in Syro-Mesopotamia (cf. Kuschke, 1970).

By now there is also evidence far to the south in Mesopotamia to show the wide distribution of faience vessels in the third quarter of the second millennium B.C. In view of the ease of transport along the line of the Euphrates such penetration from workshops in Syria is not so surprising and contemporary vessels also appear at Meskene (Emar) (Caubet 1982, 111ff.).

Parrot (1975, 7) modified his earlier dating of the "Assyrian Cemetery" at Mari after finding a faience plaque in tomb 656 bearing the name of Amenophis III of Egypt (c.1402-1364 B.C.); but, as Peltenburg (1977, 192) pointed out, there is consistent evidence from other sites, notably Tell Rimah, for dating the distinctive face-pendants, also found in the Mari cemetery, to the thirteenth century B.C. The graves in this cemetery, in courts 106 and 131 of the earlier palace, were well furnished: "caractérisé par d'élégants récipients en fritte vernissée (pl.XV.1): bonbonnières à couvercle, ampoules, coupes, jarres, godets", as well as the face-pendants (Parrot, 1937, 83; cf. Mallet, 1975). As the repertory of forms has much in common with Mari, it is broadly to this period that the faience vessels from Babylon belong (Reuther, 1926, 15, fig.10). At Babylon the excavator interpreted the contemporary "fire-pits" as kilns for the manufacture of faience objects, but the site is reported in too little detail to check this suggestion. Two sets of graves, designated as either from Alteren Kassitischen Schicht or the Jüngeren K.S., yielded in addition to vessels, a variety of beads and amulets in faience (Reuther, 1926, pls.47-8, 58).

More recent excavations at Nippur revealed in Kassite levels "a few examples of glaze on a body of siliceous paste" (McCown, 1967, 78). Within the brickwork of the altar of the Kassite Ningal Temple at Ur Woolley found an assortment of objects including two faience vessels and a faience potstand (Woolley, 1965, 103; U.3304, 3314, A, B); a bronze worshipper statuette (Woolley, 1965, pl.28:U.18628), similar to examples from Nuzi II was also included. Isolated vessels are reported from graves and buildings of the Kassite period in the

Hamrin region (Boehmer, 1982, 32; Invernizzi, 1980, figs.68-9).

That there was a faience/frit industry serving the Kassite court is evident even from the relatively meagre range of finds so far published from excavations at Aqar Quf (Dur Kurigalzu) (cf.p.205 here for glass). As early as level IV, built on virgin soil sometime in the first half of the fourteenth century B.C., room 15 produced the model head of a camel (?) and "fragments of several small objects made of frit.....some of which were the horns of small animals" (Baqir, 1945, 13-14, pl.XXVII). A finely modelled stag's head from level IB, a century or more later, was described as "in limestone or more probably frit" (ibid., pl.XXVII). Some tiles may be glazed sintered quartz rather than glazed terracotta; according to the excavator: "several tiles were found at level IC and at level II. They are made of chalky material, and some still retain traces of blue and yellow glaze" (ibid., 14).

It is by no means clear where the primary centres of faience production were located in this period of extensive production from about 1350 to 1100 B.C.; but they may not be considered in isolation from the contemporary glass and metal industries (see pp.31ff., 201ff.). Although workshops at such sites as Assur, Nuzi and Tell Rimah apparently supplied local temples and palaces with simple votive objects and personal ornaments, some production of vessels for trade is evident. Vessels, of Peltenburg's "International Western Asiatic Style", have been found in Cyprus, in Syro-Palestine and in Mesopotamia. So wide a distribution of such a relatively fragile commodity is significant. They were clearly part of a luxury trade; but how far this extended beyond gift exchange at court level, it is impossible to gauge at the moment; nor may it be said with any confidence where they were made, though workshops in Syria are most commonly favoured. Both Ras Shamra (Ugarit) and Meskene (Emar) have produced industrial debris indicating local faience production (Louvre collections). Some scholars (cf. Culican, 1971) have seen their stylistic homogeneity as indicative of a single source or a very closely related group of workshops, whilst Barrelet (1968, 108, n.3) and Peltenburg himself prefer the concept of a multiplicity of manufacturing centres; the latter arguing that, though western influences may have been crucial in initiating certain types of production, "some apparently Mesopotamian vases occur in the west during the apogee of this style, c.1400-1200 B.C." (Peltenburg, 1974, 108).

It is no easier at present to define the relative roles of many local production centres, as against a few major ones, in the distribution of much more commonplace objects. "Mitannian Common Style" cylinder seals of faience are found throughout the Near East, and into the peripheral regions, in the Late Bronze Age. Dabney (1984) argues that distinct groups, locally manufactured, are evident at Nuzi, at Alalakh in Syria (c.1550-1200 B.C.), and at sites in Palestine (c.1350-1200 B.C.) An equally extensive distribution of simple faience beads of standard design is illustrated by the occurrence at Marlik, near the Caspian Sea in Iran, and at Ur, of strikingly similar faience pomegranate and double-spiral beads in the fourteenth and thirteenth centuries B.C. (Maxwell-Hyslop, 1971, 194).

A distinctive group of faience face-pendants also demonstrates the difficulty of attribution. In a comprehensive study of them

Peltenburg (1977) showed a Mesopotamian concentration in their present distribution pattern and argued for a thirteenth century date. He cogently challenged Kühne's (1974) earlier view that they were of "Hurri-Mitannian" origin and dated to the beginning of the second half of the second millennium B.C. In Mesopotamia they have been reported from sites like Abu Ghuraib, Billa, Isin, Kish, Mari, Nippur, Tell Rimah, Shuweimiyeh in Haditha, Sippar, Tell ed-Der, Ur and Uruk; from Susa and Tchoga Zanbil in Khuzistan; from Gezer in Palestine, Hala Sultan Tekke in Cyprus and Ras Shamra in Syria. Peltenburg (1977, fig.217) pointed out the contrasting distribution of the earlier metal and moulded glass nude female figurines, mainly of the sixteenth to fourteenth centuries B.C., which at present have a Levantine concentration and are unreported from Mesopotamia (cf. Barag, 1970). He also urged a separation of the "face-pendants" from the "face goblets" in any study of faience production in the Late Bronze Age.

The most spectacular faience and glass industry of the thirteenth century, so far known, is that revealed at Tchoga Zanbil in Khuzistan by Ghirshman's excavations (1946-62). This city, focussed on a spectacular complex of shrines and funerary palaces, with underground vaulted tombs, round a monumental ziggurat, was founded by Untash-GAL in the second quarter of the thirteenth century B.C. Faience was extensively used there for personal ornaments and seals, for vessels, including small cosmetic containers shaped as female heads, for human and animal statuettes, and for votive maceheads (Ghirshman, 1966, 1968, passim; Amiet, 1966, figs.267-9, 272). More scattered evidence from earlier excavations at neighbouring Susa had indicated the existence of a Middle Elamite faience industry, but not its great range (cf. Amiet, 1966, fig.317). This industry's activity seems to have been largely extinguished by the Babylonian invasion of the later twelfth century B.C., when there is a break in recognizable production in Khuzistan until about the eighth century B.C. A contemporary glass and faience industry on the island of Failaka in the bay of Kuwait appears to have been more closely connected with this Elamite craft tradition than with that evident in Babylonia (cf. Pollard and Højlund, 1983).

(v) <u>The First Millennium B.C.</u>:

The difficulties involved in any study of the material culture of Mesopotamia between the eleventh and ninth centuries B.C. have already been considered in relation to metalwork (pp.33ff.). It is virtually impossible to tell whether certain industries passed completely into eclipse when archaeological evidence is absent, consistently so in this case, or lingered on in isolated centres whose products were not widely distributed. When specialist products reappear after an interval, albeit in fresh forms and styles, the latter is an obvious inference. In certain cases, as with glazed terracotta in architecture, there is some textual evidence to support continuity (see p.171). When material evidence becomes available again it indicates that the Neo-Assyrian repertory in faience in no way matched that of the Middle Assyrian period. The stimulus for revival may well have been western, ultimately perhaps "Phoenician", since "Egyptianizing" traits are particularly evident in "Neo-Assyrian" faience amulets and pendants (Oates, 1968, pl.XXXVIa-c).

It is to western Iran that attention must first be turned. Here, as with glass (p.206), the destruction of level IV at Hasanlu, c.800 B.C., offers a valuable terminus ante quem, earlier by almost two centuries than any yet available in Mesopotamia. Two published faience vessels from level IV at Hasanlu provide an archaeological horizon for a whole series of similar vessels reported from clandestine excavations in western Iran in the last quarter of a century (cf. Fukai, 1981): a small, blue-glazed faience box-shaped vessel, with feet and with holes for securing a lid to a round mouth, and a cone-shaped, footed goblet, of cream-glazed faience, decorated with crude geometric designs in blue and brown (Dyson, 1968, 90, centre, left; pl.XXXIX, colour). Fukai (1981) has provided a finely illustrated survey of the art market examples, many casually attributed to "Ziwiyeh" (Muscarella, 1977); but, at present, it is impossible to give them a close dating in the range ninth to the seventh centuries B.C., save for those most nearly matching the Hasanlu examples. Fukai's English text does not always clearly distinguish between glazed terracotta vessels (and tiles), which were also found both at Hasunlu and Ziwiyeh in controlled excavations, and faience. The body material of some of the latter (Fukai, 1981, 11, pl.16-17)* is given as 87.2% silica (faience). Analyses of the glazes showed that for the dark purple colour "iron, manganese and titanium compounds were used"; for the blue, no lead, "but (they) contained cupric oxide, as well as calcium and potassium, that functioned as a fluxing agent"; for the yellow, containing 24% lead, "the element imparting colour to the yellow glaze was antimony, titanium was used in order to strengthen the yellow colour"; the white glaze did not contain lead (Fukai, 1981, 21) (cf.p.166). Although west Iranian glazed tiles of the Iron Age often show Assyrian influence in their decorative motifs, the faience vessels have much more distinctive characteristics. They seem to constitute a repertory of shapes and designs largely independent both of the earlier Middle Elamite and of the Middle Assyrian industries, and are unrepresented in contemporary Mesopotamia.

Study of the Neo-Elamite faience industry is complicated by chronological problems and recurrent uncertainties about whether the vessels are indeed faience, as seems most often to have been the case, or glazed terracotta (cf. Amiet, 1966, figs.370ff.). De Miroschedji (1981, 9ff.) has argued that the cylindrical pyxides with moulded decoration are Middle Elamite in date. It is the cuboid and variously shaped containers with incised designs rather than moulded decoration, sometimes polychrome, that are Neo-Elamite, though they were probably not made after about 650 B.C. (cf. Fukai, 1981, 6-7). The local seal industry of Susiana in Neo-Elamite times also continued the earlier tradition of using faience (cf. de Miroschedji, 1982, 52-3).

In Mesopotamia in the seventh century B.C., and for an unknown length of time before that in the Neo-Assyrian period, beads, amulets and seals, were manufactured of frit and faience, as were such traditional items as eyes for inlay into statues and votive maceheads. In the Ninurta Temple at Nimrud Layard (1853, 357-8) discovered a

---

* Some of the chemical analyses, as published in the English text, are problematic.

miscellaneous collection of objects, some of which were made of faience; from the same building Mallowan (1966, 82) recovered "a glazed faience receptacle, and two banana-shaped objects made of frit". Although faience was used in "Neo-Assyrian" times for cylinder seals in a variety of styles, it is most obviously employed for seals cut in what Buchanan (1966, 106, nos.606ff.) called the "Debased Linear Style". Production of such seals may have begun as early as the mid. ninth century B.C. (Parker, 1955, 103ff.). It was to continue into the Persian period. They are well represented among the cylinder seals excavated at Assur (Moortgat, 1940, nos.689-731) and at Nimrud (Parker, 1955; 1962). The frit/faience repertory of a small provincial town of the period is illustrated by finds at Tell Rimah (Oates, 1968, 132-3) and further afield at Sultantepe (Lloyd and Gökce, 1953, 49-50; Lloyd, 1954, 107), where western influence is evident both in amulets and vessels. In Andrae's (1925) publication of the Neo-Assyrian glazed terracotta vessels found at Assur, very little reference is made to faience. It encourages the assumption that locally polychrome glazed terracotta vessels had virtually superseded faience vessels in the Neo-Assyrian period. Andrae (1925, 35, 42, pl.18c) described only one faience bowl: "a centre rosette of yellow, light blue and white petals in turn, around them a wreath of spots and yellow circle.....The glaze is different from that on the earthnware bottles, much thinner and more delicate". It might be an import from the west.

Although the Neo-Assyrian court patronized a vigorous industry producing glazed architectural terracottas (see pp.170ff.) and glazed baked clay vessels, there is no evidence that there was a comparable Neo-Assyrian faience industry producing its own distinctive repertory of vessels and votive objects. Where there is evidence of faience amulets and beads, seals and pendants, they may merely be another aspect of the "Syro-Phaenician" technological and commercial impact on Assyria, rather than an indigenous revival of the craft well known before in Middle Assyrian times. Artisans working in this tradition were resident in Assyrian centres.

The same would seem to be true of Babylonia in the first millennium B.C., where glazed terracotta was the preferred medium. Isolated finds from the Achaemenid period need not be the products of local workshops. From Ur Woolley catalogued a small vessel stand of "green faience" (U.227: Woolley, 1962, pl.31) and a "box of glazed frit, made to take a swivel lid, originally blue" (Woolley, 1962, U.16731), from a grave attributed to the Persian period. At this time, as in the previous period, faience or frit pendants and amulets, especially "eyes of Horus", in an "Egyptianizing" style were widely distributed across the Near East.

Wherever it is possible to check, Woolley's (1962, 103) summing up of the position at Ur seems to apply:

"While it would be rash to say that any new materials for beads were introduced in the Persian period it is true that one of the oldest materials, paste or glazed frit, was now employed in new shapes and with a greater richness of design. Small pendants of brown, red, yellow or blue paste in the form of miniature tools such as axe-heads and adzes are common on necklaces and there are many varieties of polychrome beads, ovals of green with yellow ends.....rosettes and spotted eye-beads far more gay in appearance than anything we have from the Neo-Babylonian age. Glass too is more commonly used, and there are examples of the inlaid or mosaic beads which are often associated with Phoenicia" (cf. Oda, 1965, for technical comments on such beads from Iran).

B: Methods of Manufacture:

As has already been pointed out (see p.133), faience bodies were made primarily from silica (sand or powdered sandstone, quartz or flint) and alkaline binding agents (natron or plant ashes). As there are no known ancient descriptions of manufacture, ancient methods have been reconstructed in modern times from a combination of evidence: from analyses of ancient faiences and glazes; from their structural examination under a microscope; and from experimental attempts, under controlled conditions, to reproduce ancient faiences. In the circumstances it is not surprising that a number of different proposals are current. Sometimes they vary over fundamentals, more often over details of procedure, since the proposed methods and recipes do not always reproduce the characteristics of the originals. Even with the ancient Egyptian faience industries, which are relatively well known, opinions vary and it is only recently that a multifacetted approach has begun to provide a broader basis for study (Kaczmarczyk and Hedges, 1983). As there is very little published information on Western Asiatic faience, technological study has to proceed by comparison with Egyptian procedures. In these circumstances it can not be stressed enough that this is only an interim solution. It is already apparent that there are aspects of the production of vitreous materials in the Near East which differ from the Egyptian traditions to a greater or lesser extent.

For the whole of the Near East there is virtually no archaeological information on the manufacture of faience. Schaeffer´s (1962, 37) suggestion that faience might have been fired in the same kiln as tablets at Ras Shamra is very unlikely from the evidence cited. The only convincing evidence for a workshop manufacturing faience, primarily beads (including wasters), has been identified at Tyre in the Late Bronze Age levels (Bikai, 1978, 7-8; Brill, 1978). In a courtyard of stratum XVI the excavators found a "work table" and a "basin" for grinding and preparing raw materials, including calcium carbonate; in the overlying stratum XV the courtyard appeared to remain in use for the same purpose. This time there was evidence for a kiln, incorporating a pithos, and a pile of selected sea pebbles.

Stray terracotta kiln setters with glaze adhering to them might indicate local glazed terracotta or faience production. When the setters are actually made of faience (if the evidence holds), they even more probably indicate local faience production, as at Assur (Andrae, 1935, 99-100, pl.39s,u,x). Wasters may be equally equivocal in their implications. Reuther (1926, 15) reported large lumps of bluish-white glassy material from "Kassite" levels at Babylon, but precisely what vitreous material was in production is not certain from such evidence. To judge by Egyptian analogies a faience worker´s tools would have been simple; moulds again may only be confidently identified when directly associated, as at Tell el-Amarna in Egypt, with the types of faience artefact made in them (Petrie, 1894).

Reisner's (1923, 134ff.) description of the faience industry of the Egyptian Middle Kingdom, as illustrated by the remarkable range of finds from Kerma in Nubia, is still the best concise report of the debris from an ancient faience industry. Until such information is available for the Near East, one paragraph of his account suffices to indicate the kind of industrial evidence likely to be yielded eventually by excavation there:

"The conclusion that the workshops for the manufacture of glazed objects were at Kerma, is borne out by the discovery of accidentally glazed setters, misfired pieces of faience, lumps of copper oxide, lumps of keriak (silicate), mortars, pestles, charcoal, and beds of wood ashes at Lower Deffufa. There were also a number of spoiled beads, as well as quantities of carnelian and quartz crystals from which beads were made. One oven was seen just north of Room C, but it was too damaged to be drawn. It was of pottery, truncated (?) cone shaped, about 80-100cm. in diameter, and apparently heated from outside. As the pottery was probably baked in open kilns, this oven may have been used for glazing faience and stone." (Reisner, 1923, 135).

Webb (1978) has usefully reviewed the scattered evidence for faience manufacture in the second quarter of the first millennium B.C. in Egypt and the Western Mediterranean.

(a) Cores and Manufacture:

Both Lucas (1962, 158) and more recently Kaczmarczyk and Hedges (1983, 123) have demonstrated that ordinary sand, sometimes washed and sifted, sometimes not, was the standard basis of Egyptian faience. Powdered quartz was exceptional; only for white faience does there seem to have been a clear preference for powdered quartz. The oldest Western Asiatic object tested so far, a bead of the later fourth millennium B.C. from Brak, was shown to have an ordinary faience body. The copper percentage suggest that it was originally glazed blue or green; a very high Na:K ratio indicates that natron, ordinary salt, or low potassium plant ashes were the alkali source (Mallowan, 1947, 254-5 cf. Stone and Thomas, 1956, no.26). Bimson (1973, 183) demonstrated that, despite its clay-coloured appearance, a lid fragment of the third quarter of the third millennium B.C., level VIII at Tell Taya, had a high quartz body with clay content of only 5%. Contemporary faience beads from the same site had "typical coarse sintered-quartz" bodies. At Nuzi, in level II about 1400 B.C., faience beads were relatively infrequent. Vandiver (1982a, 90) distinguished two main types. The cylinder beads have "a fully oxidized body and a glaze that adheres well.....similar in style and technology to the common Egyptian tubular beads". The melon beads, shaped by grinding, are different: "the glaze does not adhere well.....the interior of the body is gray-blue due to incomplete oxidization, and large oval and spherical pores and small black flakes indicate the possibility of an organic additive.....The melon bead glaze is higher in Mg and more nearly approximates the glass composition". Pollard's analyses (in Pollard and Moorey, 1982) of contemporary faience bodies from Tell Rimah showed a marked degree of purity, with the potassium and calcium contents generally less than 0.5 and 1.0% respectively, and the iron concentration less than 0.5%.

When no systematic research has been done on the body of faience objects from Western Asia, it should be borne in mind that in Egypt, although a considerable percentage of bodies were white (Kaczmarczyk and Hedges, 1983, 186, table XXX), there were many cases in which the body had been deliberately pigmented as background for the glaze. This was the difference underlying the variants of Egyptian faience first distinguished by Lucas (1962, 161ff.) and recently refined by Kaczmarczyk and Hedges (1983, 218ff.). The bodies of later third millennium B.C. faience at Tell Taya, for example, appeared to the naked eye as white, yellow or pink (Reade; pers.comm.). The question is whether the body materials here had been deliberately pigmented to form a suitable background for the glaze or whether, as may more often have been the case in Mesopotamia, the colour derived naturally from the sand of which the body was made. Similar variations are evident in faience bodies from other Mesopotamian sites.

Noble's (1969, 436-7 experiments, for making self-glazing Egyptian faience, offer the basis for a general consideration of production (cf. Tite et al., 1983):

"Ingredients are mixed with a little water and kneaded into a gritty paste. In working with the material it becomes apparent why large and complicated shapes were not made from faience by the ancient Egyptians. If the paste is too damp, it readily slumps and will not hold its shape; if it is too dry, cracks and is non-plastic. With just the right degree of moisture, it can be modelled into simple forms or pressed into one-piece terracotta molds.....The mixture is not allowed to dry in the mold; it merely serves to form the piece, which is then removed and allowed to dry freely". Throwing faience vessels on a potter's wheel, owing to the limited plasticity of the material, is extremely difficult (Allan et al., 1973, 170, pl.4). It is unlikely to have been used save for simple, and comparatively small, open shapes. Vessels were made in a number of other ways. If the Halaf period dating of the vessels from Arpachiyah is correct, and this is unlikely (see p.142), they were being made before the introduction of the potter's wheel. Vessels might be moulded either in one piece or in sections, and the pieces assembled or modelled round a "form" of straw and reeds or something similar, which would burn out on firing, but commonly leave a negative impression inside the vessel. Detail or finishing was sometimes done, after moulding, with a sharp instrument. The vessel might also be made solid and, after partial drying, the excess material scooped out before firing. Amulets, seals and some beads were mould-made. Where evidence exists, not yet in Mesopotamia, simple open moulds of baked clay sufficed. Beads could also be roughly modelled round a tube or wire, and later ground, scraped or incised (Vandiver, 1982a). Some beads show signs of having been bored after firing.

Experimental work has indicated that Egyptian faience was fired at between 800° and 950°C (Noble, 1969; 950°C; Kiefer and Allibert, 1971; 870°-920°C; Kühne, 1974). Such a range of temperature is sufficient to release the metallic oxides for colouring and to flux the surface salts for the glaze. These temperatures were well within the range of ancient Mesopotamian kilns (Tite and Maniatis, 1975). In publishing the late third millennium faience from Tell Taya Reade (1973, 184) noted that: "two of the season's blue faience vessels were also of interest: one apparently bore the marks of the tripod setter, on which it must have rested in the kiln, and another had fragments of green metallic filings, presumably copper, adhering to part of its surface, as if the colour was sometimes achieved without mixing a compound properly into the glaze". Peltenburg (1972, 133) recognized the presence of tripod-stand marks on Late Bronze Age faience vessels of his "Western Asiatic Style"; this includes vessels from Assur. They are also evident on Middle Assyrian faience bowls from Tell Mohammad Arab (M. Roaf; pers.comm.). Although terracotta tripod separators have been regularly recognized amongst pottery from Sasanian and later levels in Mesopotamia, they are not normally reported earlier. It is likely that the "setters" to separate earlier glazed pottery and faience vessels were cruder and may not be easy to recognize away from a kiln.

(b) Methods of glazing:

(i) Application of a glaze:

It was for long assumed that ancient Egyptian (and other) faience glazes had been manufactured by the almost universal method of glazing used for pottery in more recent times. For this premixed ingredients are applied to the surface by dipping or brushing to be subsequently fused by kiln firing (cf. Petrie, 1909, 107; Beck, 1934a, 69; Lucas, 1962, 172ff.). For this the powdered raw materials (quartz, calcite, natron) are mixed with water. When applied to the dried body, the porous body absorbed water from the glaze mixture leaving a fine, powdery coating which melted during firing to form a glaze. The characteristics of such a method most obvious to inspection by eye are the uneven application of the glaze and drip or flow lines. The thickness of the glaze depends in this method on body porosity and the water content of the slurry.

(ii) Self-glazing by efflorescence:

In this faience production process (Noble, 1969) water soluble alkaline salts such as sodium or potassium carbonates, sulphates, and chlorides, in the form of natron or ash, migrate during the drying from the body to the surface by capillary action as water evaporates from the surface. The alkalis effloresce, or precipitate as salts on the surface, forming a white powdery layer (referred to by Binns et al., 1932, as "wicking out"). When the dried faience object is fired in an oxidizing atmosphere the alkaline surface layer acts as a flux and the surface fuses to form a glaze. An appropriate colourant is included in the recipe to provide the required glaze colour. Binns et al. (1932) termed this method "self-glazing". In this method the drying rate is critical; a thicker coat of glaze is formed on those parts of the surface which are exposed to a greater flow of air and have a faster drying rate.

(iii) Self-glazing by cementation:

As no glaze slurry is applied prior to firing, this method is also called "self-glazing", though it is distinct from the previous one (Wulff, 1968; Kiefer and Allibert, 1971). The object is modelled from crushed quartz mixed with an adhesive, then embedded in a powder of hydrated lime, powdered quartz, alkaline plant ash and the required colourant. Smith (1972, 102) has described the ensuing reaction "(this) is heated to promote the formation of a reaction-layer of sodium silicate glass containing about 73 per cent silica. The composition is molten at 800°C and it wets the silica but does not wet the lime, so that the (objects) are virtually levitated by surface tension in the midst of the bed of powder without sticking to it. The same principle was later used in making the first hollow objects of glass, the so-called sand-core glass vessels, but the "sand", of course, was a non-wettable material probably lime, certainly not silica" (see p.214). Wulff (1968) and Tite et al. (1983) have shown, however, that firing temperatures of 1000°C are required.

Recent studies of a long chronological series of Egyptian faience objects (Vandiver, 1982a; 1983a; also Kaczmarczyk and Hedges, 1983) have shown that all these methods were at one time or another in use in Egypt, often simultaneously. After the prehistoric period of experimenting with the forming and glazing of beads and amulets, self-glazing by efflorescence was standard procedure. Already by the first part of the third millennium B.C., during the Protodynastic period (Dynasties I-II), application and efflorescence were in use, occasionally even on the same object. Cementation may not have been current until the Middle Kingdom; there is not sufficient evidence as yet to date its first appearance in Western Asia.

(c) <u>Production and nature of individual colours in glazes</u>:

Kaczmarczyk and Hedges (1983, 140ff.) have summarized the chemical nature of glass pigments in general in so far as they are relevant to faience manufacture. In a work of this type detailed scientific data is not appropriate, but certain points need to be made. A single element may be responsible for producing several different colours depending on its state of oxidization and whether it is present as a compound or in solution. When in solution in a glass, the combined effects of composition and firing conditions are crucial, as is most clearly illustrated by iron. Under oxidizing conditions, and with low concentrations of alkali, it is possible to obtain a solid solution of iron oxide giving a pink colour; under oxidizing or only mildly reducing conditions a blue or green colour is produced, but strongly reducing conditions produce gray or black glazes. Thus, whilst chemical analysis can throw light on the production of different colours, it does not of itself offer evidence of the maker's intentions, particularly with early glazes. Browns and reds were probably not deliberately distinguishable for some time; whilst green and blue are difficult, since there are so many intermediate states with little chemical difference between them. Only with deliberately manufactured polychrome faiences may the maker's intentions be more certainly inferred.

For ancient Egypt the following colour development broadly applied: green, blue, brown glaze appear in the fourth millennium B.C.; black, white, and purple appear in the First Dynasty, c.2900 B.C.; yellow was added by the early XVIIIth Dynasty (c.1500-1350 B.C.), with more intense blue, indigo and violet later in the same dynasty. In Mesopotamia blue, green and black may be observed by the end of the prehistoric period; brown/red and yellow by the end of the third millennium B.C.; and a greater variety of shades by the Kassite and Mitannian periods. Hedges (1982) has tabulated the metal oxides in a series of Mesopotamian faiences dating from prehistoric times to the

Achaemenid period.

Bimson (1973) pointed out that the glaze on a faience vessel lid of the Early Akkadian period from Tell Taya was coloured by copper with lead present in small amounts, "however, the lead content was not high enough to affect the colour given by the copper, which is "turquoise" blue as in an alkaline glaze and not grass-green as in a lead glaze". At Nuzi Vandiver (1982a) reported that premelted ground glass was used as a raw material in glazes on both clay and on sintered quartz bodies, though work is still proceeding on the technological relations between faience and glass at Nuzi. At Tell Rimah (Pollard and Moorey, 1982; Dayton, 1978, 436-7) the yellow glaze contained lead and antimony, with the lead greatly in excess and free of zinc; lead antimonate combined with copper was used as a green pigment. The black pigment in glass and faience was ferruginous manganese of highly variable iron content. Manganese had been used for black faience at Kish in the middle of the third millennium B.C. (Pollard and Moorey, 1982). The Tell Rimah faiences differ from like colours in the Egyptian repertory very little. Kaczmarczyk and Hedges (1983, 288) recorded only two clear distinctions: the absence of zinc in lead-containing glazes and the very much higher iron concentrations. Hedges (1982, 98) noted nickel in the black glaze of Western Asiatic faience, but not in Egyptian. He also reported finding antimony in Egyptian faience only when a yellow pigment (lead antimonate) was involved; in Western Asia it seemed to appear as often with, as without, lead.

Peltenburg (1974) has published unquantified analyses by McKerrell of glazes on Late Bronze Age faience vessels from Kition in Cyprus, including objects which be believed had been made in Syro-Palestine and possibly also in Mesopotamia. The white faience did not have a tin glaze; the glaze was produced by the addition of alkali to the recipe for uncoloured cores, as elsewhere at this time. The supposed existence of tin glazes in the ancient Near East appears to be a misconception, based on misreadings of early analyses. The main colourant for green was copper; sometimes a mixture of copper-blue and lead-antimonate yellow. Yellow was produced with lead-antimonate or iron oxide. In blue the colourant was commonly copper (or by the second millennium, at least, bronze i.e. oxidized scale, the corrosion products of a copper-tin alloy). Calcium antimonate was used as an opacifier, creating a matt blue, possibly a feature specifically distinguishing the "Western Asiatic" from the "Egyptian" manufacturing traditions (Peltenburg, 1974, 110; Hedges, 1982, 98).

(d) Polychromy:

In Western Asia, as in Egypt, it was only after glass workers had pioneered polychromy, in the middle of the second millennium B.C., that it entered the faience worker's repertory (Peltenburg, 1971, 9). Even then they were slow to develop techniques best suited to their own material. In Western Asia different coloured glazes on faience rested side by side, directly applied to their carriers, or were applied one on another, following closely the procedures of glass workers.

C: Architectural uses:

For much of the time in Mesopotamia the bodies of glazed bricks were made of clay, but in the Achaemenid period at Babylon there is evidence for bodies of faience. As this development is best seen in the context of glazed brick production in general, it is dealt with here on p.182.

Note on Analyses.

Isolated analyses are cross-referenced in the preceding text. So few analyses of faience from sites in Mesopotamia have yet been published, and even then in a manner difficult to correlate, that it seemed wisest just to give references here and not to attempt any correlation at this stage:

1. BRAK: Stone and Thomas, 1956, 64, 66, no.26: late prehistoric segmented bead (body only).

2. JAMDAT NASR: Hedges in Moorey, 1978, fiche 3:F05: late prehistoric beads.

3. KISH: Hedges in Moorey, 1978, fiche 3:F04-5: personal ornaments and vessels, Early Dynastic I to Achaemenid.

4. UMM EL-JIR: Hedges in Moorey, 1978, fiche 3:F05: Early Dynastic III and later beads.

5. TELL AL-RIMAH: Dayton, 1978, passim; Pollard and Moorey, 1982: personal ornaments and vessels, c.1400-1200 B.C.

6. TELL TAYA: Bimson, 1973: beads and vessel lid, Akkadian period; Neo-Assyrian polychrome sherd, whether of baked clay or sintered quartz is not stated.

7. FAILAKA: Analyses undertaken at the Oxford University Research Laboratory for Archaeology and the History of Art, 1984 (Pollard and Højlund, forthcoming). 5 later second millennium fragments ("Kassite").

IV: GLAZED TERRACOTTA:

Few aspects of the technology of the ancient Near East have led to more confusions and false assumptions than the production of glazed terracotta objects. Varied and enterprising as was the faience industry of Egypt it never seems to have been extended to glazing terracotta before the Roman period, not even when glass vessel production was introduced in the early eighteenth dynasty from Western Asia, at a time when glazed pottery was already current in Syria and Mesopotamia. Modern study of the subject has been bedevilled by recurrent failures to distinguish accurately in excavation reports between glazed bodies of terracotta and glazed bodies of sintered quartz (faience).* Even when glazed terracotta was recognized it was assumed, in the absence of unambiguous and up-to-date analyses, that the glazes in this case must be lead glazes, since it was thought the alkaline (soda-lime-silica) glazes used by the contemporary faience industry would not have adhered to an earthenware

* It should be said that this is by no means always easy by eye under field conditions and there were certain hybrid compositions yet to be properly separated by laboratory analyses.

body. Even now, when it has been shown analytically (see p.187) that lead (the flux with a compatible coefficient of expansion to fit a clay body) was not used in significant quantities in Mesopotamian glazes until an advanced date, confusions persist through popular misconceptions over what constitutes a true lead glaze. It is important to draw a sharp distinction between a true lead glaze, in which a high percentage of lead (say over 20%) is present as a flux and bonding agent in the glaze composition, and those glazes in which lead is present in relatively small percentages as a colour modifier in specific colours. It is pointless to speculate why lead glazes were not used until so late, save to remark that pristine alkaline glazes are brilliantly coloured by contrast with their duller lead glaze counterparts. They also offered a potentially greater colour range and may have had a more glossy appearance.

This observation conveniently raises a basic problem in the study of ancient glazed terracotta objects from Iraq where, even in antiquity, natural conditions were not conducive to their survival in a pristine state. Once glazed terracotta was buried the hazards were all the greater in prevailing soil conditions, particularly in the centre and south of the country. When it is possible to examine the glazes first hand by eye, it is obvious, in all but the rarest cases, that their colours are weathered or have been chemically modified. Even allowing that the following statement by the American excavators of Khorsabad turns on an extreme case, it is an important caution: "When found, the colour had largely faded or disintegrated, during which process the blue apparently first turns to green before becoming a pale, dead yellow, while the green reverts to a primary blue" (Loud and Altman, 1938, 41). Publication has, unfortunately, added another problem. Authors, since the middle of the last century, with commendable intent, have used contemporary processes of colour reproduction to illustrate some of their finds. They easily give a misleading impression, since colour-tinted drawings may restore the colours too decisively, whilst colour photography may distort, especially in reproduction, the surviving colour tones. This hazard is increased by the fact that some of the most commonly illustrated glazed-brick friezes, those in East Berlin from Babylon and those in the Louvre from Susa, have been restored in modern times in museum laboratories so that the extent and tones of the surviving ancient surface are by no means always evident. Much pottery, particularly from clandestine excavations in western Iran, has had its glazed surface "stabilized" and "improved" for sale on the antiquities market.

It is not always easy to draw an absolute separation between bodies of sintered quartz/sand and terracotta by eye. For instance, in the second quarter of the first millennium B.C., at Sanam and El Kurru in Nubia, and on the island of Samos, composite bodies of crushed quartz and terracotta are reported. Only systematic analyses of a wide range of body materials for Mesopotamia will indicate if, and when, this applied there (for bricks see p.182).

A: Pottery:

(i) Middle Assyrian (c.1550-1150 B.C.):

The earliest dated examples of glazed terracotta sherds in the Near East are those reported to be from level VI at Tell Atchana (Woolley, 1955, 299). The technical developments leading to the emergence of glazed pottery occurred at very much the same time, and probably in the same region, as the first experiments with the manufacture of core-formed glass vessels (see p.201). Level VII at Tell Atchana was destroyed within the Middle Bronze IIB period, sometime in the middle of the seventeenth century B.C., almost certainly by the armies of the Hittite king Hattusilis I. The subsequent level VI is, in ceramic terms, still largely tied to the traditions of Middle Bronze IIB-C, whereas level V is tied to the Late Bronze I period, from about 1550 B.C. Woolley described the sherds from VI as "a few miscellaneous fragments of green- or blue-glazed pottery" (AT/47/55). A complete glazed jar was found in a grave attributed to level V (Woolley, 1955, pl.LXXXIIIe); but the three associated cylinder seals indicate a later date, probably in the fifteenth century B.C. (Collon, 1981, nos.51, 63, 60). Analysis has shown these to be the ordinary alkaline glazes used by faience workers, now for the first time applied to a baked clay rather than to a siliceous core (Peltenburg, 1969, 80, n.33). They are not lead glazes as was automatically assumed at the time of discovery.

The first reported examples of glazed pottery from Mesopotamia are those found in Temple A at Nuzi destroyed about the middle of the fourteenth century B.C.; this is a context renowned for its range of glaze, glass and faience (see p.202). Starr (1939, 91, 403, pl.92Z) reported a tripod bowl "covered with an all-over coating of the familiar blue-green glaze" and several small glazed pots or bottles (Starr, 1939, 91, 391-2). In Babylonia relatively rare occurrences of glazed pottery, and rare glazed figurines, have been reported from Kassite levels at Nippur (Gibson, 1975, 16; 1978, 81), at Ur* (Woolley, 1965, 98) and on the island of Failaka, in the bay of Kuwait (Pollard and Højlund, 1983). At Tell Rimah, from a room in the Middle Assyrian temple, whence also came a collection of glass and faience, is reported "a large, pale green spouted bowl" described as being of "glazed pottery" (Carter, 1967, 286). In the north glazed sherds are also reported at Billah, Brak and Tell al Fakhar (Mahmoud, 1970).

At Assur, where the Middle Assyrian graves contained faience and glass vessels (see p.204), glazed pottery is rarely mentioned explicitly in the published reports (cf. Haller, 1954, 115: tombs 37, 51; also grave 133, which is not "Neo-Assyrian"). However, Andrae (1977, fig.164), in his general account of Assur, illustrated a range of monochrome glazed pottery jars and bottles of the kind found at Nuzi. Assur also produced a unique relief plaque described as "eine farbigemallierte Tonplatte" (6cm.x11cm.x2cm.). It is pierced for suspension at the top and is moulded in low relief with a robed Asiatic reminiscent of contemporary Egyptian representations of Asiatics on polychrome faience tiles (Andrae, 1913, 133, fig;231; 1935,

* Some, if not all, of this is faience.

81, fig.63; cf. Stevenson-Smith, 1965, 45, figs.69-70).

Although true polychromy had appeared in Western Asiatic faience, perhaps under the influence of glass making, by the beginning of the Late Bronze Age in the middle of the second millennium B.C., its earliest appearance on glazed terracotta is less easy to establish. Its use in that craft was inhibited by a technical problem, since in the case of baked clay the glazes do not so easily fuse with the carrier as in faience. A type of cloisonné had to be devised, with the polychrome set between glaze outlines, in order to prevent the glaze colours from running and mixing. This trick had certainly been mastered by the Neo-Assyrian period, both on bricks and on pottery, but how much earlier is uncertain. In a simple way it is evident in architectural glazed terracottas of the earlier thirteenth century B.C. at Tchoga Zanbil in Khuzistan (Ghirshman, 1966) where extensive use of glazed bricks is evident for the first time (cf. Peltenburg, 1971, 9).

(ii) Neo-Assyrian to Achaemenid (c.850-300 B.C.):

After the twelfth century in Assyro-Babylonia, as in Syro-Palestine, there appears to have been a break in the tradition of glazing pottery; but it is possible, especially in Babylonia, that this is more apparent than real, since excavated levels of this period are still very rare. In neither region is there explicit evidence for the time at which production was resumed. In Assyria polychrome glazed bricks (see p.170) were in use again by at least the early ninth century B.C. (see p.172) and by the eighth century, if not before, glazed pottery was once again current in Babylonia. It was now often polychrome in contrast to the monochrome glazed wares of the Late Bronze Age. Peltenburg (1969, 86) argued that a revival of glazed pottery production in Syria, which has been dated on the basis of evidence from Al Mina to the last third of the eighth century B.C. at the latest, was ultimately inspired from Babylonia: "In the second half of the eight century both Tiglath-Pileser III and Sargon II deported Babylonians to N. Syria, and it seems only natural that glazed vases or even a Babylonian glazer were included" (Peltenburg, 1969, 86). Stimulus diffusion along the Euphrates might have been a contributing factor. Much of the glazed pottery from Babylonian sites is too badly preserved, or too sparingly published, to allow for the close analysis that might permit secure isolation of local styles and production centres along the lines suggested by Strommenger (1967, 29). Any traces of continuity with Kassite craft traditions have yet to be elucidated.

In the "Neo-Assyrian" levels excavated at Babylon glazed vessels were relatively rare (Reuther, 1926, 22-3, 28, pl.75a-d, 76:13; Koldewey, 1911, 35ff., fig.54; 1913, fig.152 (colour)). They were glazed in light blue, white and yellow; polychrome decoration was only occasionally recognisable, usually in white, light and dark yellow, light blue and dark blue/black. By the Neo-Babylonian period glazed vessels were commoner, again in white or blue glaze, and with polychrome designs consisting of rows of triangles, rosettes, squares and zigzags. At Nippur (McCown, 1967, 79, pl.100-1), again in "Neo-Assyrian" levels, the base glazes were light blue, greyish white or blue-green in that order of frequency. "Polychrome designs occur in combinations of dark brown, light and dark yellow, orange, a gray and a blue which probably represent the same original colour, light green or yellowish green, and white". Gibson (1980, 202) has referred to "dwellings of the early first millennium containing extraordinary numbers of glazed vessels" at Nippur.

A remarkable group of glazed vessels was found in the pioneer excavations at Nippur. It is hard to place them with confidence chronologically, though they may be of the post-Kassite period. They were "in a large urn on Hill I, 11m. below the surface" (Peters, 1897, II, plate facing p.186). A box on four feet, with pyramidal top, is made of a spiked or knobbly ware; two smaller boxes, with lids, and three jars, one with a stopper, are more conventional. "All thickly glazed with blue and yellow.....in stripes". With the vessels were over a hundred small faience objects, bored for suspension: "some round, some oval, some crescent-shaped, and all coloured, some black and white, and some blue and yellow". A single published vessel of "knobbly ware" from the 1976 excavations at Nippur appears to be of the same type as this group. The excavator described it as a "glazed pottery incense burner". It was in a burial (no.61) cut down into level I, dated c.900 ± 200 B.C. (Gibson, 1977, 27, fig.).

The revived use of glazed wares at Ur is not securely dated, as in his "Neo-Babylonian" category, Woolley almost certainly included examples of seventh century or earlier date. He offered a detailed comparison between wares of the Neo-Babylonian and Achaemenid periods (Woolley, 1962, 90; cf. Hall, 1930, 184-5, figs. 147-8); but this also has to be used with caution in view of the difficulties of confidently attributing vessel types to one or other of these periods: "Already by the time of Nebuchadnezzar about one grave in every three can boast a glazed pottery vessel; in the Persian graves the proportion is higher, but more symptomatic is the fact that of a hundred types which are almost if not quite exclusively Persian 43 are represented by glazed examples.....we can say that half the characteristically Persian types are glazed whereas of the 59 types listed as peculiar to the Neo-Babylonian period only 12 show examples of glazed ware. The glazed forms are for the most part broad rimmed shallow bowls or plates.....; slender-necked bottles.....; small squat jars.....; or such exotic forms as the handled flasks and pilgrim bottles". In one case bowl sherds were decorated with a polychrome design of animals and a tree on a white ground (Woolley, 1962, pl.35:U.6654). Hall (1930, 184-5, fig.148) reported half an open bowl decorated with "stripes of blue and yellow outside and a design of blue and yellow spots within".

Neo-Assyrian glazed pottery of the later eight and seventh centuries B.C. is best known through the examples found at Assur, to which Andrae (1925) devoted a monograph illustrated in full colour. The predominant forms were buckets, jars and closed bowls, of various sizes, with polychrome glazes on the exterior. Some of the buckets and jars were of considerable size; the standard jars are generally 8 to 9 cm. high at the most, with larger examples reaching four or five times that size. The buckets may well have been used exclusively for cult purposes. One of the larger jars contained a cremation burial (Andrae, 1925, pl.25A; Haller, 1954, 52-3, fig.63A). "A favourite grave furniture - at any rate in late Assyrian times, that is the eighth and seventh centuries B.C. - were the little bottles manufactured in the mass, at first richly decorated, later quite simple" (Andrae, 1925, 34). Drinking cups, or open bowls, and dishes do not seem to have been glazed in this way, so the glazed jars, presumably in themselves a luxury industry, may have been containers for valuable unguents. Only a few sherds had survived with the colours preserved to something like their original intensity. On a rich light blue, or green, base glaze designs were rendered in warm golden yellow, ivory and dark blue. "On the small objects woven patterns are used, such as the chevron band, and rows of circles, rings, squares, triangles, or wreaths of leaves; in the big ones there are besides friezes of battlements, rosettes and bases of pillars, taken over from architecture" (Andrae, 1925, 34-5). Some of the more elaborate designs employed traditional motifs from the religious repertory: animals flanking "trees-of-life" with a winged disk hovering above; scenes of worship with altars, symbols and supplicants.

Similar glazed pottery has been found on other sites in Assyria, but publications are few. It is said to have been "not uncommon" at Nimrud (Oates, 1959, 138). Mallowan (1966, 120, fig.61, 187) refers to one large example decorated with an ostrich hunt, from the North-West Palace at Nimrud and to another, not described, in the main reception room of a private house. That such vessels were treasured long after they had passed from production is illustrated by a blue, white and yellow glazed "Neo-Assyrian" jar in a grave of the Hellenistic period at Nimrud (Oates and Oates, 1958, 130, 152, pl.XXVIII: 15). Comparable vessels were found at Khorsabad (Loud and Altman, 1938, pl.63:231, 244) and an isolated example has been published from site "C" at Tell Rimah (Carter, 1967, pl.9).

Westwards into Syria this Neo-Assyrian industry does not appear to have had any marked impact. Peltenburg, as has been seen (p.168), argued that a distinctive group of glazed flasks, with a wide distribution in the East Mediterranean, had come initially from Syrian workshops. If influenced from the east, it was by Babylonian not Assyrian craft traditions. They have design arrangements, shapes and certain technical traits distinct from those evident at Assur. Areas on the surface were reserved in the base white glaze for later filling-in with green and brown glazes. This "planning technique" is found in Babylonia not at Assur, where outline techniques are used for the purpose (Peltenburg, 1969, 83-4, fig.2j).

The glazed pottery traditions of western Iran were again distinct, with differing methods of glaze application and contrasting colour schemes, though some of the shapes and designs appear also on Neo-Assyrian sites in Iraq (cf. Hennessy, 1979, pls.111-2 (colour); Fukai, 1981, pls.8-9, 10-15 (probably glazed terracotta, not faience); Ghirshman, 1964, pl.398; Wilkinson, 1967, fig.14). Such vessels appear at Susa in the eighth to seventh centuries B.C. (Amiet, 1966, pl.378).

B: Architectural Uses:

(i) Bricks:

(a) Middle and Neo-Assyrian(c.1550-650 B.C.):

Among the wide range of glazed objects from stratum II at Nuzi there are only two published examples of glazed bricks: "from the scarcity of this class of object, it may be assumed that these two were used singly for some special purpose, and not as building material or architectural decoration" (Starr, 1939, 412, pl.116A). They were smaller than the standard bricks and white glazed. Apart from this exceptional case, no glazed bricks or orthostats have yet been published certainly earlier than the Neo-Assyrian period in Mesopotamia; but it is probably to the reign of Adad-Nirari I (c.1305-1274 B.C.) that the earliest textual reference to Assyrian glazed bricks belongs (Reade, 1979a, 19).

Tiglath-Pileser I (c.1114-1076 B.C.) has been credited with the glazed brick frieze on the lower facade of the Assur Temple at Assur, depicting battles and processions (Haller and Andrae, 1955, 56). But Weidner (1926) argued, on epigraphical grounds, that this decoration should be attributed to Sargon II (c.721-705 B.C.); Fridman (1969) has presented cogent reasons for identifying the monarch as Tiglath-Pileser III (744-27 B.C.); whatever the case it is Neo-, not Middle, Assyrian work. Tiglath-Pileser I, in describing the palace he built at Nineveh, speaks of:

> "bricks glazed (the colour of) obsidian, lapis lazuli, pappardilu-stone [yellow], (and) parutu-alabaster [white]. I installed on its towers replicas, in obsidian-coloured glazed brick, of date palms....." (Grayson, 1976, 33; modified after Oppenheim, 1970, 17).

This indicates, whatever the true date of the Assur frieze, that by the end of the Middle Assyrian Period architectural details, if not "panoramas", were already executed in glazed brick. Although moulded brick friezes were a feature of Kassite architecture, the known examples show no evidence of having been glazed. Throughout the Neo-Assyrian period it seems likely that "panoramas" in glazed brick were to be found most commonly in palaces.

Reade (1979a, 19ff.) has concisely summarized the main characteristics of Neo-Assyrian glazed brickwork. He isolates three main categories of glazed bricks or tiles: bricks glazed on the square side like a modern tile; bricks glazed on one edge, and sometimes also on part of an adjoining edge or side; bricks with one edge moulded and then glazed. The history of this craft is rather unevenly represented in the surviving material evidence, with a heavy concentration on the ninth and later eighth centuries B.C. The more extended designs, not very common, are usually on "tiles", whilst edge-glazed bricks were used to frame architectural units. At Assur some of the glazed bricks were so large that Andrae described them as "brick orthostats" (Andrae, 1925, pl.7-10). Some older references to "painted bricks" are to glazed ones; the outer surface decays leaving the effect of matt paint.

A fragmentary orthostat (66.5x46.5cm.) (Andrae, 1925, pl.7; Reade, 1983, pl.18) found in the Temple of Anu and Adad, near the east corner of the Adad Ziggurat at Assur, dated by its inscription to Tukulti-Ninurta II (c.890-884 B.C.) shows a chariot scene, with little but parts of the horse and charioteer preserved, framed between chevron bands. The scene is rendered in dark line on a light blue ground in a combination of yellow, blue, white and black. On another inscribed fragment (Andrae, 1925, pl.8) a god, with drawn bow, set within a disk, hovers over another charioteer.

In the text carved on the so-called "Banquet Stela", from Nimrud, Assurnasirpal II (c.883-59 B.C.) described, for the first time, what may have been a more extensive use of glazed brick friezes, on the exterior courtyard walls of the North-West Palace:

> "I depicted in greenish glaze (?) on their walls my heroic praises, in that I had gone right across highlands, lands, (and) seas, (and) the conquest of all lands. I glazed bricks with lapis lazuli (and) laid (them) above their doorways" (Grayson, 1976, 172ff.).

It is unfortunate that vestiges of these decorations found by Layard are now largely lost, since it may be that the phrase here translated as "greenish glaze" describes narrative scenes painted rather than rendered in glazed brick. The south side of the great outer courtyard of this palace was occupied by the triple entrances to the state apartments:

> "Between the bulls and lions (i.e. of carved stone), forming the entrances in different parts of the palace, were invariably found a large collection of baked bricks, elaborately painted * with figures of animals, and flowers, and with cuneiform characters. It is remarkable, that on the back of these bricks, or on one of the sides not coloured, are rude designs, in black paint or ink, of men and animals, and marks having the appearance of numbers. They appear to have been built into a wall above the sculptures" (Layard, 1849, II, 13-14; 1853a, pl.84; cf. Dayton, 1978, 374, fig.337).

In excavating the Temple of (Ishtar) Šarrat-niphi at Nimrud, built by Assurnasirpal II, Layard found that the exterior walls had been adorned with glazed bricks as were the battlemented "altars" on either side of the entrance (Layard, 1853, 359; cf. Reade, 1983, watercolour on contents page); such was also the case with the Ninurta Temple (Reade, 1983, fig.3) and the Kidmuru Temple, excavated by Rassam in 1878 (Rassam, 1897, 225).

Fragmentary glazed brick friezes, from the royal palace at Nineveh (Campbell Thompson and Hutchinson, 1931, 80-1, pl.XXVIII-XXXII), included within framing bands of guilloches and chevrons, military scenes and scenes of worship, comparable to those on the better known carved stone orthostats of Assurnasirpal II's reign, though the bricks might be early in the reign of Shalmaneser III.

---

* Layard's standard word for "glazed".

"The colours most commonly used were white, yellow, green, black, more rarely brown, and very rarely red" (ibid., 83). Scattered glazed bricks were also published from "above or near" the Nabu Temple site at Nineveh, again using blue, green, yellow, white and brown, the latter usually for outlines (Campbell Thompson and Hutchinson, 1929, pl.LVII.350-8).

The most grandiose architectural monument of the reign of Shalmaneser III (c.858-824 B.C.) was the ekal mašarti (Fort Shalmaneser), which occupied the south-east corner of the city enclosure at Nimrud. Oates (1962, 9, pl.II-III), in justifying his restoration of the West Gate, noted that:

"The proportions.....and some idea of the overall size of the battlements is given by the rows of impressed rosettes in glazed brick which were commonly used to decorate the upper parts of the walls. No glazed bricks have been found among the ruins of the outer wall of Fort Shalmaneser, but they were employed on the north face of the South-east corner tower, and from components recovered there we know that the height of each rosette was three courses".

A semi-elliptical panel, over four metres high, surmounted the courtyard doorway of room T3 in Fort Shalmaneser, and on the outer wall of the throne room (T1) there had been a glazed brick frieze of rosettes, found scattered in the courtyard. Reade (1963; cf. Dayton, 1978, pl.24 (colour)) restored and published the panel, which bears an inscription of Shalmaneser III. In the lower centre panel two kings stand facing one another below a winged disk containing "Assur", then comes the inscription, and above it rampant bulls, heads turned back, flanking a tree. The whole design is framed first by rosettes and guilloches, then by a triple border: from inside to out, of alternating pomegranates and buds; of kneeling goats and palmettes. The glaze colours used were green, yellow, white and black on an entirely (originally ultramarine) blue background.

Individual panels are less commonly reported. One, measuring 12"x9", from Layard's excavations at Nimrud, of the ninth century B.C., shows a king holding a libation bowl, followed by attendants (Layard, 1853, 167-8; Perrot and Chipiez, 1884, 307-8, pl.14; Andrae, 1925, 29; BM Guide, 1922, 170, pl.XLIII; Reade, 1983, fig.41 (colour)). In a dwelling house (no.69) at Assur, on the outer line of the Neo-Assyrian city wall, an "orthostat" (56x27.5cm.) was set into the lower part of an inside wall. This may be a secondary use (Preusser, 1954, 54-5, pl.25b; Andrae, 1925, pl.10 (colour)) for an object made in the ninth century; but its dating is not certain. A deity, or his statue, set on a podium, is worshipped by a man standing before it. In the field above his head is a locust; in a line at the top: a sun disk; a winged disk; a crescent; and a rosette.

If the attribution to Tiglath-Pileser III (c. 744-727 B.C.) is correct (Fridman, 1969), some idea of the more elaborate figured designs on glazed bricks of the middle or later eighth century B.C., more closely matching the sculptured friezes, may be gained from those on the podium in front of the temple of Assur in the city of that name. They showed a king on campaign and a fragmentary part of a king enthroned (see p.171).

The best known Neo-Assyrian glazed brick friezes, since they were excavated in the nineteenth century and coherently published at an early date, are those at Khorsabad set up by Sargon II (c.721-05 B.C.). As Layard (1849, II, 13) recognized Assyrian builders seem to have reserved glazed brick decoration particularly for panels or friezes set above the true voussoirs of the arches of gates and doorways (cf.Pillet, 1962, pl.VII; Reade, 1983, fig.7). Place (1867-70, I, 234; III, pl.9-17; cf.Loud and Altman, 1938, 15) discovered almost intact the glazed brick decoration of town gate 3 at Khorsabad. Within a border of small rosettes, typical Neo-Assyrian winged genii with pails and cones, alternating with large rosettes, provided an apotropaic frieze. The figures are in yellow; the rosettes in white on a blue ground. More spectacular were the tableaux flanking entrances in that part of the palace designated "Harem" by Place, but shown by the subsequent American excavators to be a temple complex (Loud, 1936, 80ff.). In these palace shrines the glazed brick friezes flanked the doorways leading from the courtyard into the temple itself, where they took the place of the sculptured winged bulls and genii flanking palace portals. They rose about 1.50m. in height, from ground level, and extended the width of the buttress with a projection of 1.00m. from the buttress face.

On the Sin and Shamash Temples at Khorsabad (Loud, 1936, figs.99, 104-6, 110, 115) single figures of a lion, an eagle, a bull, a fig-tree, and a seeder-plough are set in sequence on the front face, with the king and his chief minister (?) at either end. In the Ningal Temple space forced the omission of the eagle and the bull. The figures were outlined in black on a vivid ultramarine blue ground; they were of yellow with black hair and beards. The leaves of the fig tree were green; the border rosettes white-petalled with yellow centres and stripes. In the Nabu Tample (Loud and Altman, 1938, 56ff.) a pair of friezes flank the doorway through which access was gained to the central court, whilst in the central court they adorn the facade of the inner temple exactly as round court XXVII in the palace. The friezes here were less well preserved; but did not, in what is recognizable, substantially differ from the others (ibid., pl. 17D, 44).

This by no means exhausts the role of glazed bricks at Khorsabad, or presumably elsewhere in the later eighth century B.C., since they are regularly found, if not always in quantity, in the debris of rooms, where they had served in some minor way to decorate the interior, perhaps as narrow rows, commonly of rosettes, between wall top and ceiling. A more individual use is their presence, as a single band of rosettes, in alternate horizontal joints of the stone facing of the arched bridge connecting the palace and the Nabu Temple (Loud and Altman, 1938, 56, pl.12). Their only occurrence as part of a free-standing structure was in the central court of the Nabu Temple, where they formed part of an altar (ibid.,42, pl.22C-F). This stood 1.50m. square, but both its original height and its surface designs were lost.

Outside Assyria, at Tell Halaf in Syria, what may have been an altar of comparable type, with glazed designs on its sides, stood before the ninth century palace portal. It was in much better condition and its geometric designs were still evident (Naumann, 1950, 71ff., colour plates I-III). In this case the design consisted of registers of rosettes in green and yellow, a scale pattern, in yellow and white, and a guilloche in green, yellow and white. In some cases here the bricks had been cut to shape to make up the pattern rather than it being painted on the surface of rectangular bricks.

Amiet (1966, pl.L) has reconstructed a similarly decorated podium of the eighth century B.C. in a temple at Susa. It is possible that it is from this same building, or a related structure, that two pieces of decorated glazed terracotta, now in the British Museum, originally came. They were published by Smith (1928, pl.XIV i, XXII d: BM 9196, 117981) as from the excavations by Loftus at Warka; but the protome, at least, is much more likely to be from his excavations at Susa.

In his inscriptions Sennacherib (c.704-681 B.C.) refers to his palace as embellished with glazed bricks the colour of obsidian and blue (Luckenbill, 1924, 107 vi 42); he also provides what may be the first textual reference to "glazed" bricks for a temple (ibid., 148). There is no archaeological material certainly relevant to these references. To the reign of his successor Esarhaddon (c.680-669 B.C.) may be attributed a series of decorated glazed bricks ("tiles") found by Layard, in what is now known as "Fort Shalmaneser" at Nimrud, re-used in a pavement whose exact location is unknown (Layard, 1853, 165-6; 1853b, pl. 53-4). They show Assyrian soldiers in seventh century uniforms campaigning in Egypt (Postgate and Reade, 1976-80, 317). The technique is distinctive; Layard described them as follows: "the colours have faded.....The outlines are white, and the ground a pale blue and olive green. The only other colour used is a dull yellow" (Layard, 1853, 166). Although Ashurbanipal (c.665-627 B.C.) boasts of building an entire New Year's Chapel of blue and red glazed bricks, appropriate material evidence for his reign has yet to be identified.

Before passing to the glazed brickwork of the Neo-Babylonian period it is important to notice that moulded or shaped glazed-bricks were relatively rare in Assyria. They have been reported, and then unusually, at Nineveh (Rassam, 1897, 222; Campbell Thompson and Hutchinson, 1931, pl.26:2, 4) and at Nimrud (Postgate, 1973, 193), from a panel above an Ashurnasirpal door.

(b) <u>Neo-Babylonian to Seleucid (c.600-300 B.C.)</u>:

Both the Old Testament and classical authors preserved memories of the glazed brick decoration of Babylon. Ezekiel (23:14) recounts how Oholibah "saw male figures carved on the wall, sculptured forms of Chaldaeans, picked out in vermilion.... "Whilst Diodorus Siculus, writing in the first century B.C., included a description of two major palaces, one on either side of the Euphrates, given by Ctesias (early fourth century B.C.):

"And within this (the outer palace wall) she (Semiramis) built a second, circular in form (Koldewey preferred "annular"), in the bricks of which, before they were baked, wild animals of every kind had been engraved, and by the ingenious use of colours these figures reproduced the actual appearance of the animals themselves ..... "(II.8.3-6). Then:

"On both the towers and the walls there were again animals of every kind, ingeniously executed by the use of colours as well as by the realistic imitation of several types; and the whole had been made to represent a hunt, complete in every detail, of all sorts of wild animals, and their size was more than four cubits. Among the animals, moreover Semiramis had also been portrayed, on horseback and in the act of hurling a javelin at a leopard, and nearby her husband Ninus, in the act of thrusting his spear into a lion at close quarters" (II.8.6-9).

From the eighteenth century travellers commented on the prevalence of glazed bricks at Babylon and other sites in the south. But it was not until the German excavations at Babylon that the account given by Diodorus was actually illustrated: "The discovery of these (glazed) bricks formed one of the motives for choosing Babylon as a site for excavation. As early as June 1887 I came across brightly coloured fragments...." (Koldewey, 1914, 26). Koldewey (1914, 129-131) later commented on the description left by Ctesias in the light of his own discoveries. The first passage appeared to describe the Neo-Babylonian decorations of the "Processional Way" and the "Ishtar Gate" (see below), whilst the latter recalled the fragmentary glazed bricks of the "Persian Building", for only there were human figures discovered (cf. Haerinck, 1973). Nebuchadnezzar II (c.604-562 B.C.) refers in his inscriptions to the blue bricks of the temple tower Esagila at Babylon, especially those of the sanctuary which crowned it (Langdon, 1912, 114 i 43). The "silver" coloured bricks decorating the approach to another chapel were probably also glazed (Langdon, 1912, 128 iii 56; cf. 158 vi 36).

At Babylon two complete Neo-Babylonian schemes of decoration in glazed brick may be reconstructed. In the great royal palace, the "Southern Citadel", the throne-room lay on the south side of the main courtyard approached through three doorways, one in the centre flanked by one toward each end of the facade. Although details in the restoration of this decoration are debatable (Heinrich and Seidl, 1968, foldout plate), its general scheme is agreed. The entire wall was decorated with panels of slender columns with volute capitals, connected by a series of palmettes, rendered in yellow, white, light blue and green, on a dark blue background. Above was a frieze of white double palmettes, bordered below by a band of squares, alternately yellow, black and white. White borders, against the dark blue background, heightened the effect. Below was a procession of lions passant on the central archway (Koldewey, 1914, 104ff; 1931-2, 84ff., pl.37-8, colour; Meyer, 1965, pl.174 in colour = Moortgat, 1969, pl.292, not in colour). Better known, primarily from the modern reconstruction in the Vorderasiatischen Museum in East Berlin, is the so-called "Processional Way" and the "Ishtar Gate" through which it passed. Along the lower walls of the "Way" were lions passant framed above and below by rosettes; blue glazed brick battlements ran along the top. On the walls of the "Gate", from the top to deep in the foundations, where they were not glazed, were separate rows of bulls and "snake-dragons". They were widely spaced, within patterned borders, on battlemented facades entirely surfaced with deep blue glazed brick (Koldewey, 1914, 38ff.; 1918; Meyer, 1965, pl.168-71, in colour; Moortgat, 1969, pl.289-91, not in colour).

The scattered evidence for glazed bricks, usually monochrome, from other sites in Babylonia offers no coherent designs like those recovered at Babylon, only indication of the use of glazed bricks quite commonly in religious, if not also in secular, structures. At Ur Woolley reported that "in the rubbish lying against the sides of the Ziggurat there were found quantities of bricks of Nabonidus of which one side, or one side and one end, were covered with a bright blue vitreous glaze" (Woolley, 1939, 133). Examples now in the British Museum (BM 116981) are blue, blue-green, and deep red in colour. None were found in situ and Woolley therefore assumed that they had fallen from high on the structure, perhaps from the sanctuary on top, by analogy with Nebuchadnezzar's Esagila description.

The famous inscription in which Darius I described the building of his palace at Susa clearly informs us that "the men who wrought the baked bricks these were Babylonians" (DSf: Kent, 1950, 142ff.; Vallat, 1971). In Mesopotamia itself only at Babylon are there vestiges of glazed brick decoration certainly of the Achaemenid period. The so-called "Persian Building", adjoining the Southern Citadel, contained fragments of polychrome glazed bricks, probably from friezes of guards (Koldewey, 1931, 121ff., pl.28 in colour; Haerinck, 1973). The date of this building is uncertain. The excavators attributed it to Artaxerxes II, but Olmstead, followed by Schmidt (1953, 28) and Haerinck (1973, 129) have preferred Darius I. Certainly it has the same type of layered, re-surfaced, floor as the buildings at Persepolis and Susa most certainly dated to his reign. As might be anticipated, such decoration was relatively rare at Persepolis, where stone was readily available, and restricted in its use of motifs (Schmidt, 1953, 70-2). But it was well represented at lowland Susa, associated with various rulers through the Achaemenid Dynasty. The repertory, whether of individuals or groups of people, of animals and monsters, and of floral designs (Haerinck, 1973, 118ff.), has much in common with that carved in stone at Persepolis, where much of it was originally picked out in paint of comparable colours (Roaf, 1983, 8-9).

The persistence of the glazed brick tradition in Mesopotamia is clear from the decoration of the Seleucid Anu-Antum Temple at Uruk (Jordan, 1928, 18ff., pl.48ff.).

(ii) Baked clay wall "nails" (sikkati) and plaques:

The idea of decorating walls with terracotta cones arranged in patterns, their end dipped in paint, goes back to the prehistoric period in Mesopotamia. The custom of securing a square plaque to a wall by a peg driven through a hole in the centre is almost as old (Boese, 1971, 155ff.). During a long history many variants of the "peg" or "nail" appeared (Andrae, 1925, fig.38; 1977, fig.187); but it was not until after the middle of the second millennium B.C. that they were manufactured in glazed terracotta. In the fifteenth century B.C. at Nuzi (with at least one exception), at Tell Rimah a century or so later, and at Assur, the heads of terracotta wall knobs, some with low relief floral or geometric decoration, were covered with green glaze (Starr, 1939, 407-9, pl.97-8; Carter, 1964, 40, top; Andrae, 1935, pl.41). In general wall pegs were associated with such royal buildings as fortifications, palaces and temples, and their heads sometimes carry royal inscriptions (cf. at Assur: Andrae, 1913, pl.LXXXVI-II, XCIII, XCV, CI-III); but the glazed examples of the middle and later second millennium B.C. are primarily found in religious buildings. There were also "wall-pegs", sometimes glazed, as in the Middle Assyrian Ishtar Temple at Nineveh, that had open ("trumpet") mouths (Reade, 1979a, 21).

Some of the most explicit information on the setting of wall-pegs in the second millennium B.C. was recovered at Nuzi: "In many cases the shanks were not of sufficient length to have borne any great weight; consequently, their main purpose must have been decoration. Since they were found only in the Ishtar section of Temple A, and in L5 and L8, the palace chapel, it is evident that this form of decoration was restricted to the cult. In L5 they were found in position, inserted in the wall 178cm. above the pavement, 95cm. apart, the first being 130cm. from the corner (Fig.21). Elaborations of this same type were found in Temple A (pl.98, E-H), used apparently in the same manner" (Starr, 1939, 59, cf.151-2). In Mesopotamia at this time there is no evidence for plaques secured to the wall by such wall-pegs; whatever their decorative role it was self-sufficient. A terracotta wall knob from Tell Rimah was glazed green with an alkaline glaze coloured by copper (Pollard and Moorey, 1982, 45ff.).

The most spectacular find of glazed wall-pegs was made by Ghirshman at Tchoga-Zanbil in Khuzistan, where some of the rooms in the first stage of the Ziggurat, built in the thirteenth century B.C., had been used to store them in quantity (Ghirshman, 1966, 12, pl.XVII-XIX, XXV). Fragments of comparable wall-pegs, inscribed for Silhak-Inshushinak, were found at Susa (Amiet, 1967, 28-9). At Tchoga-Zanbil the pegs were associated with square plaques which they had secured to walls through central holes. The plaques are decorated in low relief with quarter rosettes, their petals blue/green, white or yellow; the projecting pegs also have floral decoration, sometimes with petals rendered in white and yellow. According to Fukai (1981, 5) blue, green and white glazes were used; two occasionally in combination, but never all three. He also published the analysis of a blue-green glaze from one of the knobs, showing it to be a non-lead copper glaze (ibid., 6).

No information is at present available on the use of such glazed knobs between the twelfth and ninth centuries B.C. in Mesopotamia or Western Iran. Most of the known Neo-Assyrian examples seem to date to the ninth century B.C. Andrae's study of 1925 remains the basic treatment of this series, though it is based on only one site, Assur. "The head of the knob was given a tapering neck and one or more rings at the foot, at the place where it enters the wall. The long stem which must remain firmly in the wall is left unglazed, while the foot, neck and head were often enamelled (i.e."glazed"). If the stem was left out the hollow knobs had to be fixed to the wall with wooden pegs, and to attach it immovably to the pegs it was made fast with a pin for which a hole is found in the neck. The ring at the base is sometimes widened to a disk....But usually and especially from the ninth century onwards, the stem is left out, and the neck and head of the knob are made hollow. In this way arise the <u>knob plate</u>, in which the disk at the base preponderated and became the chief part of the decoration" (Andrae, 1925, 65).

With the information now available from Tchoga Zanbil, the

plaque may not be seen as a typological development of the ninth century, though how its revival then relates to the Middle Elamite prototypes seen at Tchoga Zanbil remains obscure.

Clear evidence for the arrangement of these lozenge or cushion-shaped, sometimes round, wall plaques in Neo-Assyrian buildings is scarce, perhaps because of the height at which they were normally set on a wall. On one wall of the Palace of Assurnasirpal II at Assur (Preusser, 1955, 21, pl.14) were the "ghosts" of seventeen square wall plaques, at about the height of a man, set in a row. Andrae (1925, 66) thought that this was a later arrangement, " as if in late Assyrian times someone had had the pleasure of collecting and bringing together again, to some extent suitably, well-preserved knob-plates from the palace ruins, perhaps at the order of the Babylonian satrap resident in Assur". The reason for such an hypothesis is not evident. Most of the plaque decorations restored at Assur were the familiar floral friezes of the period, known also on sculptures and rare mural paintings. A few bore geometrical petterns and there were even fragmentary examples carrying scenes of lion-hunting (Andrae, 1925, pl.9:31-6; fig.39, 41, 45, 47-8). Similar plaques have been found on the other major Neo-Assyrian palace sites, as also in temples, though they have rarely been well published (Khorsabad: Loud and Altman, 1938, 43; Nimrud: Layard, 1853b, pl.55.3; Rassam, 1897, 225-6 (BM 91679-82, 131661); Perrot and Chipiez, 1884, fig.127-8; Balawat: Dayton, 1978, fig.338; Nineveh (British Museum)). The series begins in the ninth century, at least at Assur, and endured through into the seventh century there. Andrae detected some modifications through time in the decoration. In commenting on the colour schemes he remarked that "we have to deal only with yellow and light blue, only in a few cases with any other colour - for instance dark blue. White and black give the essentials of the design" (Andrae, 1925, 73).

The fashion for glazed polychrome wall plaques secured by pegs with decorated heads is also found in western Iran in the first millennium B.C. In the ninth century they seem to have been more immediately under the influence of Neo-Assyrian designs, though some local traits are already evident. At Hasanlu, for example, in the late ninth century debris of level IV, plaques comparable to those from Assur appear to have been secured to walls in certain cases with pegs having anthropomorphic or zoomorphic terminals in distinctively local styles (Dyson, 1959, 14, fig.13, lower right; Fukai, 1981, fig.9). Wall tiles of this type have also been reported from trenches cut into the citadel at Ziwiyeh, perhaps a century or more later than those from Hasanlu IV (Dyson, 1963, 35). At Susa the Middle Elamite tradition of glazed terracotta endured, though undocumented until the eighth century B.C.,when Shutruk-Nahunte II built a small temple at Susa. Amiet (1967) has restored the archaeological context of its scattered terracotta decoration. The actual scheme of decoration is impossible to reconstruct with certainty. At some height on the walls zoomorphic pegs were used (Amiet, 1966, figs.368, 380ff.; 1967), but whether to secure plaques is not evident from the publications.

Pegs and plaques do not appear to have survived into the Neo-Babylonian period; perhaps they were an "Assyrian" phenomenon.

(iii) Glazed Fists ("Hands of Ishtar"):

Among the baked clay architectural fittings used in the Neo-Assyrian period there is a group of life-sized, and slightly larger than life-sized, shanked human fists, whose exact function has yet to be established (van Buren, 1930, 270-2; 1945, 59-9; Peltenburg, 1968). In the most recent treatment Peltenburg proposed that they were used primarily "as window-ledge fixtures of a decorative and apotropaic nature", largely on the evidence of a basalt frame from Hama, level E, carved with a series of alternating, projecting talons and recessed rosettes (Peltenburg, 1968, pl.XIV). As these fists were either glazed or covered with bitumen, sometimes overlaid with sheet metal, an external or an internal use is possible, though the glazes would not have withstood water effectively. In the North-West Palace at Nimrud Smith found one "planted upright in the wall, embedded in mortar between the bricks" (Smith, 1875, 75ff., 429). At Assur they were also reported in a vertical position, though Andrae, probably correctly, took this to be a re-use (Andrae, 1925, 63, fig.36). He preferred to see their role as decorative: inserted inside rooms under the roof beams or as supports for glazed friezes (Andrae, 1913, 7; 1925, 67). A role in ceiling or roof construction was almost certainly their most common use (cf. Preusser, 1955, pl.14b).

Fists have been reported loose in debris from all major Neo-Assyrian sites (Nimrud: Layard, 1853a, pl.84, 14, 16; Smith, 1875, 75ff., 429; Mallowan, 1953, 11; 1954, 121, 124, 143; Nineveh: Campbell Thompson and Hutchinson, 1931, 99, pl.XX.55, 57-8; Khorsabad: Place, 1867-70, I, 86; Assur: Andrae, 1913, 8, pl.8, fig.5, 36, pl.CII.9464, 43, pl.LXXIX; Preusser, 1955, 21, pl.17). In Syria a variant form is known from Zinjirli, where they resembled cupped hands; they are glazed, but uninscribed (Andrae, 1943, 60ff., fig.73, pl.31d-e). They may also have been used in Urartu (Peltenburg, 1968, 58, n.6). They are first recorded in Assyria with inscriptions of Assurnasirpal II and appear to be most popular in the ninth century B.C.; none have seventh century inscriptions. The inscriptions refer to palaces, to temples, and possibly to gates. They have also been found loose in private houses at Nimrud, and by the city wall at Assur, where thirteen were also found in a tomb (Haller, 1954, 155: tomb 53).

No Middle Assyrian examples have been reported, nor are any precursors evident outside Assyria before their earliest appearance there. They seem to be a local phenomenon. As has been noted, their structural role is obscure and may have been various. None is made to carry any great weight. The only constant in their role may have been apotropaic (van Buren, 1945, 57ff.). As intimate association with a particular deity is not evident in the available archaeological and textual evidence, the popular description "Hand of Ishtar" is best avoided as a generic term.

C: Terracotta Statuary:

The earliest examples of glazed statuary from Mesopotamia are fragments from the fifteenth century temples B and C in G50 in Stratum II at Nuzi. According to the excavator: "these objects, broken and out of place, give the impression that between the time of

Temple C or B (which are identical in this building) and Temple A, a destruction took place as thorough in the case of the objects, if not the walls, as that which obliterated Temple A" (Starr, 1939, 110), the last in the series. In these levels were "the haunches of a lion figure identical with the red and yellow ones but covered with the familiar green glaze.....and incomplete yellow-glazed Ishtar figure with black on the genitals and breasts (pl.99R)" (Starr, 1939, 416). The latter was the only Ishtar figurine treated in the round found in this excavation.

The range of glazed terracotta from the later Temple A was greater. In the north corner of G29 was "a green-glazed terracotta representation of a woman's breast (pl.102, E) and fragments similar to it which may have been pieces of another" (Starr, 1939, 90, cf.423). These were taken, though smaller than life-size, to be parts of a composite, large-scale deity statue. In the courtyard was a blue-green glazed terracotta statuette of a standing female worshipper (Starr, 1939, 96, cf.419-20). Traces of small glazed animal figurines, of types found unglazed in the private houses, were also recovered from Temple A (Starr, 1939, 97, pl.110B, 108); but they in no way compare with the larger animal sculptures in terracotta for which this shrine is best known. There were a pair of lions (47cm. long) with an even coating of green glaze, standing on an unglazed base (Starr, 1939, 97, 430ff., pl.110A, 111A); and another pair, this time couchant, "covered with a thick yellow glaze which streaks across their red-painted bodies" (Starr, 1939, 97, pl.111B, 430ff.). Only one was found in the Temple A Cella, significantly complete; the others were in fragments in the courtyard outside. Also in fragments within the cella was a green-glazed boar's head (Starr, 1939, 98, pl.112B). It had the appearance of having been a wall knob projection; two tusk-like objects of blue frit may belong to it. There was also evidence for a comparable ram's head (Starr, 1939, pl.110C). More unusual was "an ear of a lion in terracotta, covered with a deep yellow glaze of exceptional thickness. The rim and veins within the ear were shown in relief and executed in green glass" (Starr, 1939, 98). Starr believed that glazing was restricted solely to objects for temple use or to cult vessels.

The standing lions were made in five parts: body and head, with open mouth leading to a hollow interior; front legs; two hind-legs made separately; the base, unglazed save for drips (Starr, 1939, 430ff.). The assembly of these objects was distinctive:

"The only adhesive used in joining these separate parts is the glaze itself. By it alone the front legs - dipped all over in glaze - were held in place both above and below. The two hind-legs were similarly cemented to the base and body as well as to each other. Not until the body - also with an over-all coating of glaze - was set in its final position (the hindquarters are not exactly centered over the legs), was the assembled object placed in the kiln. The vitrification of the glaze firmly cemented the sections together, and the excess which oozed from the joints served to make an unbroken surface between the separate parts which were lost when the object was broken up in the destruction of the temple" (Starr, 1939, 431). The glaze was in very poor condition and the body was a very coarse ware, dull brick-red in colour and friable, with temper of sand and straw. The surface was not smoothed down before glazing and this has affected the lie of the glaze. The glaze provides the adhesive also in the boar's head (Starr, 1939, 435), which is made from two perfectly symmetrical sections.

The pair of couchant lions are technically interesting in other ways. They are decorated with a comparatively rare yellow glaze and a type of red paint only found on these objects at Nuzi: "a foundation of red colour applied there before the application of glaze. The colour is a brilliant, deep red, loosing much of its intensity on exposure to the air. It runs over the body. The gaudy contrast of deep red ground and intentionally streaked yellow glaze must have been striking in pristine state".

No glazed terracotta statuary has yet been found in Mesopotamia that matches the guardians of temple entrances found at Tchoga Zanbil in Khuzistan in the great thirteenth century ziggurat complex. As reconstructed these bulls and griffins are almost a metre and a half high and over a metre long (Ghirshman, 1966, 49-50, pl.XXXII-V; 1968, pl.LXVIII). Such terracotta temple guardians had been known in Mesopotamia since at least the earlier second millennium B.C., but no glazed examples have yet been recorded.

As with other aspects of the glaze and faience industries, there is a hiatus in the evidence after 1200 B.C. for between four and five hundred years. Even then, although glaze came to be widely used on architectural fittings and pottery, it is very rarely reported on Neo-Assyrian baked clay statuary. From seventh century Assur Andrae (1925, 59ff., pl.30) reported a horse's head, the hollow body of a pack animal and the major part of a figure restored as a female worshipper. All were polychrome glazed, the woman in a garment richly patterned in green, yellow and blue/black (Klengel-Brandt, 1978, no.288, cf.289). By the Neo-Assyrian period, if not earlier, in central and southern Iraq small baked clay figures were more commonly glazed, as was contemporary pottery. The most notable examples are a series of tiny horsemen which remained popular through the Achaemenid period, glazed and unglazed (Barrelet, 1968, 128-9). As Babylon has yielded at least one glazed Pazuzu head, and even at Assur there may be glazed horsemen, this practice certainly runs back into the seventh century (Reuther, 1926, 24, fig.23; Klengel-Brandt, 1978, 60).

D: Techniques:

(i) Bricks:

(a) Bodies and glaze colourants:

The body of glazed bricks until the Achaemenid Persian period, does not appear to have differed significantly from that used for unglazed bricks. In the Achaemenid period a body of sintered quartz or faience appears. In describing the fabric of the glazed bricks found in the Apadana at Persepolis Schmidt described them as "a tan, rather porous, frit-like mass, apparently made of lime and sand" (Schmidt, 1953, 71). This is precisely how Koldewey (1931, 122) had earlier described the glazed bricks of the "Persian Building" at Babylon, following analyses by Rathgen. This indicates a conscious attempt to provide a body fabric to which the glaze would adhere better. Certainly the surface of surviving Achaemenid glazed bricks is commonly less "crazed" than that of Neo-Babylonian or Neo-Assyrian ones. This might demonstrate either that the Neo-Elamite faience industry had had a greater impact in Iran on those producing bricks for glazing than did its Mesopotamian counterpart (whose existence at this time is anyway doubtful (see p.157)), or that Egyptian technology had already, in this respect, influenced Achaemenid craftsmen. Dayton (1978, 383) has also published an analysis of a Persepolis brick commenting that "the body was very similar to a modern sand-lime brick; no clay was present as the absence of alumina showed. The calcium oxide proportion was about 9 per cent, and the amount of $Si\ O_2$ about 90%." By contrast his examination of a Neo-Assyrian glazed brick, from Fort Shalmaneser at Nimrud (Dayton, 1978, 375), revealed that it was a vegetable tempered clay brick, which did not appear to have been fired much above $900^{\circ}C$. It was wedge-shaped, not rectangular in section, with a glazed face 110mm. long; the back face tapered to 102mm. It had been fired face up in the kiln, as the glaze had run in streaks down the unglazed side in firing. The glaze showed the poor adhesion commonly encountered on Neo-Assyrian glazed bricks. Whether such bricks, when used on outside walls, were given a special surface treatment has not yet been established.

Some glazed bricks, probably of the Neo-Babylonian period, found by Layard at Borsippa were analysed by de la Beche and Percy (Layard, 1853, 166): "The yellow is an antimonate of lead, from which tin has also been extracted, called Naples Yellow.....The white is an enamel or glaze of oxide of tin.... The blue glaze is copper, contains no cobalt, but some lead;....the red is a sub-oxide of copper". These results have recently been corrected and modified by analyses done by Hedges and Werner (cited by Dayton, 1978, 49; cf. also 378ff.). Hedges analysed three different colour areas, white, yellow and green, on some weathered Neo-Assyrian bricks from the site of the Nabu Temple at Nineveh (Campbell Thompson and Hutchinson, 1929, pl.LVII):

"These were estimated for their content of Cu, Zn, Pb, Sn, Sb. Pb was present above the trace level only in the yellow areas, together with Sb, and X-ray diffraction work on a sample of the material showed that it contained lead antimonate (with very probably an excess of lead in the glaze). Antimony was present both in the white and yellow areas. Presumably the white of the glaze is due to the presence of calcium antimonate. Tin was only present in traces on the green areas, which were the only areas containing a substantial amount of copper (probably originally up to 5% of the oxide). Zinc was present in all areas at the 0.1-0.2% level; it was not associated with the presence of copper in the 9 samples examined" (Hedges and Moorey, 1975, 31).

Fitz (1982) has studied the Neo-Babylonian glazed bricks, now in East Berlin, from the Processional Way and the "Throne-room" at Babylon. Cobalt and cupric ions in trace amounts produced the blue glazes; cupric ions contributed to the turquoise coloured glazes. Lead antimonate was the yellow colourant, whilst calcium antimonate was the opacifier for the white and some of the turquoise glazes.

Fukai (1981, 25ff.) has published analyses of four colours on a glazed brick of the Achaemenid period from Susa. The yellow glaze contained 44% lead, with lead antimonate as colourant. Fukai (1981, 21) has found comparable yellow, high-lead glazes on Iron Age objects from Azerbaijan. For the white no lead was reported. The green glaze, coloured by copper, had lead in it. The black/purplish colour was low in lead with copper oxide, iron oxide and cobalt oxide as colourants. In the analyses of Susa brick glazes published by Dayton (1978, 378-9) blue is attributed to cobalt modified by antimony, the orange/yellow to iron oxide rather than to reduced copper. The dark lines on a brick from Persepolis were said to be of iron oxide rather than manganese; but the character of these separators in polychrome designs requires more research (see p.185 below).

(b) Manufacture: design and layout techniques:

(1) Neo-Assyrian period (c.1000-600 B.C.):

In the late eighth century at Khorsabad the glazed bricks were the standard fire-baked bricks. They were "never employed singly, they were first set up as units, perhaps in the "paint-shop", where the design was painted upon the face of the temporarily erected panel or structure. The painting finished, the bricks were carefully removed for the firing required to effect the glaze. At the time of removal for glazing they were marked in such a way as to expedite their final re-assembly" (Loud, 1936, 93). "Such a process is evidenced by the fact that, while an entire decorative design may be repeated over and over again without apparent variation, one example is never exactly the same, brick by brick, as another" (Loud and Altman, 1938, 14).

Reade (1963), in publishing his reconstruction of a large ninth century panel from Nimrud (see p.173 here), has provided a more elaborate description of the procedures used in manufacture and assembly. Fitters' marks were painted in thin glaze on the upper side of almost all bricks, one to denote the course to which the brick belonged, the other its position in that course. Significantly there were omissions when the shape or decoration of a brick was itself distinctive; distinctions were also made by using different glaze colours. The bricks were slightly baked and, when necessary, shaved down prior to painting. "The design in all its details was then drawn on them in black; these lines have sometimes survived where the glaze has perished. The fitters' marks were added while the bricks were being put into or taken out of position; for the marking was certainly done before the glazing, and it is quite clear that, for glazing, the bricks were picked up individually. This is indicated by several slight inconsistencies of line, where the painter did not conscientiously follow his preliminary outlines, and some of the colour, as where a single chevron, black on one brick, becomes white on the next; moreover the driblets of glaze, which continually obscure the fitters' marks, never coincide on adjoining bricks" (Reade, 1963, 40). They were baked again after decoration, with the painted side upwards; on many bricks the melted glaze has run over and trickled down the sides.

In the case of Assyria it is possible to make direct comparisons with mural paintings. As might be expected the techniques of draughtsmanship were close. The Assyrian, by contrast to the Neo-Babylonian and Achaemenid, glazed brick friezes were murals in every sense save that glaze replaced pigment as the medium and surface treatments were different. Evidence, where available, from Neo-Assyrian sites other than Khorsabad and Nimrud, indicates that procedures were uniform in court workshops.

The range of colours found in Neo-Assyrian glazed friezes is well documented. The colours used for geometric designs were green, yellow, dark brown and white, with less commonly an intense blue and more rarely a strong red. On the figured designs it is again yellow, blue, green, white and brownish-black. Caution is needed in commenting on the colours used, since modern reproductions often give a richer impression than may be justified. In a single composition the range of colours is only from three to five: brown, green, blue, dark and pale yellow (Layard, 1853a, pl.84.12); orange, lilac, white, yellow and olive-green (Botta, 1849-50, pl.155, fig.3). Combinations of four colours are more common: red, white, yellow and black (Botta, 1849-50, pl.155, fig.2); deep and pale yellow, white and brownish/black (Layard, 1853b, pl.55, fig.6); yellow, blue, white and olive-green (Layard, 1853b, pl.53, figs,3 and 4; pl.54:12-14). When as few as three colours are used, the background is generally also the colour of the figures. Yellow, blue and white on a blue ground (Layard, 1853b, pl.53, figs.2 and 5; pl.54, fig.9); the same colours on a yellow ground (ibid., pl.53, fig.1); white and yellow on a blue ground (ibid., pl.54, fig.7); white and yellow on olive-green (ibid., pl.53, fig.1). The strength of outline is marked. It is almost always coloured differently from the object depicted, either in black, white, yellow or brown. The figures, by contrast with the sculptured reliefs, are generally slimmer and less muscular; size varies considerably, from brick-size figures to others standing at least three feet in height.

(2) Neo-Babylonian and Persian periods (c.600-330 B.C.):

Various methods have been proposed for the construction of the Neo-Babylonian glazed reliefs, the earliest being that of Thomas in 1863 (cf. Oppert, 1863, I, 145) and the most comprehensive Koldewey's (1914, 28ff.). A method of mass production is evident from the constant reproduction of the same few motifs. Presumably a master relief was made for each motif and then moulds taken from it for forming the individual bricks into which the master pattern had been divided up. "The edges of the figures do not project more or less squarely as they do in Assyrian alabaster reliefs, but in an obtuse angle. Also there are no even upper surfaces as there are on Assyrian stone carvings. Both peculiarities would considerably facilitate the withdrawal of the tile from the mould" (Koldewey, 1914, 29). The brick was fired before glazing, "the contours were then drawn on it with black lines of a readily fusible vitreous composition, leaving clearly marked fields. These were filled with liquid coloured (glazes), the whole dried and then fused, this time apparently in a gentler fire. As the black lines had the same fusing-point as the coloured portions they often mixed with the colours themselves ....With the Persian glazes.....these black lines have a higher melting point and therefore remain distinct and project above the coloured glazes after firing" (Koldewey, 1914, 29-30). Fukai (1981, 25) has explained these prominent outlines as indicative of a squeeze technique, like the use of a bag for icing a cake. Built-up lines were made by squeezing out the glaze through a nozzle from the bag.

As this technique is particularly evident on an unusually elaborate glazed jar from the excavations at Babylon (Koldewey, 1914, fig.152 (frontispiece); Mayer, 1965, pl.157: both colour), dating to the seventh or sixth century B.C., Fukai argued that the technique passed directly from Neo-Babylonian to Achaemenid craftsmen. As the brick-makers at Susa (see p.177) are known to have been Babylonians this makes good sense; but the matter still needs detailed investigation.

Some of the surviving glazed bricks from Persepolis and Susa (cf. colour plates in Dayton, 1978, pl.27-8) appear to the eye to have cloisons carefully made from short, thin strips of some metal, possibly sheet copper or bronze. This method of laying out the design is in marked contrast to the black lines, whether flat or projecting, just commented on and traditional before the Achaemenid period. Such "metal" outlines seem also to be used on at least some parts of the glazed brick decoration of the Achaemenid palace at Babylon (cf. Koldewey, 1931, pl.39f, t). This technique has a close affinity with that used for the polychrome inlays so popular in the "Court Style" of jewellery produced under the Achaemenids.

In the Neo-Babylonian period green and brown do not seem to have been fashionable colours on glazed friezes (cf. Koldewey, 1918, 27; 1931, 85), though they reappear in Achaemenid times at such major sites as Babylon, Persepolis and Susa (Haerinck, 1973, 120). Blue and yellow, black and white, retained their popularity throughout; red (deceptively changed to green, see p.187 here) appeared at Babylon in Neo-Babylonian times (Koldewey, 1918, 27; 85), but not apparently in the Achaemenid period.

In order to facilitate the assembly of a design, as in Assyria (see p.184 here), the bricks were marked on the upper edge in rough glaze with a series of simple signs and numerals. Sign was intended to match with sign in the laying. This system was best seen at Babylon in he "Southern Citadel", where more coherent sections of glazed brickwork were recovered. Here each colour was outlined in black and the position marks used again as on the "Ishtar Gate". Koldewey described it as follows: "Here (i.e. on the capitals) the markings consist of numerals combined with dots. They are marked on the upper edge of the bricks with a poor, somewhat blackened, glaze. The signs that distinguish the courses are in the centre, those for the lateral arrangement are close to the vertical joints. Each of the latter signs is a counterpart of the sign near the vertical joint of the brick adjoining it. On the central signs that mark the courses the top course of the upper row of volutes has one stroke, the second has two, and so on up to seven. The seven courses of the lower row of volutes are numbered in the same way, but the groups of strokes are preceded by a dot to distinguish them from those of the upper series. For the sequence of the bricks one of the intermediate ornaments forms a single unit with the capital adjoining it to the right. All the bricks that belong to the same unit bear the same number of strokes. The counting runs from left to right. The numerals are crossed by a transverse stroke, which, in order to mark the direction of the signs, has a dot attached to it. This direction line is parallel to the vertical joint on the central ornament, and parallel with the front of the brick on the volutes. It is quite probable that the separate groups were first provisionally built together, at any rate for the purpose of drawing the design, which is still visible in red colour under the (glaze), so as to secure that boldness and freedom of outline which delights us with its beauty at the present time. But when once the process of (glazing) began - the transportation of the bricks, the drying, the burning, and all the unavoidable processes that had to be carried through before the bricks could be placed in the wall - it would be impossible to keep them together. The marks would then afford the only means of placing them correctly on the walls, and rendering it easy to deliver them in groups to the respective masons.

In order to close the joints completely the bricks are slightly wedge-shaped. The joints between the courses are laid in mud over asphalt, which, as we observe in other careful building, does not extend to the front of the building but stops at a distance of half a brick, thus avoiding any blotching of the face of the wall.

In addition to the black outline and the dark blue ground, the colours employed are white, light blue, yellow and red. The red now has everywhere the appearance of green, but where this colour is thickened, as for instance where drops have trickled down, a core of brilliant red is found coated with green, which must be the result of a superficial change of colour that has occurred during the course of ages. We have also some large pieces of glaze from ancient breakages in which we can observe this same fact. The green coating extends to a depth of 2 to 3 millimetres, which in the ordinary glaze on the brick would entirely supersede the original red colouring. This is an important point, because the manufacture of opaque red glaze has been attended with considerable difficulty even in recent times, while transparent red glaze is made with ease at the present time. Thus in forming a judgement on the sense of colour of the ancient Babylonian it must not be forgotten that this fine red was included in his scheme. We can all imagine a red-haired, but not a green-haired lion" (Koldewey, 1914, 30). This chemical change is typical of a red glaze (or glass) coloured by reduced copper. This effect would not have been observed had the colourant been iron.

(ii) Pottery:

None of the glazes on pre-Islamic Mesopotamian pottery or bricks analysed, in recent times, has been found to be a lead or a tin glaze in the strict sense of the word; they are all alkaline glazes. This was first established by Salvetat in the nineteenth century; but obscured by later inaccuracies and false interpretations (cf. Birch, 1858, 148). In a series of analyses done in Oxford Hedges (with Moorey, 1975; Hedges, 1976; 1982) established a series of points about ancient Mesopotamian pottery glazes. Compared with modern alkaline glaze (Hedges, 1982, table 2) there is a little more Si and a little less Al or Al + Fe. It is beneficial if the alkali, especially the soda content, is kept as low as possible, since it is the main contributor to the expansion coefficient, and thus the lower the percentage, the better the chances of adhesion. Vandiver (1982a) has suggested that the adhesion of an alkaline glaze to a clay body was promoted by the presence of soluble salts deposited on the surface as efflorescent products during drying.

In the vessels so far examined there is no major difference between the body fabrics of glazed and unglazed vessels, though one might have expected the former to contain more quartz. Thin section microscopy discerned no differences (Hedges, 1982, 99). Nor is there any significant change in glaze composition through time. In the Sasanian period at Kish there was a slight increase in Al and a corresponding decrease in Si, perhaps allowing for a lower maturing temperature and a greater control over the glaze.

Although glazes on pottery from sites as far apart as Failaka and Nineveh, dating from the Kassite to the Sasanian period, revealed comparatively little variation in the glaze composition (Pollard and Højlund, 1983), there are distinctions that may be indicative of local traditions. At Nippur the glazes were higher in potassium than at Kish and Nineveh. At the same site tin was a common occurrence, but not lead or zinc; zinc was ubiquitous at Nineveh, and at Kish tin was only found with any frequency in Parthian and later samples, and then lead is commonly found as well. It is also possible that barium in the glaze or body might well be indicative of an origin in the Nippur region. Glazes on clay from Kassite levels at Failaka (Kuwait) have a remarkable excess of magnesium over lime in the glazing. This is probably to be explained by the fact that the alkali consisted of salt derived from sea water. The magnesium calcium ratio is consistent with the use of salt from such a source. This may also have some connection with the high-magnesia glasses characteristic of Mesopotamia in the second millennium B.C. (Pollard and Højlund, 1983), since craftsmen may have been using salt from local salt pans.

The whole question of colourants in ceramic glazes is discussed here in the section on faience (p.162).

V; Egyptian Blue:

(a) The material:

The blue frit referred to as Egyptian Blue was produced by firing an intimate mixture of quartz, calcite and a copper compound together with a small amount of alkali. The resulting frit consists of unreacted quartz and the mineral Egyptian Blue together with varying amounts of glass phase. The Egyptian Blue mineral is a silicate of copper and calcium ($CuO.CaO.4SiO_2$), which is identical to the natural mineral cuprorivarite. The positive identification of Egyptian Blue frit is not possible by visual examination but requires X-ray diffraction analysis. As a result, there is regular confusion in the literature between Egyptian Blue and other blue frits, sometimes even, on occasion, between Egyptian Blue and blue faience or blue glass.

As well as providing a pigment, when powdered, Egyptian Blue may be mixed with a binding agent, moulded and refined, to make artefacts. It has the distinction of being one of the very few materials studied here that survived into the Graeco-Roman period, so that it is not only referred to in classical authors, but appeared in the pioneer eighteenth century excavations at Pompeii in Italy, where it rapidly attracted the attention of chemists. The secret of its manufacture appears to have been lost between about the fourth and the nineteenth centuries A.D. In the fourth century B.C. Theophrastus (On Stones, 55, ed. Caley and Richards, 1956) commented as follows on kyanos (cf. Foster, 1979, 10ff.; Hittite: kuwanna; Mycenaean Greek: ku-wa-no):

"Just as there is a natural and an artificial red ochre, so there is a native kyanos (cf. sections 39 and 51: lapis lazuli and azurite) and a manufactured kind such as the one from Egypt. There are three kinds of kyanos, the Egyptian, the Scythian and the Cyprian. The Egyptian is the best for making pure pigments, the Scythian for those that are more dilute. The Egyptian variety is manufactured, and those who write the history of the kings of Egypt state which king it was who first made fused kyanos in imitation of the natural kind; and they add that kyanos was sent as tribute from Phoenicia and as gifts from other quarters, and some of it was natural and some had been produced by fire. Those who grind colouring materials say that kyanos itself makes four colours; the first is formed of the finest particles and is very pale, and the second consists of the largest ones and is very dark. These are prepared artificially...."

In the first century B.C. Vitruvius (VII:I,1) recorded that Egyptian Blue (caeruleum), the invention of which he credits erroneously to Alexandria, was made by fusing together sand (which in the case of Egypt is a natural mixture of quartz and calcium carbonate), copper filings and natron. Pliny (XXXIII:57-8), about a century later, called caeruleum a kind of sand.

After the pioneering analyses, among them those of Sir Humphrey Davy (1778-1829), the most significant of the earlier studies is that by Foqué in 1884. He identified the compound as $CuO.CaO.4SiO_2$ and defined the optical characteristics of the crystals. Various scientists then attempted to synthesize this ancient pigment, most comprehensively Laurie and his colleagues in 1914. They systematically investigated the effect of the firing temperature and the concentration of alkalis. They reproduced Egyptian Blue by heating together fine sand, copper carbonate, calcium carbonate and sodium carbonate (fusion mixture); the blue crystals were formed at about 830° - 850^oC. Amongst various more recent experimental reproductions and analyses, three are of particular importance. Chase (1971) investigated the firing temperature range over which Egyptian Blue could be produced and the effect of different firing atmospheres. He included a sample from Nuzi. Bayer and Wiedemann (1976) used differential thermal analysis to investigate the formation and stability of the pigment. Tite and his colleagues (1981; forthcoming) aimed to obtain information on the ancient methods for producing the wide range of Egyptian Blue fabrics which vary from soft and friable to hard and semi-vitrified as well as from light to dark blue. Their samples included material from Egypt (late Old Kingdom onwards), and sites of the Late Bronze and Iron Ages in Syro-Palestine and Mesopotamia. Their results showed that Egyptian Blue can be satisfactorily synthesised by firing a mixture of quartz, calcite, malachite (copper carbonate) and sodium carbonate (alkali) to $900-1000^oC$ for a few hours. It was also demonstrated that small objects of Egyptian Blue were most probably produced using a two-stage firing cycle with grinding and moulding to the required shape between the first and second firings. The colour depends on the size of the Egyptian Blue crystals and thus the coarse-textured material produced by the first firing tends to be a darker blue than the fine-textured material produced after grinding and firing for a second time. The hardness, which depends on the extent of the interconnecting glass phase present in the material, varies according to the firing temperature and alkali content. Thus, the Neo-Assyrian objects from Nimrud and Nineveh which have a low alkali content (0.5 per cent) tend to be softer and more friable than earlier ones from Egypt, the majority of which have a higher alkali content (1-5 per cent). Kaczmarczyk's (personal information) analyses of Late Bronze Age examples from Ras Shamra indicate the use of both copper and bronze "scrap" as colourants and no marked chemical distinction in composition between the Syrian and Egyptian material at this time.

(b) Historical survey of use:

In view of its name something may appropriately be said first about the appearance of Egyptian Blue in Egypt, although there is at present no certainty that it originated there rather than somewhere in Western Asia. Lucas (1962, 340ff.) summarized the matter, but rather more detail is needed to establish the possible date of its earliest appearance in Egypt. Samples attributed to the Fourth Dynasty have yet to be identified by analysis. The more important occurrences of the Fifth Dynasty, from the ruined Sun Temple at Abu Gurob and in the decoration of Per-Neb's tomb at Giza have been analyzed (Ransom Williams, 1932, 29ff.). Stevenson Smith (1946, 256) extended the history of Egyptian Blue back into the Fourth Dynasty with three major cases of its use, identified by eye, attributed respectively to the reigns of Cheops, Chephren and Shepseskaf. On the absolute chronologies at present used for the Old Kingdom in Egypt, this would indicate well authenticated use by about 2400 B.C., with more or less certain extension back to about 2550 B.C., that is the range of the Early Dynastic III period in Mesopotamia. The possibility of its even earlier occurrence in Egypt has been raised by at least one "blue paste" vessel attributed to the Protodynastic royal tombs at Abydos of the earlier third millennium B.C. (Petrie, 1920, pl.XXIII:14); but this is a vessel of faience with a blue core.

Blue frit has been identified in a cavity in a Late Prehistoric limestone inlay fragment of typical Mesopotamian type, now in a private collection (Kozloff, 1981, no.2); but the analysis was only a preliminary one. As with so much else in the history of materials in Mesopotamia the numerous graves of the "Royal Cemetery" at Ur offer the first real check on its appearance, but here the looseness of Woolley's terminology creates its own problems. "Glass paste" beads, as distinct from frit or faience, are first listed for two graves of the Early Akkadian period (PGs 1517, 1601: U.13531, 13598; Woolley, 1934, 587-8). There is then a long interval until the third quarter of the second millennium B.C. before it can be traced again (at Nuzi), although there is no reason to believe that it ceased to be used, if only sporadically, for beads, amulets, etc. Stratum II at Nuzi yielded numerous blue frit beads (Vandiver, 1982a; see p.150 here) and also half a moulded macehead, found in a private house, which Gettens identified as Egyptian Blue (in Starr, 1939, 460). It is reported as inlays in a gold bracelet of the Kassite period from Dur Kurigalzu (Maxwell-Hyslop, 1971, 164, pl.127). Both scaraboid seals and small "ingots" of Egyptian Blue were reported from Late Bronze Age levels at Ras Shamra (Louvre display, 1983) and the material was widely used for jewellery in the Late Bronze Age.

After the technological Dark Age, from about 1200-900 B.C., the best information on artefacts of Egyptian Blue comes from the destruction debris of level IV at Hasanlu in north-west Iran c.800 B.C. As this site is known to have had diverse commercial contacts with Syria and Mesopotamia it is always difficult to assess which of the objects found there were produced in or near Hasanlu and which were imported from countries to the west, across the Zagros. These finds include complete and fragmentary vessels, some moulded with low-relief designs, some modelled three-dimensionally in zoomorphic shapes: a beaker (Chase, 1971, fig.5:1; Dayton, 1978, fig.334A); a beaker fragment (Porada, 1965, pl.33); and a "lion vessel" with gold overlays (van Loon, 1962). A solid lion-head fitting (Dayton, 1978, pl.22:2) and a so-called "sistrum handle" (Chase, 1971, fig.5:2) indicate the existence of a wide repertory of objects. A sheet-bronze ram's head rhyton has blue frit inlays in its eyes, eyebrows and nostrils (Dyson, 1960, 128, upper left) and there are comparable blue inlays in a gold knife-handle (Porada, 1965, pl.31). Art historical criteria indicate both local production for the vessel fragment decorated with rampant goats and for the knife-handle, whilst the "lion vessel" might well be an import.

It is usually assumed that the main stimulus for the production of objects in Egyptian Blue at this time came from somewhere in Syro-Phoenicia, though local evidence is extremely rare and virtually confined to coloured inlays in carved ivories, predominantly those in the so-called "Phoenician Style" (Barnett, 1975, 240), only occasionally in "Syrian" ivories. Both are best known from sites outside their homelands (see p.207 here). Inlaid "Syrian ivories" were found in Hasanlu IV (see p.208). From seventh century levels at Nimrud has come evidence which might be taken to indicate a trade in blocks or ingots of Egyptian Blue. In Room ZT 15 of the north wing in the North West Palace "one large amphora, ND 1975, of a type familiar from Palestine, contained inside it lumps of Egyptian Blue, which was the substance principally used for the incrustation of ivories" (Mallowan, 1966, 180). In room SW 6 of Fort Shalmaneser "there were large lumps of bright Egyptian Blue, ND 6448, which was certainly stored for the purpose of incrusting ivories, for the material used in the 'cloisons' is identical... The discovery of these raw materials implies that there were repair shops in the arsenal where, as we have already seen, there was presumed evidence of shell carvers in SW 3" (Mallowan, 1966, 408). Here Mallowan is assuming that craftsmen, working in the "Phoenician" style, perhaps themselves expatriates, were exercising their skills in the palace workshops using raw materials brought from elsewhere. Both Dayton (1978, 32) and Tite and others (1981) have published analyses of the Nimrud ingots. Dayton (1978, 32) also published from the same site two other raw materials which he described respectively as "a stick of very bright 'hyacinth blue' chalky material", which gave a strong tin reading, perhaps indicative of bronze scrap as a colourant, and a trace of cobalt; another, similar stick, was paler blue.

Although in referring to glass fragments from the throne-room of the "Burnt Palace" at Nimrud Mallowan noted that they "are associated with several specimens of Egyptian Blue" (Mallowan, 1966, 209), few have yet been published. He illustrated an Egyptian-style headdress, presumably from a composite statue, of faience and Egyptian Blue (Ibid., fig.144: N.D.1075(i)). From the Ninurta Temple at Nimrud

Layard (1853, 357-8) had reported:

"Figures of winged deities, etc., of clay, coloured in the mass with a blue derived from copper; eyes, beards, hair and ornaments..." Surviving examples are now in the British Museum, including one fragment of a winged goddess (Layard, 1853, 357, left = BM 118785); a beard fragment was also illustrated by Layard (1853b, pl.55:1). In carved ivories the inlays are sometimes of glass over a backing of Egyptian Blue (Barnett, 1975, 240).

That vessels, of forms also current in metal, were being made in Egyptian Blue is evident at this time from a phiale found at Tell Halaf (Hrouda, 1962, 66, pl.49: 1-3; 50:1) and from two vessels found in Tumulus P at Gordion in Turkey, where both were regarded by the excavators as luxury imports (Young, 1981, 30ff., pl.14). One was a juglet, the other a lotus-handled dish. Comparable vessels, in stone and in Egyptian Blue, are reported from Iron Age Cyprus. Despite "Egyptianizing" traits, there is little reason to postulate an Egyptian source, for there, if anything, green frit was more popular than blue in the Late Period (Kaczmarczyk, personal information).

It has long been recognized that Egyptian Blue was a material particularly characteristic of collections of scarabs and small objects in "Egyptianizing" style found throughout the eastern Mediterranean during the eighth to sixth centuries B.C. (James in Payne, 1962, 467-8), though, as has been noted, it was not a material then much used in Egypt. When found in Egypt it is usually on Delta sites, such as Naucratis and Tanis, renowned for their communities of foreigners, artisans, traders and mercenaries. Throughout Western Asia, including such sites as Assur, Nimrud and Nineveh in Mesopotamia, it is also current for seals, amulets and beads.

The German excavators at Babylon revealed a variety of Egyptian Blue artefacts. In the "Principal Citadel" of the Kasr Koldewey reported finding "remains...of large reliefs consisting of a beautiful blue paste, similar to lapis lazuli. The figures were made up of separate pieces, each of which only contained a small part, such as a lock of hair. On the back of these separate pieces there was a prismatic addition, by which they were affixed to some background of which we know nothing" (Koldewey, 1914, 158-9). A miscellaneous collection of precious and semi-precious stones, largely objects or fragments of them, were found in what had originally been two baskets in a mud-brick house under a building of the Parthian period ("Amran Perlen-Depotfund"). Koldewey (1914, 221) believed them to be the raw materials of a bead manufacturer; inscriptions indicate a date for at least some of the objects in the Neo-Assyrian and Neo-Babylonian periods. Among them were objects of imitation stone, including large, blank cylinder seals of "blue paste" (Koldewey, 1914, 48) and various frit seals. They belong to a distinct class of cylinder seal, dated between the ninth and seventh centuries B.C., assembled by Moortgat from examples excavated at Assur, Tell Halaf and Babylon (Moortgat, 1940, nos.689-731); in some cases the material "blue frit" is specifically identified. Egyptian Blue was also used at Babylon for mould-made vessels. Reuther reported a fragmentary relief vessel showing a nude female clasping her breasts "in blue paste (imitation lapis)" (Reuther, 1926, 140-1, fig.90). He also described a small vessel as being of "imitation lapis" (Ibid., 223, pl.75A).

The range of Egyptian Blue production in the Achaemenid period is best represented in the excavations at Persepolis in Iran. Uninscribed and inscribed wall pegs were found in "Xerxes' Harem" (Schmidt, 1957, 50, fig.4, pl.42:27). Inlay beards made of this material continue directly an earlier tradition (see p.191) (Schmidt, 1957, 72, pl.41:14-15). Some other Egyptian Blue objects were distinctively Egyptian (Ibid., 68, pl.31:3). In the chart of the "Royal Tableware", 14 (of 317) footless plates, 1 (of 7) footed bowls, and 6 (of 182) sherds, are described as "composition" (Ibid., 91, table VIII). A buff coloured pottery sherd, from a store-room (21) of the Apadana, had blue pigment incrusted on part of its interior surface to a maximum thickness of 0.05cm., perhaps indicative of local manufacture (Ibid., 133). By far the finest object in Egyptian Blue, found more recently by the Iranian Antiquities Service at Persepolis, is the head of a young prince or princess (Porada, 1965, pl.45; Spycket, 1980), wearing a mural crown. It is 6.5cm. high and was probably once the head of a statuette.

In his report on the Egyptian Blue at Persepolis (in Schmidt, 1957, 133ff.) Matson proposed a process of manufacture, which subsequent research has indicated is almost certainly incorrect.

(c) Textual references:

There is no direct textual information in Akkadian on Egyptian Blue. Oppenheim (1970, 13, 7b) believed that it might be the material referred to in certain cases where artificial lapis lazuli was implied; but there is no certainty about the contexts he cites.

NOTE

The Problem of Origins

The literature on the earliest occurrence of glazed materials in Mesopotamia is particularly hazardous to use. This is notably so at Arpachiyah, where a variety of glazed materials were encountered. Here both recording of context and description of materials is often obscure or ambiguous. As Mallowan distinguishes 'frit' from 'glazed frit' he may mean the specific distinction between relatively unweathered and badly weathered faience. He also mentions bitumen ornaments inlaid with "vitreous paste", but it is not clear whether they were of the Halaf or Ubaid periods. If there is no true frit at Arpachiyah, then the 'bluish frit' (see p. 143 here) from Eridu might be the earliest reference to blue frit or Egyptian Blue; but the matter remains entirely open.

Chapter 3: GLASS

A: The Material:

Glass has a homogenous, non-crystalline structure, quite different from faience or frit. In a glass, at some point in the manufacture, the ingredients become completely fused into a liquid melt, which is cooled and then sets as glass. In its softened form, glass is remarkably tractable. It can be rolled into sheets, drawn into tubes, rods or threads; it can be cut with shears or teased into shape. It can be pressed into moulds or, as was discovered, probably somewhere in Syro-Palestine, in the first half of the first century B.C., blown like a bubble from the end of a pipe.

B: Mesopotamian Glass and its study:

Finds of early glass in Mesopotamia are relatively rare and nearly always ill-preserved (cf. von Saldern, 1970, 204), so they have attracted much less attention than has been accorded to the well-preserved, brilliantly coloured polychrome glass objects from New Kingdom tombs in Egypt. Although one of the first books ever written on ancient Babylonia refers to glass and glazed bricks (Rich, 1816, 29), and finds were sporadically made throughout the nineteenth century (cf. Peters, 1898, II, 134ff., Koldewey, 1914, 255ff.), pioneer students of the history of glass in this century, like Kisa and von Bissing, dismissed the claims of Mesopotamia. Meissner (1920, 235) thought otherwise, arguing that there was a local industry as old as that in Egypt. Starr (1939, 457-9) opened a new epoch for glass studies in Mesopotamia, when in his final report on the excavations at Nuzi he devoted special attention to his remarkable glass finds. Then, in the following year, Fossing's Glass Vessels before Glass Blowing (Copenhagen, 1940) brought all the Mesopotamian evidence sharply to the fore. He isolated a group of early "Mesopotamian" glass vessels, recognized a local industry as early as 1300 B.C., and defined its relationship to Egypt, though making clear an independent Mesopotamian trend. In the years following the Second World War Harden (1968), Barag (1962; 1970) and von Saldern (1966; 1970) have set the study of Mesopotamian glass on a firm basis of assembled examples and comparative typology. Oppenheim (1970) gave the subject a fresh dimension by detailed investigation of textual sources, whilst Brill (1970; 1978) and others have pursued complementary studies of composition and technology.

C: Historical Survey:

(i) The prehistory of glass production in the Near East (to about 1500 B.C.):

Since Beck's fundamental review of the evidence in 1934, no

substantial progress has been made in study of the earliest glass objects in the Near East. Beck was able to list only 17 objects, not all of them of certain date, and not one of them a vessel, made before about 1500 B.C. in Egypt or the Near East. Although the list might be slightly lengthened fifty years later, there is still no certain evidence for vessel production before the middle of the second millennium B.C. Most of the specimens were, and remain, beads, found singly or in relatively small groups. None of Beck's examples, nor any of the subsequent finds, may confidently be dated before the middle of the third millennium B.C. Beck's main conclusion, that it was probably in Western Asia rather than Egypt, that regular glass production originated, has also stood the test of time (Barag, 1970, 132).

Beck's caution about the correct dating of two glass beads attributed to level 4 of the Great Pit MM at Nineveh has been sustained by subsequent research (Beck in Campbell Thompson and Mallowan, 1933, 180-1, pl.LXXIX, 25-6). One bead was pale blue, spherical in shape; the other "a hexagonal cylindrical bead, which has been cut from a hexagonal rod", had been a brilliant green and from its specific gravity Beck considered it to be a lead glass. As Beck cogently remarked "it seems to me most improbable that these beads can be as early as stated". If these two beads are put to one side, the earliest glass bead from Western Asia is that from Tell Judeideh in the Amuq Plain of Syria, attributed to phase "G", dated to the earlier third millennium B.C. (Braidwood, 1960, 341, fig.258). It is a well-preserved, short, oblate spherical bead, pale yellow-green in colour. Matson (ibid., 341) reported that:

"The bead was formed around a straight rod (d.3.6mm) which gave the core a uniformly cylindrical shape, but the outer surface, which was probably shaped by rolling the soft glass on a flat surface after it had been gathered on the rod, is misshapen and pock-marked. The presence of so many seeds and cords in the glass indicates that it was melted at a fairly low temperature at which the viscosity was still quite high, so that the glass had the stiffness of thick molasses. It could not be stirred or worked easily in such a state, so that it was not cleared of seeds and cords. Had it been possible to use a higher melting temperature, a better quality of glass could have been made".

In grave 5A of pit L4 at Nuzi (attributed to Stratum IV: Starr, 1939, 32, 515) was "a plain, straight copper pin of even diameter, 14cm. long, with a large glass bead, 15mm. in diameter, serving as a head... It is, in fact, in much sounder condition than the majority of the Nuzi glass beads (i.e. of Stratum II)" (Starr, 1939, 380). In the same grave was a cylinder seal, still on its copper pin, assigned by Boehmer to the later Akkadian Period (c.2350-2150 B.C.) (Starr, 1939, pl.55 F, P; Boehmer, 1965, no.1566); broadly the chronological horizon to which this grave was attributed by the excavators, who found a concentration of "Old Akkadian" texts in Stratum IV (Starr, 1939, 516). Since such "bead-heads" on pins had earlier been made of lapis-lazuli, the association of the earliest glass, as the earliest faience, with the cheaper reproduction of semi-precious stones is evident. Unfortunately Starr did not illustrate this find.

Of greater technical significance is a pale blue-green "chipped glass rod" from the "Main level of the Northern Palace" at Tell Asmar in the fill of room E 16:16 (Delougaz, 1967, 189-190; 246; As.31:671; cf. Gibson, 1982). Objects littered about the floor of this room included a small upright-handled jar (B.526:471); an important hoard of jewellery (including faience objects) was found buried below the floor of the room (ibid., 190, 245: As.31: 868-926). Frankfort (et al., 1934, 56) provided the following description of the context for Beck's technical report:

"It was found above 16 E 16 in the layer of débris corresponding to the period of the desertion of the Akkadian palace, which, as we have seen, contained almost exclusively objects of Sargonid times. It was found definitely beneath walls of a ruined building which had contained tablets of the 38th year of Shugli. The foundation of the walls of the building stop half a meter above the level at which it lay". Frankfort himself dated it to the Guti period or "more probably to that of the dynasty of Akkad". Gibson's (1982, 534) revisions in the dating of this area would not invalidate this conclusion; though the tablet (sic) would now be dated Shulgi 40 or 41. Beck published two technical comments on this find (Frankfort et al., 1934, 56ff., fig.51; Beck, 1934, no.7, figs.2 and 3); he reported that "it appears to have been modelled or moulded to its present shape, and had not been cut out of a solid block. The glass is very pure; it has a few small bubbles but is surprisingly free from striae or inclusions of quartz or dirt". The object was not analysed. It looks as if it might be a type of glass tube regularly used in manufacture.

Perhaps the most famous single piece of raw glass from Mesopotamia is the small lump (BM 115474) excavated at Abu Shahrein (Eridu) by Hall in 1919: "In the rubbish but not in the floor of the house, immediately beneath the Bur-Sin (Amar-Sin) pavement, was found a lump of opaque blue vitreous paste which I recognised as true glass" (Hall, 1930, 213). This would date it to the dynasty of Akkad or, more probably, the early Ur III period. Beck (1934, no.8, figs.4 and 5) concluded:

"This specimen may have been meant to be carved or to be remelted and moulded to the desired shape; or even, it may have been intended to grind it up and use it as a glaze. In any case it was probably a manufacturer's piece of material and the probability is that it was made in the immediate neighbourhood of where it was found".

The glass is pure ultramarine in colour; it has been analysed by Garner (1956, 147ff.), who identified the colourant as cobalt (0.15%). "Although the amount of copper is 0.40%, three times that of cobalt, it has a much weaker colouring property and in fact would hardly influence the colour at all". The cobalt was identified as one of the arsenical ones, either cobaltite or erythrite, and attributed to an Iranian source, perhaps the mines at Khemsar, near Kashan, an erythrite of beautiful amethyst-pink crystals, whereas the cobaltite of Azerbaijan is a metallic grey in colour and much less likely to attract attention. The cobalt is free of manganese, zinc and nickel, strengthening the case for an Iranian source.

Oppenheim (1970, 19-20) has called attention to the very rare appearance in texts of the Ur III period of a material known as *anzahhu* (Sumerian AN.ZAH), once as a bowl, once as a raw material. He believed it to be a "primary glass"; that is a key intermediate material in the production of a finished glass. The C.A.D. renders it "an imperfectly fused, crude, frit-like glass"; but Foster (1979, 17; cf. Loding, 1981, 12) has taken it to be a mineral, perhaps a type of quartzite. Whatever the case, and the available evidence is indecisive, it does not provide certain evidence for glass vessels at this time. At most, if an artificial and not a natural substance, it complements the archaeologically attested production of faience vessels (see p.147 here) and imitation of coloured stones in glass, as on the Nuzi pinhead.

Oppenheim (1970, 85) also argued that the Tell Asmar and Eridu fragments "indicate a technical turning point in the history of Mesopotamian glass making. We may read here the level at which the anonymous craftsman who produced the "primary glasses" (see p.36) was supplanted by the chemist. The new alkali-silicate glasses were combined with the "primary glasses" in order to produce imitation of coloured stones". This goes far beyond the present evidence of archaeology. What the known material culture indicates is a continuous tradition in the manufacture of imitation stones, whose composition varied considerably; but before the sixteenth century B.C. little, or none of it, was of glass in the accepted modern sense (ingredients fused into a liquid melt). It was various types of sintered quartz, glazed and unglazed, i.e. frit and faience, particularly when used for anything other than beads.

In his analyses of the "Old Babylonian" word-lists of the earlier second millennium B.C. Oppenheim isolated certain substances as what he called "primary glasses" (see above): *anzahhu*, *kutpû*, *huluhhu* and *huhû* (Oppenheim, 1970, 19ff.). Virtually nothing is known of these substances; but, in what Oppenheim saw as a significant, undated, sub-period, *anzahhu* played a major role in the production of imitation coloured stones. "It may be assumed that their composition varied greatly and that the mentioned designations refer not only to uses and colours, but to specific still unknown properties as well. These materials were used to make beads, inlays, cylinder seals and probably bowls..." (Oppenheim, 1970, 84). Oppenheim (1970, 84-5) extended his philological analyses into the area of cultural history by arguing that all the early designations of his "primary glasses" were not Akkadian words and that even the Sumerian AN.ZAH, from which *anzahhu* is derived, could well be a loan word from some unknown source. Although he admitted that the ultimate origin of these words had not been established, he hazarded the guess that they came "into Mesopotamia from the northwest together with the substances they denote". Such a view is hard to substantiate from the available archaeological evidence, even if philologically sound; but it is not impossible.

In the long series of third millennium graves at Ur, commonly known as the "Royal Cemetery", Woolley catalogued two instances of glass beads, one from a Late Akkadian grave (PG 1213): "some glass ringbeads" (U.12003); the other from a Neo-Sumerian grave (PG 973): "beads...glass" (U.11427). From only one of the 198 graves attributed to the Isin-Larsa period (LG 121), of the earlier second millennium B.C., does he report "beads 18 chiefly carnelian and some glass" (Woolley, 1976, 205). Both the identification and the archaeological context of the "glass" beads from Enunmah are equivocal; they were near a drain in room 9 (Woolley, 1974, 50). In the main text they (U.152-3) appear as "minute crystal and pebble beads and a few larger paste and stone beads", (ibid., 50). In the catalogue U.153 is put in room 13 and listed as "Beads, minute, of transparent white glass and opaque glass paste" (ibid., 84). This would appear to be an example of the common visual confusion between rock crystal and glass; the supposed early glass beads from Tell Farah (Shuruppak) were also finally said to be rock crystal (Beck, 1934, 10). Although Beck was almost certainly right to reject as "before 1600 B.C." some red beads from Ur given him by Woolley for examination, the others from the site he examined may not be independently checked. Beck (1934, 18) reported that "they were found with a series of faience beads definitely stated by Woolley to date before 1600 B.C. The dust consists largely of broken down glass of at least two sorts, one practically colourless, the other yellow glass so common in Egypt during the XVIII dynasty".

At Assur various votive deposits were placed in the corners of the Ziggurat. Two sets were excavated. The lower, on bedrock, included some glass beads. On the comparative dating of associated brickwork the excavator dated this bedrock deposit to Shamshi-Adad I (c.1813-1781 B.C.), though it is not certain that it was not contaminated by Shalmaneser III´s workmen a millennium later (Beck, 1934, 17; Andrae, 1938, 90, pls.45-6; Haller and Andrae, 1955, 2ff.). The upper deposit also contained glass beads. It included silver and gold disks inscribed for Shalmaneser III (c.858-824 B.C.) and a stone tablet of Shalmaneser I (c.1274-45 B.C.). However, an inscription of Adad-nirari I (c.1307-1275 B.C.) from Assur speaks of "the great new Ziggurat built by my father Arik-den-ili" (c.1319-1308 B.C.) (Edzard, 1964), so these beads might be fourteenth century in date. Glass beads, with others of agate and faience, were found in an old Assyrian grave at Assur (no.486) (Haller, 1954, 39). At Tell Asmar from houses below the "Southern Building", which are attributed to the Early Larsa Period, in the first century or so of the second millennium B.C., the excavators reported a "glass tubular fragment" and a "glass bowl fragment" (Delougaz, 1967, 263, no.31:547; 264, no.31:58). No details have yet been published about either of these finds; if correctly reported, the latter would be of great significance at this date.

From such meagre and scattered evidence little of significance may be concluded. The sample is so small (particularly when potentially misdated examples are removed) that the remaining beads might be interpreted as no more than compositions intended as faience which turned completely vitreous when overfired. The "glass tubes" and the Eridu lump would then have to be seen as by-products of glaze production for faience manufacture and not as evidence for an independent glass industry. However, if a more positive line is taken about the earliest glass in Mesopotamia, closer to Oppenheim´s position (Oppenheim, 1970, 76, 83, 85), it is necessary to lay emphasis on one point. As the scribes who compiled the lexical lists upon which Oppenheim drew for his analysis were cataloguing the names of stones they were naturally interested in including those coloured stones produced artificially. Although the meagre material evidence continues to show that before 1500 B.C. such imitations were predominantly of faience or frit, it does at least indicate the emergence of some production in glass: the seeds of a repertory that was to remain a significant aspect of Mesopotamian glass manufacture for long after 1500 B.C. Indeed the very existence of the "glass texts" (see p.224) in later Middle Babylonian and Neo-Assyrian times is evidence for its continuing importance. Since the modern student of early glass technology tends to see the production of glass vessels as the key factor, it is easy to forget that this was only one aspect of the industry after 1500 B.C. and not, as it happens, the one of interest to contemporary scribes.

In this instance, particularly, a broader perspective is instructive. In the Near East, outside Mesopotamia, there still appears to be no certain evidence for glass beads before the Middle Bronze Age, save for the isolated Amuq find (p.195). In Anatolia Schliemann reported three small glass spheres and a glass bead from the Burnt City of Troy (IIg); but there is firstly some doubt about their context (Schliemann, 1884, 478, 480) and then much debate over the absolute dating of this level. Both at Boğazköy and at Alishar there is evidence for glass beads in the Middle Bronze Age levels (Boehmer, 1972, 175, no.1809; von der Osten, 1937, 284-5, fig.309). In Syria Woolley (1955, 269) reported polychrome glass beads from the temple of level VII at Tel Atshana. In Iran blue glass beads, coloured with cobalt, have been reported from graves at Geoy Tepe, usually assigned to the period between 2000-1500 B.C. and probably earlier rather than later in it (Burton-Brown, 1951, 127ff.; Crawford, 1975, 12). Thus, within the broad network of Old Assyrian trade contact, there is some firm evidence for glass bead production that seems more than accidental.

The evidence for Egypt has been admirably summarised by Lucas (1962, 181; cf. Cooney, 1960, 11):

"There is, therefore, no doubt whatsoever that glass beads and tiny glass amulets date from as early as about the Fifth Dynasty (c. 2465-2323 B.C.), and it is exceedingly probable that they were all of Egyptian manufacture and that they were the outcome of the use of glass as a glazing material for steatite and quartz (both solid and powdered). Some of these early beads, however, are not normal glass, but what elsewhere I have called 'imperfect glass'... Though possibly intended as faience, which is glazed quartz frit, they are not, since the material is of similar composition throughout without any coating of glaze, and, therefore, they must be classed as glass. This imperfect glass consists of glass matrix in which a considerable proportion of uncombined quartz is embedded."

Two udjat eyes of black and white glass found at Sedment, dated to the Tenth Dynasty, are the only other objects that may date before the Middle Kingdom (Lucas, 1962, 182). None of the other "early glass" objects from Egypt has survived critical examination in recent years. The occasional production of glass in the Middle Kingdom and through into the Second Intermediate period includes beads (Lucas, 1962, 181) and some inscribed scarabs (Martin, 1971, nos.441, 1198). The Boston Museum of Fine Arts has a tiny glass plaque, inscribed with the names of Ahmose (c.1550-25 B.C.) and Amenophis I (c.1525-1504 B.C.), which is taken to be contemporary with the later part of

Ahmose's reign (Gordon, 1982). "An elemental analysis (in the Boston Museum laboratory by Van Zelst and Sayre) by means of energy dispersive X-Ray fluorescence spectrometry showed the presence of calcium, a main element in glass production, copper as a colorant, antimony as an opacifier, strontium as an impurity coming in with calcium, and iron as an impurity" (Gordon, 1982, 294).

The first regular production of core-formed glass vessels in Egypt is still hard to date. If two glass sherds found in the debris of the tomb of Tuthmosis I (c.1504-1492 B.C.) really belong to his reign, then production may have begun by the end of the sixteenth century B.C.; but this remains doubtful (cf. Barag, 1970, 181). It is to the time of Tuthmosis III (c.1479-1425 B.C.) that the earliest reliable evidence belongs. Vessels of this reign represent an important initial stage of development, since shapes and designs were not yet standardized. In Barag's (1970, 183) view some show no evidence of foreign influence at all, whilst one has adopted decoration familiar in Mesopotamia and two may be actual imports from that area. It is possible that following the Asiatic campaigns of Tuthmosis III the skills of glass vessel production, if not actual glass workers from Syria, were first established in Egypt; but it is not until the reign of Tuthmosis IV (c. 1401-1391 B.C.) that a standard local repertory (most evident a generation or so later at Tell al-Amarna) is first evident.

Whatever the case there is no question, as is sometimes stated, of the "invention" of glass somewhere in the Near East in about the sixteenth century B.C. There was simply a major change in glass technology. Where this took place, and under what circumstances, is still unknown. As the older artificial substances, the frits and faiences, are chemically very similar to the opaque, richly coloured glasses of the initial Late Bronze Age in Syro-Palestine, it is likely that the change took place in an area with a well established frit/faience industry. Such an innovation is likely to have emerged in a well patronized court or temple workshop. The moulding of faience nude-female figurines in the Middle Bronze Age, as at Alishar and Kültepe (Dayton, 1978, fig.94) in Turkey, is exactly matched amongst the earliest evidence for glass production (Kühne, 1969; Peltenburg, 1971, 8; see below). There are also frit or faience precursors, in the same general area, for the earliest disk-pendants and spacer-beads of moulded glass (Haevernick, 1965; Barag, 1970, 192ff.; Spaer, 1984).

Smith (1981, 202, 248) has argued that the cementation process for faience manufacture ("self-glazing"; see p.161 here), involving "very subtle behaviour of alkalies and silicates in differential contact with lime and silica surfaces (cf. Wulff, 1968)" gave rise directly to the manufacture of the earliest so-called "sand-core" glass vessels in which the "sand" was a non wettable material, probably lime, not silica. If so, it would be interesting to know whether this technique was first widely used for faience production in Syro-Palestine in the mature Middle Bronze Age as it seems to be a Middle Kingdom innovation in Egypt (Vandiver, 1983a). The simultaneous appearance (see below) of evidence for the earliest glass vessels and the first vessels of glazed terracotta (see p.167) at Tell Atshana in Syria, at the very end of the Middle Bronze Age, is certainly not coincidence. They are two aspects of a single major technological breakthrough in the faience workshops of the Levant at this time rapidly taken up in the court workshops of the region.

(ii) The Mitannian, Kassite and Middle Assyrian traditions (c.1500-1150 B.C.):

As has just been argued, somewhere in the Near East in the sixteenth century B.C. there was a major technological change in the use of glass, requiring new techniques, the use of special tools, and a good knowledge of the particular properties of glass. For the first time this hot, viscous substance was coiled around a core to make vessels. The earliest evidence for such vessels is still extremely meagre. Woolley found isolated sherds of polychrome, core-formed glass vessels in level VI at Tell Atshana (Alalakh) in Syria, for which the lowest accepted dating is the later sixteenth century B.C. (Woolley, 1955, 300 AT/39/225; Barag, 1970, 157, no.4). By contrast, no other vessel sherds outside Atshana (level V was also reported to have yielded glass sherds) may safely be dated before the second half of the fifteenth century B.C.; so caution may still be in order in any assessment of the context of these pieces at Alalakh.*

Cast or moulded glass objects were also reported from Atshana VI, notably a nude-female plaque (Woolley, 1955, 247, pl.LVIb; Barag, 1970, 188ff.); another was in a grave attributed to level V, (Woolley, 1955, 220:AT/39/106). Stratum IX at Megiddo (c.1550-1475 B.C.) yielded a glass nude-female plaque (Loud, 1948, 182, pl.241:4: as "fayence") as well as a complete, moulded, plain glass disk-pendant and moulded, blue glass spacer-beads (Loud, 1948, pl.210:39, 41, cf.Barag, 1970, 192). Examples of moulded blue glass star-shaped disk-pendants, nude-female plaques and spacer-beads in the Aegean, in archaeological contexts of the sixteenth or fifteenth centuries B.C., have been identified as oriental imports (cf. Barag, 1970, 189ff.). As such they are further support for the argument that there were active glass workshops in Syro-Palestine by the earliest phase of the local Late Bronze Age (c.1550-1500 B.C.).

No site excavated in Mesopotamia has levels securely dated to the crucial period, c.1600/1550-1450 B.C., so no glass certainly of this early period is yet known there.

Oppenheim (1973) investigated the direction in which the invention of glass vessel manufacture spread across the Near East in the third quarter of the second millennium B.C. through two key Akkadian words. A text from Ugarit indicated that they were synonymous: the expression ehlipakku, perhaps of Hurrian origin, and the word mekku, of West Semitic origin. The former is attested at Boghazköy in Anatolia, in Alalakh (Atshana), Ugarit, Qatna and Nuzi but does not appear in the "glass texts", the latter was used in Tyre, Ugarit, and in Assyria. This commodity, by one or other name, is requested by thg Egyptian pharaoh, in the Amarna letters, from a number of rulers of city-states in Palestine and Lebanon, including Ascalon and Tyre. Mekku is translated in the standard dictionaries as "a kind of

---

* There are sherds of pottery and faience vessels attributed by Woolley to levels VI-V at Tell Atshana that are demonstrably of a later date (in the local Late Bronze Age).

precious stone" (C.A.D. s.v.: cf.AHw, s.v.: "stone"), whilst Foster (1979, 21) rendered it "finely powdered quartz". On the basis of his study of the Mesopotamian glass texts, Oppenheim (1970, 57ff.; 1973, 261) argued that it was the "raw material used by craftsmen who fashion the glass containers"; that is an artificial glassy material. Even in the key text (Oppenheim, 1970, 55-6) mekku does not have to be glass; it could well be a natural mineral ingredient used in the manufacture of a frit or glass. On his translation Oppenheim postulated an export trade in raw glass from Syro-Palestine.

Although Oppenheim appreciated the anomaly of an Egyptian pharaoh keenly requesting raw glass "in spite of the fact that all ingredients necessary to produce glass objects were abundantly available in Egypt" (1973, 263), this did not shake his faith in his translation. The existence of flourishing glass workshops at Amarna (Lucas, 1962, 191ff.) urges even greater caution over Oppenheim's translation. Although the analytical evidence is sparse, information on two colourants also tends against Oppenheim's argument. The cobalt blues analysed to date from Egypt and Mesopotamia "clearly indicate that different sources of cobalt have to be sought for the Egyptian and Mesopotamian brands of cobalt" (Kaczmarczyk and Hedges, 1983, 290). Brill's results on lead isotopes are equally distinct "the results establish beyond reasonable question that the yellow pigments were manufactured at different places, the Egyptian in Egypt and the Mesopotamian elsewhere...so the yellow pigment itself was not - in this instance - an article of trade" (Brill et al., 1974).

Syrian glass craftsmen may well have been instrumental in accelerating the development of true glass production in both Egypt and Mesopotamia, but this equivocal lexicographic argument is better left out of the debate. By the time evidence for glass is widely available, in the early fourteenth century B.C., international contacts at court level were well established and far reaching, quite sufficient to diffuse luxury objects and relevant technologies between Egypt, the Mitanni and Babylonia, as Oppenheim (1973, 263ff.) himself well appreciated. Whether there was a complementary trade in raw glass over long distances has yet to be established.

The most remarkable collection of early glass from Mesopotamia is that recovered from stratum II at Nuzi, equally well-known for the associated faience and glazed terracotta found there (cf. pp.150). Barag (1970, 136ff.) has provided the basic typological study of the vessels, but much remains to be done on the other glass objects. Vandiver (1982a; 1983) has begun to examine the technology of the industry and her work has already shown the variety of techniques in use. This glass was concentrated in the destruction level of Stratum II (Starr, 1939, passim) which may not date much before 1350 B.C. (cf. Wilhelm, 1976). As there is a general tendency to date Nuzi Stratum II to the fifteenth century B.C., it is important to stress the possibility that many of the excavated objects may be a generation or so after the end of that century and so more nearly contemporary with the Amarna glass factories in Egypt than is often appreciated. No cemeteries were excavated at Nuzi, so all the glass came from the town, where it was concentrated in Temple A (Cella G29, Court G50, Room H7 and Street 8) and in the Palace (Rooms L8 ("shrine"), M79 and Court M100), with further finds in the "House of Shilwi-Teshup" (Room 44, Court 15 and Court 7) and in Room C22.

One of the most remarkable discoveries about the beads found in such quantity in Temple A at Nuzi was the probability that a significant number (especially unpierced ones) had been part of the architectural decoration, though no wall was preserved to a great enough height to show exactly how (Starr, 1939, 92ff.). This was "substantiated by the discovery of an unbaked brick in the northwest end of the room, in which had been set a row of elliptical and circular eye-beads" (Starr, 1939, pl.119, K1). Two plaques, one of limestone, the other of baked clay, which had been "overlaid with coloured glass in which geometrical and conventionalized tree-of-life designs are inlaid" (Starr, 1939, 93, pl.131, A, B), were assumed also to have been attached to the mudbrick walls. The excavator thought that even a substantial number of pierced beads had been strung as wall hangings and "were not special offerings to, or decorations for, the statue of the goddess", as would commonly be assumed of beads found in temple contexts.

Thousands of beads were recovered (Starr, 1939, 94, 445ff.) with many fragments of core-formed glass vessels, whose shapes (goblets, piriform bottles, shallow bowls and stands) were those of contemporary pottery, some of which was glazed. Apart from a shallow monochrome glass bowl, the reported Nuzi glass vessels were polychrome, with blue as the primary base colour; thread decoration was in white, yellow, orange (amber) and pale blue; rim twists were of blue and white or brown and white. A variety of geometric patterns, meanders, festoons, feather-patterns, chevrons and twists were used. A bowl from the Palace (Starr, 1939, 461, pl.119J; Barag, 1970, 140, no.14, fig.13) resembles the mosaic glass better represented at Assur. Barag described it as "made of red-brown, opaque composition mixed with thin yellow and white threads". Equally unusual is a goblet-fragment, again from the Palace, published as "faience" by Starr (1939, 461, pl.119I). Barag (1970, 140, no.15) is more cautious. It is of a mixed opaque composition, red-brown, yellow, dark blue and turquoise (?) faded white in which dark blue spots are glass. Further sherds from similar vessels were found.

This uncertainty usefully serves to illustrate the dangers of trying to categorize by eye the products of an industry that was both inventive and various (cf. Vandiver, 1982a; 1983) producing a number of closely related artificial materials. Several lumps of deep blue, raw glass were found, but in a locus (C62) whose exact position on the site is obscure (Barag, 1970 ,140-1, fig.16). Glass was used for ceremonial maceheads and "staff-heads" (Starr, 1939, 459-60). Among pendants, round ones with or without moulded stars, are the best known (Barag, 1970, 40) and have been associated with the cult of Ishtar, whose devotees are perhaps also represented by the moulded nude female glass plaques found at Nuzi and elsewhere (Barag, 1970, 140).

The glass finds from Nuzi, now in the Semitic Museum at Harvard, have been the subject of a special technical study by Vandiver (1983) and what follows draws heavily on her account. Accepting that most of the finds date from the later fifteenth or earlier fourteenth century B.C. she argues that they do "not represent the incipient stages of an emerging technology"; much had already happened. Artistic skill was matched by technical expertise; the range of chemical compositions was relatively limited, though the number of colourants was large; the related soda-lime-silicate technology at Nuzi was also sophisticated and various (see p.150). The glass repertory included artefacts whose find spots suggested that they were luxury objects for the rich; but glass beads were very widely distributed on the site. Glass, like faience before it, was used to imitate semi-precious stones and shells. Patterns drawn from nature, as well as amulets shaped like birds, flies, frogs, flowers, and stars, were characteristic of this industry. The technical repertory was equally diverse: rod-forming for beads; moulding for "Ishtar" amulets; disks of molten glass impressed with designs; rare examples of rotary grinding of cold-glass blocks; core-formed polychrome vessels; free-formed amulets.

Although at this stage, as with expanded faience production (see p.150), this is an industry primarily concentrated in the "Great Organisations", the palaces and temples of major cities, it is not strictly correct to say that "nearly all" early glass is found in temples (as Peltenburg, 1971, 8). As is noted above at Nuzi, personal ornaments have a wider distribution. More elaborate artefacts of glass are certainly concentrated in palaces, temples and in rich graves, as would be expected of a luxury material; but this is not incompatible with simpler production in "cottage industries", away from the palace and temple workshops, at least by the later fourteenth and thirteenth centuries, when glass is more widely evident and was much used for beads, and even cylinder seals, as in ordinary graves at Tell Mohammad Arab (M. Roaf; pers.comm.).

The glass finds at Assur, another site with an active faience industry at the same time (cf. p.152), are mostly dated to the thirteenth century and again display considerable technical variety. Oppenheim (1970) suggested that the "glass" texts assembled six hundred years later in Ashurbanipal's library at Nineveh had first been codified at Assur about this time. The archaeological evidence certainly indicates a local industry making glass of all kinds (mosaic and core-formed vessels, inlays and probably plaques) as well as a complementary production of artefacts in "glassy" (sintered quartz) materials. Core-formed glass vessels and sherds of them were largely found in graves, notably in tombs 37 and 133, which held glass contemporary with that from Nuzi II, and in various parts of the Ishtar Temple (Barag, 1970, 141ff.), erected by Tukulti-Ninurta I (c.1243-1207 B.C.). This glass has not yet been fully published, though much of the [illegible]sociated "faience" has (see p.152). It is usually dated to the [illegible]eenth century B.C., though some of it might be slightly earlier. [illegible]ar as it is possible to judge at present, the variety of shapes and [illegible]oration is greater than at Nuzi, though the colour range is compara[illegible]. The mosaic glass fragments (Haevernick, 1968; von Saldern, 19[illegible]15, no.7) are of outstanding importance and constitute the most [illegible]rtant second millennium B.C. group of such glass yet published. The fragments are from vessels as well as inlay plaques for walls and for furniture, using red, blue, yellow, green (?) and white glass. Cullett was also found, reinforcing the case for local production. Figured mosaic designs include human beings, animals, and floral devices reminiscent of contemporary motifs on seals.

Polychrome sherds from glass vessels found at Brak and Chagar Bazar probably date to the later fifteenth or to the fourteenth century B.C., (Barag, 1970, 146; Mallowan, 1947, 77ff., 84). But the other important collection of glass from a northern site is that found at Tell Rimah, where, once again, there was evident indication of an active contemporary faience and frit industry. The majority of the fragments are "from Middle Assyrian levels with a terminal date of c.1250 B.C. and a mean date of 1300 $\pm$ 50 years" (Oates, cited by Harden, 1969, 29). They were found in the late phases of Level 2 (cf. Barag, 1970, 145). Core-formed vessels were represented by complete examples and by sherds; moulded glass by pendants, nude-female plaques and demon masks; beads by a variety of polychrome types. Scholarly attention has, however, concentrated particularly on the fragments of a series of straight-sided, button-based goblets made of opaque mosaic glass arranged in chevrons of green, white, blue and yellow canes (von Saldern, 1966a; 1970, 206, 213ff.; Harden, 1969). One almost flat fragment may be part of a plaque or a very large vessel. No industrial debris or raw glass has been published from this site. Two mosaic glass beakers from Marlik in Iran may be exports from workshops in Assyria specializing in the production of such mosaic vessels (von Saldern, 1966a; 1970, 208); they are from graves likely to belong to Iron I (c.1350-1000 B.C.), though this cemetery was in use for some considerable time into the first millennium B.C.

An important collection of glass fragments was recovered from the royal palace in the Kassite capital at Aqar Quf (Dur Kurigalzu), near Baghdad, in contexts the majority of which are dated to the reign of Kurigalzu II (c.1332-1308 B.C.). A text from level II at this site, dated in the 3rd year of Kashtiliashu IV (c.1232-1[illegible] B.C.), refers to glass given to artisans for decoration in the "Palace of the Stag" (Gurney, 1953, no.22). Glass was also found in the final destruction levels (Ic) at the palace, in the late thirteenth or twelfth century B.C. The published information on the glass finds (cf. Barag, 1970, 148ff; von Saldern, 1970, 207-8) is limited to the more unusual types and little or no reference is made to sherds of core-formed polychrome vessels. Mosaic glass was especially plentiful. Glass plaques (with relatives at Tell Rimah) of two main types were reported; some were plain (or variegated?), otherwise of red glass inlaid with white, turquoise-blue, red and perhaps yellow glass in designs of stars, eye-like disks, lozenges, and birds-of-prey (von Saldern, 1970, nos.3-4). Mosaic glass was also used in the manufacture of bowls and other vessels (von Saldern, 1970, nos.5-6; Barag, 1970, 148). Texts indicate that workmen at Dur Kurigalzu were being issued with a variety of precious materials for palace decoration, so here again, as seems to be the case at Assur, Nuzi and Tell Rimah, royal and temple workshops are likely to have included those of glassmakers.

Further to the south, in Babylonia, glass of early date has rarely been recorded. The pioneer excavation of Peters at Nippur produced a group of fragmentary, moulded blue glass votive axeheads, with dedicatory inscriptions for the Kassite king Nazi-Maruttash* (c.1307-1282 B.C.) (Peters, 1898, 134ff., 143ff., 373ff., with an analysis cf. Neumann, 1927). They had been hoarded by a "jeweller", with comparable objects of lapis lazuli and other semi-precious

---

* And other rulers, cf. Brinkman, 1976, 71, n.289.

stones, in the Parthian settlement at Nippur (Barag, 1970, 148; von Saldern, 1970, 215, no.8). A grave of the "Kassite Period" at Ur contained a piriform glass bottle (Woolley, 1965, 78; Barag, 1970, 147, for dating) and what may be a contemporary black glass rod of nine threads, marked at one end by pincers (Woolley, 1965, 79, n.1, 105:U.7593). On the island of Failaka glass, faience and glazed pottery were recorded in Kassite levels, with some evidence for local workshops (Pollard and Høјlund, 1983).

Evidence of an unusually interesting glass industry, of the thirteenth century B.C., was evident in the Ziggurat area at Tchoga Zanbil in Khuzistan. Here long rod-formed tubes made of black (or very dark blue) glass decorated with a white thread were used as decoration on wooden doors (Ghirshman, 1966, 30, 65, 74, fig.20, pls.XXV:1,4). Fragments of similar tubes are also known from Susa (Barag, 1970, 149, n.88a). Convex disks and "eye roundels" of glass were used at Tchoga Zanbil to decorate bricks (Barag, 1970, 149; von Saldern, 1970, 216). At this site the seals of poorer quality were made of faience; but the high quality ones, in which Kassite stylistic traits and Elamite iconography are blended, were carved on cylinders of dark blue glass (Porada, 1970), presumably in imitation of lapis lazuli. Glass seals are reported outside Susiana and Failaka at this time relatively rarely and then usually in Syria (Buchanan, 1966, no.1008; Parker, 1949, no.155) or northern Mesopotamia. To what extent Babylonia supplied the initial technical "know-how" for Elamite workshops is an open question (cf. von Saldern, 1970, 208); but, in view of the flourishing Middle Elamite metal and faience industries evident at Susa, such an indebtedness should not be too easily assumed. The impetus may be older and have come through commercial activity linking Khuzistan with the kingdom of Mitanni at a time when crucial early developments in glass production were happening there.

In view of the rarity of glass finds in Babylonia as a whole, and in the period between about 1200 and 750 B.C. in particular, the scattered glass inlays found in the main temple complex at Larsa are of particular interest. At some time during the reign of Adad-apla-iddina (c.1068-1047 B.C.), when rooms 15 and 13, and courtyard I, were paved, various inlays, predominantly of blue glass, one a beard fragment (Huot et al., 1978, 193-4, fig.27[b]), another an animal's horn (Huot, 1983, 209), were being used in the decoration of statuary. If these pieces were indeed manufactured at this time, they go some way to bridging the gap between the flourishing glass industries of the thirteenth century B.C. and the known return of glass production sometime in the ninth century B.C. A published analysis (Huot, 1976, 21-2, 28) indicates a standard Mesopotamian glass, with copper (or possibly cobalt) as the blue colourant.

(iii) The Neo-Assyrian period (ninth to seventh centuries B.C.):

(a) Revival of core-formed glass production:

Towards the end of the second millennium B.C., in Egypt and the Near East, glass production subsided so much that it appears from present evidence to have all but ceased. Thereafter, for the period before the fall of the Assyrian Empire in western Asia, only two sites at the moment provide archaeological *termini ante quem* for significantly varied finds of glass: level IV at Hasanlu, in north-west Iran, destroyed about 800 B.C., and the main destruction levels in the various major buildings at Nimrud dated to about 614/612 B.C. Both sites present problems of their own. The glass from Hasanlu is only partially published. Two fragmentary beakers and the remains of a third, are outstanding examples of the mosaic glass tradition established in Mesopotamia in the Kassite and Middle Assyrian periods (von Saldern, 1966, 9ff; 1970, 209ff., 216-17). Scholarly opinion is divided upon whether they were indeed made many centuries before the destruction level from which they were retrieved, or whether they are, for the present, a completely isolated indication that this craft tradition survived into the early first millennium B.C., possibly only in western Iran. No glasses of this specific type are known from a Mesopotamian or Syrian site after the second millennium B.C. Little has been reported of the other glass found in Hasanlu IV. It is known that no "Phoenician" ivories inlaid with glass were found there, but some in the "Syrian style" had inlays, though none has been scientifically identified as glass (Muscarella, 1980, nos.222, 229, 267). Barag (1970, 194, no.216) reported square moulded shafts of blue glass, possibly parts of furniture, and a bead with a head of an animal (?), made of black glass with white and yellow (?) twisted canes marvered in.

The date of the re-appearance of core-formed polychrome glass vessels in Mesopotamia has not yet been established with any precision. Barag (1970, 194) cites grave 119 in the Merkes at Babylon as a fixed point. It contained a pointed bottle and a globular bottle and was below the debris of a house said to have been destroyed by Sennacherib's sack of Babylon in 689 B.C. (ibid., 159-60). Comparable vessels from such sites as Assur, Kish and Nippur are not so firmly fixed in time and some may date as late as the fifth century B.C. (Barag, 1970, 156ff.; on a lower dating for the Kish vessels, Moorey, 1978, 52). Barag (1970, 155) and von Saldern (1966, 632) catalogued only a few sherds of core-formed glass from Nimrud. Two were from the shoulders of cylindrical alabastra, decorated with a wavy thread in white. Core-formed vessels are rarer in Neo-Assyrian palace excavations than in the private graves of this period so far excavated.

There are certain marked typological distinctions between the Late Bronze Age core-formed glass vessels, when the repertory of shapes was relatively limited, and their Iron Age successors. The early craftsmen had excelled particularly in the use of a variety of colours and in skilled treatment of thread decoration. The shapes of the later group show no direct links with the earlier; no open shapes occur and the vessels are smaller than before. The quality of the glass and the skill of the workmanship is also less impressive. The primary glass shapes, pointed bottles, globular jars and relatives of the alabastron, are those most popular in other materials. Handles, unknown in the Late Bronze Age glass repertory, now followed the fashion of stone alabastra in being "duck-shaped" vertical tabs or small loops. So far, outside their homeland, these distinctively "Mesopotamian" shapes in glass are best known from examples found on the island of Rhodes in seventh century archaeological contexts (Harden, 1981, 52ff., nos.78-81).

In the continuing absence of local evidence for Phoenician glass manufacture before the earliest re-appearance of core-formed vessels in Mesopotamia, the source of the revived glass industry remains obscure. With a few isolated exceptions glass disappears in Egypt from late in the XXIst Dynasty (c.1070-945 B.C.) until the XXVIth (c.664-525 B.C.), when stray examples are again known. Even so, well documented and securely dated glass of the Late Period is rare. The craft does not seem to have been revived locally on an extensive scale until the XXXth Dynasty (c.380-343 B.C.) (Cooney, 1976, XV-XVI). As there was a persistent tradition of glaze manufacture for baked clay objects in Assyria, as also probably in Babylonia (see p.168), a local impetus is not wholly out of the question. This view would be strengthened if the glass fragments from Larsa (see p.206) really were of eleventh century manufacture.

So long as the date of the earliest inlaid "Phoenician" carved ivories is uncertain, evidence for glass and faience industries in either Syria or Phoenicia in the ninth century remains oblique. Isolated imports of glass and faience in the Aegean world at this time have been taken to indicate the activity of Syro-Phoenician production centres. Conspicuous amongst such finds are a variegated glass bead, threaded with over 1000 faience disk beads, in a rich female grave in the Areopagus at Athens (Smithson, 1968, pl.33:78) and faience objects from Lefkandi in Euboea (Popham et al., 1982, 242ff., pl.20). The inlaid "Syrian" carved ivories from Hasanlu IV may also indicate glass production in the ninth century in Syria (Muscarella, 1980, 220).

(b) <u>The emergence of "clear" glass</u>:

At Nimrud moulded cut-glass fragments were overwhelmingly more common than polychrome core-formed glass sherds (von Saldern, 1966; 1970). They mark a number of crucial changes. Traditionally glass in Mesopotamia had imitated the appearance of opaque, precious and semi-precious stones, not only the blues of lapis lazuli and turquoise, but also the varied colours of banded agates. Now fresh chemical discoveries (see p.223 below) allowed for the imitation of rock crystal, which was popular at the time and worked in shapes and decorated by techniques exactly replicated in glass (Mallowan, 1966, fig.143; Oliver, 1973). Glass alabastra, notably a well-known example inscribed for Sargon II (von Saldern, 1970, 218, no.17), small handled jars, hemispherical or similarly shaped bowls, and variously shaped cups, were mould-made in glass, varying from extremely thin to very thick, from practically clear to a pale green in colour. These vessels were ground to shape and polished; some were decorated with cut designs, usually geometric; or, less frequently, incrusted with mosaic glass, or painted. The technical skill required to produce such bowls with thin walls led both Turner (1955) and von Saldern (1966, 627-8) to speculate on the possibility that the blow pipe, or another unrecognized process, different from regular moulding, had been familiar to the glass makers. It is possible that a <u>lost-wax</u> casting technique was used.

As there is still no evidence that Egypt had an industry producing cast and cut glass vessels from the XXIV to XXVI Dynasties (c.724-525 B.C.), the initial manufacture of these vessels is commonly assigned either to Phoenicia, where, as with core-formed glass, there is again at present no local evidence, or to Assyria itself.

Barag (1982, 17ff.) has cogently argued the case for Phoenicia, albeit on circumstantial evidence, attributing the concentration of finds at Nimrud to Phoenician craftsmen (or men they had trained) active in Assyrian court workshops. He isolated a series of pertinent arguments: glass inlays were used extensively in "Phoenician" carved ivories and have been found in greater quantities at Nimrud than would be expected to result from their falling out of imported carved ivories (von Saldern, 1966, 632-3; Barnett, 1975, 240); a unique cut glass bowl from Nimrud is inlaid with a comparable type of mosaic glass (von Saldern, 1970, nos.29-30); at least one of the glass bowls from Nimrud has cut designs in a "Phoenician" style (Lehrer, 1974); there are glass plaques from Nimrud painted in the "Phoenician" style (Orchard and Brill, 1978); the "Sargon Vase" is shaped like Egyptian rock crystal alabastra, more likely in a Phoenician than in an Assyrian craft tradition; the glass jug from Aliseda in Spain, a key artefact in this group of glass, is a typical "Phoenician" shape, has a debased Egyptian hieroglyphic inscription, and is from an area of known Phoenician contact; and the industry persisted after the fall of Nineveh in 612 B.C., when its products are better known from Cyprus and Italy than from Mesopotamia (von Saldern, 1970, 212).

To what extent the chemical change to clear glass may also be attributed to the Phoenicians or Syrians is at present an open question. In their study of Egyptian faience Kaczmarczyk and Hedges (1983) have argued that if one were to look for a time when antimony oxide might also have been added as a clarifier (bleaching agent) and not solely as a pigment, as was previously the case, it would be the XXVIth Dynasty (c.664-525 B.C.). The clear glass from Nimrud was made before about 614 B.C.; but a moulded glass omphalos bowl placed in tumulus P at Gordion before about 700 B.C. indicates that this industry, wherever located, was already active in the later eighth century B.C. (von Saldern, 1959; Young, 1981, 32). Both Young and von Saldern believed the Gordion bowl to be of Mesopotamian manufacture. No such glass has yet been reported from Egypt of this date. Although evidence for a local Neo-Assyrian faience industry is sparse (see p.157), there was extensive production in Assyria of glazed pottery and bricks (see pp.168ff.). This involved the use of lead antimonate to create yellow glazes or possibly (in combination with copper) green ones, so the existence of a local clear glass industry in Assyria by the reign of Sargon II (c.721-705 B.C.), if not before, is plausible. But still captured Syrian or Phoenician craftsmen might initially have brought the technical knowledge with them from the west and themselves, or their pupils, practised it at the Assyrian court. Fresh evidence is needed to elucidate the question.

(c) Coloured glass:

Moulded glass in rich colours was of enduring popularity. Ivories inlaid with monochrome pieces of glass, in imitation of precious stones, were already produced by craftsmen in Syrian workshops active during the ninth century (Barag, 1983; Barnett, 1982, 14, 44). In this case the glass inlays were secured in a distinctive way with a small depression drilled into the centre of the background, presumably to take an adhesive. Glass inlays are, however, much more intimately associated with carved ivories in the "Phoenician" style (Barag, 1983). These inlaid ivories and a series of small square polychrome mosaic glass tiles decorated with rosettes, have been found particularly at Nimrud (von Saldern, 1966, 623ff.), at Arslan Tash (Thureau-Dangin et al., 1931, 138) and at Samaria (Crowfoot and Crowfoot, 1938, 56ff.). As there is no hard evidence for dating any ivories in the "Phoenician" style (or related glass) before about 800 B.C., they are all at present dated after the end of Hasanlu IV, where none was found. Barag (1983) has argued that they may indeed be an innovation of the first half of the eighth century B.C. Nor is it clear, once again, whether they were only manufactured in a series of closely associated workshops in Phoenicia, and exported thence to Syro-Palestine and Mesopotamia, or whether, as may well have been the case in Assyrian palace workshops, expatriate craftsmen produced them in foreign centres.

A dark blue glass mace or staff-head, perhaps the attribute of a deity statue, was found at the bottom of a well in the Nabu Temple at Nimrud (von Saldern, 1970, 224-5, no.41a). Another of sky-blue glass was found at Assur (Haevernick and Kühne, 1982). They illustrate the continued use of opaque, coloured glass to make objects in direct imitation of comparable artefacts carved in semi-precious stones. It was this particular type of glass manufacture that most concerned the writers of the "glass texts" found in the "Library of Ashurbanipal" at Nineveh (Oppenheim, 1970; see p.224 here).

(iv) The Neo-Babylonian and Achaemenid periods (c.600-330 B.C.):

Glass finds from sites like Ur and Babylon (von Saldern, 1970, 226ff.), rare as they still are, indicate that luxury cast and cut glass, as well as core-formed glass, was used and probably produced to some extent in Mesopotamia in the Neo-Babylonian period (c.612-550 B.C.). The core-formed polychrome vessels, for instance, are typically Mesopotamian in shape and style, so were presumably manufactured locally.

In Mesopotamia during the Achaemenid Period (c.550-320 B.C.) the isolated finds of clear or engraved glass (cf. Nippur: Barag, 1968), still very much a luxury product, are known to be a standard product of workshops serving the Achaemenid court and aristocracy. They produced a standard repertory of forms, usually copying the shapes of tableware manufactured in precious metals (Oliver, 1970). Finds from Persepolis illustrate the range of this internationally available glasswork (Schmidt, 1957, pls.66-7). Whether it was only produced in a few centres closely associated with royal residences, or more widely, is unknown.

Core-formed polychrome glass vessels (Barag, 1970) were probably still being made in Mesopotamia in the later sixth and fifth centuries B.C. This is also the period when workshops somewhere in the East Mediterranean region emerged as producers of a distinctive range of core-formed polychrome vessels that were widely traded (Harden, 1981, 52ff.). There is only one type of vessel made at this time that appears likely to be from glass workshops in the eastern rather than the western empire. This is a group of polychrome glass, rod-formed, kohl tubes, of a distinctive square-faced, tapering form (Barag, 1975). Also in this period personal ornaments, amulets, and some seals, were made of glass to standard patterns. They are so widely distributed throughout the Achaemenid Empire, and beyond its frontiers, that it is impossible to isolate regional traditions. At this time distinctive "eye-beads", produced in Western Asia, reached as far as China (cf. Taniichi, 1983). Only in the case of cylinder seals and conoid stamp seals is it possible to define a concentration in Syro-Phoenicia and suggest that they were a local phenomenon.

<u>Summary</u>:

It is clear from the data assembled in this historical survey that glass was intentionally manufactured in Western Asia from at least about 1500 B.C., in a number of shapes and by a variety of techniques, using various colours. A trend is broadly discernible, moving from opacity, in imitation of semi-precious stones in the later third and second millennia B.C., to translucency and transparency, by the second quarter of the first millennium B.C., though the earlier tradition also persisted for some time longer. Techniques, whether in core-formed or mosaic glass, in the earlier, polychrome stage, emphasized variegated colour effects, whilst it is only in the Neo-Assyrian period that the idea of clear, translucent glass, imitating cut stones like rock crystal, became the primary goal in the workshops of the Assyrian and Achaemenid courts. The colours included blue, notably a turquoise blue, red, yellow, green, brown, black and white; the transparent colours were blue and purple. The early clear glass is, unintentionally, tinted green or yellow.

The shapes of all glass vessels in the period covered here, that is before the invention of glass-blowing, derive from ceramic, stone or metal prototypes. Beakers, goblets and piriform bottles were the favourite shapes among the early core-formed, polychrome vessels, whilst various types of pointed bottle, of alabastra, of jars, and finally of juglets, emerged in the first millennium. In moulded glass the bowl became the most popular form in the first millennium B.C.

Without direct evidence for places of manufacture, sources of glass found in Mesopotamia remain controversial. The present pattern of distribution indicates that in the second half of the second millennium B.C. there were various glass production centres in northern Mesopotamia and in Khuzistan (Susiana), whether this was also true for Babylonia is not quite so clear, though evidence from Aqar Quf, Nippur and Failaka suggest that it was. Evidence from cities like Assur, Aqar Quf, Nuzi and Susa indicates that local glass workshops, within the orbit of major royal or temple building enterprises, served local needs. To what extent there was also an extensive commerce in glass vessels, or more significantly in ingots of raw glass, from more minor factories is not yet established. If, as seems likely, Syria had its own very active glass industries, Mesopotamian export may only have been into western Iran. The mosaic glass vessels found at Hasanlu and Marlik may be examples of such an export trade. Archaeological finds of second millennium B.C. glass in Mesopotamia are still bewilderingly sporadic and random: cane mosaics at Aqar Quf and Tell Rimah; opaque glass maceheads and axeheads at Nippur and Nuzi.

In the first millennium B.C., to judge particularly from finds at Nimrud, royal palaces once again had their own glass workshops, employing craftsmen in some cases brought from the west not only to exercise their special skills, but also to train on the spot a new generation of artisans. There are numerous indications that, as in the second half of the second millennium B.C., the innovative region in glass technology was to the west of Mesopotamia. But many key questions about innovation and diffusion remain wide open. Even if it is possible to suggest the possibility of two key glassmaking traditions in the earlier first millennium B.C., which might be distinguished crudely as "Syrian" and "Phoenician" (the former more closely linked to Babylonia by the Euphrates), it is still not possible to isolate with confidence their respective roles. The chronological distinctions in the activity of ivory and metal workshops in these two regions, argued by Winter (1977), is in itself debatable. It is possible that glass manufacture revived earlier, and ceased sooner, in Syria than in Phoenicia; but hard evidence for such a conclusion is still awaited.

Although it falls outside the chronological range of this survey, evidence adduced by Adams (1981, 211-13) for glass production in the Sasanian period has implications for what might be expected in earlier periods when minor glass production centres are sought:

"...let me call attention to the series of sites specializing in glass manufacture along a newly dug Sasanian canal northwest and north of Uruk-Warka... The scale of what can only be described as industrial production there is suggested by mounds hundreds of metres long that apparently are composed mainly of glass slag. Numerous glass furnaces also can still be seen in place. To understand the resource base that led to such specialization, it is worth noting that in late Sasanian times the area south of this canal line probably had begun to include extensive swamps. Here would have been found large accumulations of snails as a source of calcium carbonate... Suitable water-laid sand might also have been supplied locally. Finally, naturally occurring plants in the same area are reported to be a source of sodium carbonate. There is little to suggest that these sites were extensively involved in fashioning glass vessels from the raw material they produced in such quantities. But the important point is that, in difficult terrain at a great distance from the main centres of administration and consumption, a large-scale industrial operation went on (probably into the Early Islamic period) that implies a fairly high degre of economic integration not merely in the cities, but extending throughout the countryside". Even on a marginal site near Ur in this period there were "glass sherds and glass cullet" (H.T. Wright in Adams, 1981, 335).

D: Technology and Composition:

(i) Glass workshops and furnaces:

Only one possible glass workshop has been identified in a Mesopotamian excavation, at Nimrud, and that appears to date to the Achaemenid period. Insufficient information has been published for any appropriate technical description. Mallowan withdrew his earlier dating to the later Neo-Assyrian period (1954, 77, 82-3; 1966, 209-10) and commented thus "...on the south side of room 47 (of the Burnt Palace) there were traces of kilns and of a glassmaker's kit, including one specimen of sealing wax red glass, probably from a crucible (cf. Stern, 1981, 44-5, fig.5). This red glass was of the lead silicate type, coloured with cuprous oxide, and contained 22.8 per cent of lead oxide. The kilns with which this fragment were associated had been sunk from a level containing Hellenistic pottery which could be dated about 200 B.C." More recent research (in Barag, forthcoming) on this type of glass, including a Carbon 14 dating, indicates that the Nimrud glass workshop probably belonged to the Achaemenid rather than to the Seleucid period. There is other industrial debris from glass making, in the Neo-Assyrian levels at Nimrud, but none of it offers coherent information on working methods. Two disks of opaque blue glass had probably been formed in rough moulds, whilst at least two pieces of cullet, or unshaped fused glass, again blue,to be used as raw material, were also recovered (von Saldern, 1966, 633; 1970, 223, no.37; cf. analysis in Turner, 1955, 67).

There is scattered information of this kind from a number of sites in Mesopotamia; but it now does little more than indicate local production, since, without the archaeological context, no evidence of workshop practices may be retrieved. "Segmental" ingot fragments, like those from Nimrud just cited, have been found at Kar-Tukulti-Ninurta, in dark blue glass (Barag, forthcoming, no.171) and at Nuzi (see p.203 here) from second millennium B.C. contexts. From Rassam's excavations at Dilbat (Dulaim) came fragments of ingots of dark red to orange glass (Barag, forthcoming, nos.168-170). They are parts of large, heavy disks with straight edges, probably cast in dish-like crucibles sometime between the eighth and fourth centuries B.C. A translucent dark blue rectangular glass ingot from Kuyunjik has been dated to the Achaemenid period by analogy with an example from Persepolis (Barag, forthcoming, no.172). Other lumps of glass from Kuyunjik and Abu Shahrein (Eridu) may be pieces of ingots or at least debris from local glass working (Barag, forthcoming, nos.182-5). Glass canes from Ur (Woolley, 1965, 105) and Babylon (Barag, forthcoming nos.173-6) increase the geographical range of evidence for local production.

In the absence of the evidence of early glass factories, it is generally assumed, as at Tell el-Amarna in Egypt, that an open hearth system was used, with shallow crucibles supported on heat resistant drums. As no really high temperatures would thus have been possible, the glass would generally have been worked in a pasty state (Petrie, 1894, 25ff.; Vandiver, 1982a, 77). But in the "glass" texts (see pp.224ff.) studied by Oppenheim (1970, 69ff.), of the later Neo-Assyrian period (though probably incorporating sources going back at least to the Middle Assyrian period), three types of kiln-like structures are mentioned. They are never explicitly described. The *kūru* kiln, with four fire openings ("eyes"), was used for fritting and sintering. It had a lower part or "fireplace", possibly constructed so that flames might rise through a pierced floor to the main furnace. This seems to be a special type of kiln, because the introduction to the texts refers to "kiln for glass(making)"; its Sumerian designation indicates that it was provided with bellows to obtain the required high temperatures with a wood fire. There was an alternative type of *kūru* kiln "with a chamber", which could be closed with a "door". Such kilns were evidently used as melting furnaces, for founding glass, perhaps a reverberatory furnace in which the flame is forced back upon the material exposed to it. The *atānu* kiln was brought into use for prolonged firings of up to a week and involved use of "moulds" rather than the standard "crucibles". The texts speak of "stands" within the kilns and slabs of baked clay upon which primary glass was cooled and then ground up for further processing. Tongs ("lifters"), rakes ("turners") and tools for handling the hot crucibles are listed. The texts mention consistently high temperatures to be held over long periods; they indicate how the boiling glass should be tested for viscosity; and the complex stages of grinding and firing required to produce the desired opaque, coloured glass. In common with glass makers until modern times, when chemical reactions could finally be analysed and controlled, they appear to use redundant materials and excessive care, simply because they could not distinguish precisely the effective from the ineffective.

(ii) <u>Mixing and Melting</u>:

There is no ancient textual evidence from Mesopotamia relative to the manufacture of glass artefacts. Artists (Labino, 1966), scientists (Schuler, 1959, 1959a, 1962) and, regrettably, forgers (cf. Goldstein, 1977), have attempted to reproduce ancient forms and techniques, sometimes just for the end product itself, sometimes as a means of elucidating ancient technology. To this the comparative study of contemporary primitive glass-making has also made significant contributions in the hands of Wulff (1966) and Charlesworth (1967).

The mineral raw materials used in glassmaking are generally in the form of oxides. The "batch", from which the manufacturing process begins, normally contains a glass former (the sand), a modifier (soda or potash) and a stabilizer (lime). In many industries it may also contain "cullet", i.e. e[illegible] glass from previous production. Then, as required, other ingredients are added: colourants; opacifiers; decolourants. The ingredients are mixed together and heated in a furnace (see p.213). At about 1400°C the entire mixture fuses into a viscous liquid. The temperature is then held steady for some hours; it is lowered (to 1100°-1300°C) towards the end of the operation to raise the viscosity of the batch so that it is ready for manufacture into artefacts. This has to be a relatively quick process, at most a matter of minutes, as the cooling glass changes rapidly from a highly viscous liquid to a solid.

(iii) <u>Forming Processes</u>:

(a) <u>Free-modelling</u>:

This, the simplest of techniques, does not appear to have been used very much in the ancient Near East, as free-modelled and tooled objects are rarely encountered in Mesopotamian excavations. Small birds from Nuzi stratum II (Vandiver, 1982a, 77ff., fig.4) are good examples of the technique.

(b) <u>Core-formed</u>:

So far as present evidence goes, core-forming was the earliest method for the manufacture of glass vessels (Barag, 1970) and may have derived directly either from techniques for forming sintered quartz vessels on a core or simply from glazing pottery vessels, though the priority of that technique over glass vessel-making has yet to be firmly established. With the discovery that sand had not been a major constituent for the core, the traditional term "sand-core" was dropped (Bimson and Werner, 1969). A core of clay and dung was made first and round this the craftsman trailed the glass (cf. Goldstein, 1979, figs.1-4), again perhaps following, if only broadly, techniques by then standard in faience vessel-making. As the ancient craftsman was probably working with temperatures lower than those available under modern conditions, the viscosity of the glass would have been higher and he is unlikely to have dipped his core into a pot of glass as has often been suggested. However, any modern reconstruction of ancient work-procedures must allow for many variations. The need for constant reheating would have required many stages in the production (Harden, 1969, 50; Goldstein, 1979, 27-9). Wall thickness and the smoothness of the vessel's surface were controlled by rolling the glass-covered core across a flat surface (marvering). Various techniques were used in decoration. The most common was the application of different coloured trails, usually marvered into the surface, but sometimes left in relief. Handles and foot-rings were then added; if appropriate, rims were trailed on and then fashioned to shape. Preformed rods or canes of glass were also applied to core-formed objects.

(c) Rod-formed:

Beads were formed on a rod, at times coated with a clay core, trail decorated, if required, and then tooled. This is well illustrated by finds at Nuzi (Vandiver, 1983, fig.3). The glass tubes for architectural inlays from Tchoga Zanbil in Khuzistan in the thirteenth century B.C. were also made in this way. One still retains the metal rod, presumably because the maker was unable to extract it (Ghirshman, 1968, pl.L:2). Some kind of separating agent would have been needed on the metal rod. Goldstein discounts the suggestion that it was lime (Goldstein, 1979, 28); Vandiver has proposed a "parting layer of calcite and clay" (Vandiver, 1983, 242). A series of rod-formed kohl tubes was manufactured in the eastern Achaemenid Empire (Barag, 1975).

(d) Mosaic Glass:

It has been suggested that the mosaic glass technique was evolved sometime in the middle of the second millennium B.C. from experiments with the rods and trails used by contemporary core-glass makers. In so far as the chronology of the process is at present understood the earliest vessels were made from sections of monochrome cane, though a fragment from Aqar Quf, which may be older than the Tell Rimah vessels, is made of a cane consisting of three concentric colours: a blue centre with red surround in a white border (Harden, 1969, 50, fig.2). Goldstein (1979, 30-31, figs.5-7) admirably summarizes and illustrates the probable procedures of manufacture. He explains how preformed canes of glass would be cut into the required lengths for the mosaic pattern. The cane-sections would then be assembled over a solid core, presumably with a binder of adhesive, and then enveloped in an outer mould (cf. von Saldern, 1966a). The mould would then have been heated slowly to fuse the canes without allowing them to become so hot that the design was lost. This would have required a sophisticated knowledge of the properties of various glasses. Once the canes had been fused, the mould would be removed, the glass surface cleaned and maybe polished. There was much scope for damage and error; but only the excavation of a glass factory will reveal the details of this procedure.

(e) Moulding and Casting Glass:

Moulding techniques had been developed over centuries for the production of small faience objects and the same methods were used for glass pendants and amulets from early in its history. Pendants and inlays were cast in simple open moulds; ground and polished after removal from the mould, which would have been put into the kiln for the glass to be fired in it. Casting in piece-moulds would have been a much more complex process, for which analogies with metal casting are misleading. At the temperatures used in antiquity the glass would not have flowed into a complex mould as did molten metal. Schuler (1959) has proposed methods for mould-casting glass that would have overcome this problem. In his reconstruction the mould, as it was heated in a furnace, was continually filled with ground or powdered glass.

For more complex vessel shapes and for elaborately decorated inlays, such as beards and wigs, the lost-wax technique was probably used to make the moulds (cf. Schuler, 1959, 49-50).

(f) Cold-cutting:

This was a technique in which glass was treated in the same manner as stone had been for millennia. As glass is slightly softer than the harder stones like quartz or flint, cold cutting or grinding was possible both for finishing glass objects, first roughed out by casting, or for working a vessel from a block of glass just as a stone vessel would have been made. Examples of the first process are known in Mesopotamia from the later second millennium B.C.; but the latter is not evident until the next millennium in Assyria. It has been suggested that the famous "Sargon vase", of the late eighth century B.C., was produced by cutting, grinding and polishing. A lump of glass, approximately 10cm. high by 7cm. in diameter, was first prepared. The block was of reasonable uniformity, obtained from well-melted glass and subsequently slow-cooled to avoid shattering. It was then worked exactly as a stone alabastron would have been (von Saldern, 1970, no.17, fig.17). However, it has also been argued that a lost-wax method could have been used for the basic shaping and only the finishing done by grinding and polishing.

None of the excavated examples of Neo-Assyrian cast, cut and polished glass has yet been the subject of a thorough technological examination; but Gwinnett and Gorelick (1983) have published their investigation of an unprovenanced vessel of this type in the Corning Museum of Glass. This vessel is at present attributed to a workshop in Syria or Assyria in the eighth or seventh century B.C., though it might be of the Achaemenid period. They demonstrated that the drill holes in the two lug-handles were probably made with a copper drill pre-bonded with an abrasive; an instrument evident considerably earlier in the manufacture of cylinder seals. No evidence survived to show whether a solid core drill or a hollow core bit had been used; this may only be determined "when the results of the leading edge of the drill can be seen" (ibid., 254). The inner surface showed that original horizontal ribs produced by drilling had been polished in a vertical direction, perhaps with an abrasive used with cloth or leather. There was no trace of the use of a lathe; the regular grooves were probably made with abrasives manipulated by hand.

(iv) Composition:

Today the commonest commercial glass is made by heating together a mixture of such materials as sand (about 75%), limestone (about 10%) and soda (about 15%), at a very high temperature, to form a liquid. When this liquid is taken from the furnace it stiffens rapidly as it cools, until at about 500°C it resembles glass as generally understood. The soda acts as a flux and the limestone as a stabilizing agent to give a durable glass. Other oxides, such as aluminium and magnesium (Al_2O_3; MgO) are also usually present; the former improves the chemical durability and reduces the tendency of the mix to crystallize, whilst the latter decreases the solubility of the glass. Many ancient glasses were frequently of this soda-lime-silica composition; the other common composition has potash replacing soda as the fluxing agent. Potash (K_2O) assists in making a glass that hardens more rapidly and at a higher temperature; but is less suitable for the moulding techniques used in antiquity. Lead oxide (PbO) is also a good modifier providing glass of low softening temperature and high density.

There are two sources of information on the composition of the glass excavated from ancient sites in Mesopotamia: that provided by modern methods of scientific analysis, which is still meagre, and that derived from the cuneiform records, assembled and critically studied by Oppenheim (1970), which are full of obscure terminology.

(a) Modern Analyses:

(1) General:

All the analyses of ancient glass undertaken between 1790 and 1957 have been assembled and critically assessed by Caley (1962). He pointed out a number of general conclusions emerging from the narrow data-base available to him for Mesopotamian glasses: the magnesia content was so high that these may properly be called soda-lime-magnesia glasses; differences in the ratio of alumina to silica might be due to a contrast between the quality of quartz sand used in Assyria and Babylonia; the ratios of magnesia to lime were decidely higher than in contemporary Egyptian glass; generally the same colouring agents appear in Mesopotamian as in Egyptian glass; a mixed soda-lime-lead type of glass had appeared in Mesopotamia by at least the eight century B.C. Caution is necessary with some early (and current) analyses, since weathered specimens were analysed, faute de mieux, and errors thereby introduced.

The flow of information in recent years has much increased; but not as significantly as in other aspects of glass studies, since suitable samples of Mesopotamian glass are not readily accessible to scientists for analysis. Garner's (1956) analysis of the Early Eridu fragment (see p.196) was an isolated piece of research. In 1969, Brill, in reviewing scientific investigations of ancient glasses, brought forward his own work on glass from Tell Rimah and from Nimrud as part of the survey. He pointed out the potential of such techniques as lead and oxygen isotope analysis for characterising early glasses and, more particularly, for grouping glasses (see p.113). This had been the intention of the pioneers using chemical analysis; but for the moment the available information is too meagre, uneven and scattered to make any real and reliable contribution to source studies.

Most progress has been made in the study of composition, notably in establishing the identity of colourants, opacifiers and clarifiers. In 1963 Sayre published a major study demonstrating that both antimony and manganese had been used intentionally as colourants or opacifying agents in glass manufacture long before they were used as decolourants. He produced the first indication that in the time of the Neo-Assyrian Empire, by at least the seventh century B.C., the role of antimony as an oxidizing agent was well appreciated (Sayre, 1963, table III). Antimony changes the state of oxidization of the iron in the glass mix from the ferrous state (a pale green tint) to the ferric state, in which iron causes only a pale yellow shade and yields an apparently colour-free glass. More recently Vandiver (1982a, 1983) has published analyses of glass from Nuzi; Brill (1969) and Pollard (1982) of that from Rimah; Pollard (1983) of that from Failaka; and Brill has contributed particularly to studies of glass from the Neo-Assyrian levels at Nimrud (1969, 1970, 1978). In 1970, when contributing to Oppenheim's philological study of glass, Brill assembled the analyses of Mesopotamian glasses, by then about twenty of them; but their variety was such that in a comparative table he sought only to describe "a hypothetical base glass to which additional colorants and colorant-opacifiers as listed separately in the table, would have been added" (see tables on pp.219-20).

(2) The Mitannian and Middle Assyrian periods: (c. 1550-1150 B.C.):

Vandiver (1983) has published a technical study of the Nuzi glass. Her representative compositions (see pp.221-2) fall within the second millennium B.C. category of relatively high K_2O and MgO established by Sayre and Brill; a tendency to higher MgO:Ca ratios (0.3-0.8) may be explained by the dolomitic lime in local Nuzi clays. The glazes on clay vessels at Nuzi have the same characteristic (MgO:Ca: 0.5-0.8). Vandiver (1983, 240) identified three mechanisms for colour formation at Nuzi. In one, solution of ions into the base glass applied: translucent turquoise blue was formed by absorption of colours other than blue by copper ions, requiring melting in an oxidizing atmosphere. In another, precipitation processes operated: red formed by the scattering of light from submicron size, droplet-shaped particles of cuprite, which precipitate in the glass melt in a reducing atmosphere over a range of temperature. Finally, there was the inclusion of unmelted particles: lead and calcium antimonates for yellow and white which require oxidization and relatively low temperature melting. Vandiver (1983, 245) suggested that the black and amber glasses were coloured by the sulphur-iron chromophore used today to colour beer-bottles brown.

COMPOSITIONS OF MESOPOTAMIAN GLASSES (After Brill, 1970, 122)

	Typical Values (and Ranges) Found Upon Analysis[1]			Hypothetical "Base Glasses"[2]			Calculated Composition For zukû[3]
SiO_2	55	64%	68	56	66%	70	56.0%
Na_2O	13	16	19	14	17	20	23.8
CaO	4	6.5	8	4	7	10	6.6
K_2O	1	1.5	3	1	1.5	3	3.8
MgO	3	4.5	6	3	4.5	6	5.6
Al_2O_3	1.5	2	3	1.5	2	3	2.2
Fe_2O_3	0.5	1	1.5	0.5	1	1.5	0.7
TiO_2		0.1			0.1		0.05
MnO		var.			0.02		0.05
Sb_2O_5		var.			0.000X(?)		nil
PbO		var.			0.000X(?)		nil
BaO		0.1			0.1		0.1
SrO		0.1			0.1		0.1
P_2O_5		0.3			0.3		1.0
SO_3		0.3			0.3		0.5(?)
Cl		0.3			0.3		0.5(?)

Calculated Semi-Quantitative Composition of A Red Opaque Glass	
SiO_2	40%
Na_2O	18
CaO	5
Cu_2O	13
PbO	12
Sb_2O_5	1
"anzaḫḫu"	1
Other[4]	10

Other elements sometimes found in the range of 0.000X-0.2 in ancient glasses from various sources: B_2O_3, Li_2O, Rb_2O, V_2O_5, Cr_2O_3, ZrO_2, NiO, ZnO, Ag_2O, SnO_2, Bi_2O_3.

1. Excluding red opaques with high PbO contents. Values in italics show approximate maxima and minima.

2. To be modified substitutionally by colorants and colorant-opacifiers.

3. SO_3 and Cl uncertain due to unknown extent of retention.

4. Includes K_2O, MgO, Al_2O_3, Fe_2O_3, etc.

Var. = variable, depending upon colour of glass.

COLORANTS AND COLORANT-OPACIFIERS
(after Brill, 1970, 122)

CuO	1-2%	blue transparent
CoO	0.05-0.2%	blue transparent
MnO	1.5%	purple transparent
Fe_2O_3	2-3%	green transparent
$Fe(S)_x$	2-3%	amber transparent
Mixture of above	5%	black "transparent"
Cu_2O	~5, ~10, ~15%	red opaque
PbO	0.0X, ~3, ~10, ~20%	
SnO	nil, .X%	
Sb_2O_5	.X, 2%	
$Ca_2Sb_2O_7$	(2-5% Sb_2O_5)	white opaque
$Pb_2Sb_2O_7$	(2-5% Sb_2O_5)	yellow opaque
$Ca_2Sb_2O_7$ + CuO/CoO		light blue opaque (turquoise)
$Pb_2Sb_2O_7$ + CuO/CoO		green opaque

Kaczmarczyk and Hedges (1983, 287) cited Sayre's complete analyses of five thirteenth century B.C. glass samples (2 white; 2 black; 1 blue), and a glazed brick, from Tchoga Zanbil. The white glass was a very pure soda-lime-silicate material without any identifiable opacifier. As the blue glass contained less than 0.02% CoO and less than 0.04% MnO, the colour was produced by copper and iron. Both of the black glasses were coloured with reduced iron and not manganese, as their iron oxide contents were far in excess of the trace amounts of manganese. None of the glasses contained much lead; 0.07% PbO in the blue glaze was the highest. Only one, the blue glass, had a detectable opacifier, 2.2% Sb_2O_5 (cf. Brill, 1970, 116ff.). Neumann (1927; cf. Caley, 1962) analysed two specimens of blue glass from "Kassite" Nippur. One was turquoise coloured and contained no lead or cobalt; but its copper was accompanied by some tin (2.6% CuO; 0.32% SnO_2). The darker blue piece was coloured by cobalt, which was accompanied by some manganese (0.93% CoO; 0.65% MnO). As no nickel or zinc was reported, and cobalt exceeded manganese, this was probably Iranian cobalt comparable with that in the Eridu glass fragment (p.196).

Pollard (Pollard and Moorey, 1982) has analysed some of the "Mitannian/Middle Assyrian" glasses from Tell Rimah. These samples showed calcium concentrations in the range 2-10%, slightly higher than the 4-8% previously reported for Mesopotamian glass (Brill, 1970, 122); but this might be explained by corrosion products on the surface. Antimony was present in all objects tested. In the fragment of a mosaic glass vessel it was present in all colours, with copper in the opaque green areas, manganese and copper in the black. This does not support Dayton's view (1978) that the copper was associated with black glass, antimony with white. The black areas of the mosaic glass had relatively less Mn and more Fe and Cu than the colouring of black faience.

The use of cobalt in blue glass in Mesopotamia is fully documented as early as the Ur III period at Eridu (Garner, 1956). In published analyses of Mesopotamian glasses cobalt is uncorrelated

Representative Nuzi Glass Compositions Determined by Microprobe (after Vandiver, 1983, 243)

Oxide	Corning Glass Standard A	Average and Range of five Analyses		Black Glass 30-2-195		Blue Opaque Glass 30-2-7-1		Blue Translucent Glass M79/1	
SiO_2	66.40	67.23	(64.97-70.55)	69.71	(68.71-72.10)	71.41	(70.27-72.47)	69.95	(67.51-70.95)
Al_2O_3	1.14	2.21	(1.81-2.93)	1.45	(1.51-1.65)	.58	(0.51-0.65)	.62	(0.53-0.72)
Na_2O	14.40	12.53	(11.67-13.01)	15.50	(15.49-15.99)	10.40	(7.74-12.08)	15.08	(14.03-15.28)
K_2O	2.88	2.75	(2.30-2.92)	2.34	(2.0-2.59)	1.31	(1.16-1.60)	2.32	(2.34-2.46)
CaO	5.29	5.35	(4.92-5.96)	3.42	(2.92-3.76)	4.82	(4.56-5.29)	3.22	(3.13-3.40)
MgO	2.75	2.65	(2.57-2.75)	3.82	(3.71-3.76)	2.90	(2.82-2.97)	6.11	(6.09-6.22)
Cu_2O									
CuO	1.22	1.75	(1.74-1.76)	.01	(0.00-0.05)	2.13	(2.05-2.20)	1.63	(1.51-1.72)
TiO_2	.80	1.00	(0.82-1.67)	.38	(0.12-0.83)	.00	(0.00)	.00	(0.00)
Fe_2O_3	1.07	1.17	(1.09-1.23)	.88	(0.82-0.94)	.33	(0.30-0.36)	.35	(0.30-0.37)
Sb_2O5	1.76	1.74	(1.59-1.86)	.00	(0.00)	1.55	(0.56-2.73)	.00	(0.00)
PbO	.05	.06	(0.00-0.11)	.02	(0.01-0.03)	.05	(0.00-0.03)	.00	(0.00-0.01)
TOTAL	97.76	98.44	(97.32-100.11)	97.62	(97.13-98.85)	96.12	(95.39-97.43)	99.28	(97.99-99.57)
OTHER ELEMENTS				S and Cl present, could not detect Cu, Mn, Co, or Zn in excess of 0.01%		Cl present Ca-Sb Inclusions 0.01-1.0 micron		Cl present	

By emission spectroscopy, the following elements were found in quantities of less than 0.1%: Ag, B, Cr, Mn, Ni, Pb, Sn, Sr, and Ti.

Representative Nuzi Glass Compositions Determined by Microprobe Analysis (After Vandiver, 1983, 244)

Oxide	Red Opaque Glass 30-2-414		Yellow Opaque Glass 30-2-414		White Opaque Glass H 12-13		Blue Translucent Glaze on Clay Body 31-1-162	
SiO_2	63.50	(62.20-67.62)	63.60	(62.33-64.51)	68.36	(68.26-69.66)	51.46	(49.00-53.36)
Al_2O_3	1.63	(1.38-1.82)	1.35	(1.25-1.45)	.55	(0.53-0.58)	10.10	(9.27-10.90)
Na_2O	15.04	(14.79-15.14)	14.55	(14.38-14.95)	13.21	(12.64-13.75)	12.37	(11.67-13.34)
K_2O	4.15	(4.07-4.19)	4.17	(4.15-4.19)	2.75	(2.66-2.92)	3.31	(3.18-3.62)
CaO	5.60	(5.68-5.72)	5.48	(5.44-5.50)	8.42	(7.79-8.56)	8.93	(8.20-9.91)
MgO	3.61	(3.57-3.63)	3.48	(3.41-3.56)	4.55	(4.38-4.92)	4.23	(4.08-4.36)
Cu_2O	3.58	(3.37-3.82)						
CuO			.09	(0.08-0.10)	.14	(0.11-0.17)	1.37	(1.24-1.47)
TiO_2	.16	(0.15-0.17)	.00	(0.00)	.00	(0.00)	.80	(0.70-1.01)
Fe_2O_3	3.32	(3.37-3.82)	.96	(0.80-1.04)	.22	(0.20-0.25)	4.35	(4.17-4.38)
Sb_2O_5	.00	(0.00)	1.95	(1.75-2.06)	2.05	(1.90-2.68)	Not measured	
PbO	.40	(0.40-0.45)	4.35	(4.20-4.53)	.00	(0.00-0.02)	.15	(0.11-0.20)
TOTAL	100.99	(99.32-101-59)	99.98	(97.83-100.70)	100.25	(99.46-100.88)	97.07)	(95.21-99.98)
OTHER ELEMENTS	S and Cl present, Cu rich inclusions		Cl and Mn present, Pb-Sb inclusions		Cl present. Ca-Sb inclusions			

with MnO or Fe_2O_2 and is seldom accompanied by more than trace amounts of nickel or zinc. This is in marked contrast to the cobalt used in Egyptian faience, glass and pottery pigments in the New Kingdom. Kaczmarczyk (pers.comm.) believes that cobalt in the alum from the Dakhla Oasis is invariably accompanied by all of these elements and is the likely source for Egyptian glass makers in the New Kingdom. Mesopotamia (and Egypt in the first millennium) is likely to have drawn on the rich deposits of high grade cobalt in Iran. When manganese appears with the cobalt in Mesopotamian glass it is likely to have been introduced independently as a clarifier (cf. Sayre, 1963). On Failaka in the Kassite Period two types of blue glass were current, one coloured only by copper, the other with 0.10% cobalt in addition, to develop a deep blue colour (Pollard; pers.comm.).

(3) The Neo-Assyrian Period (c.1000-600 B.C.):

Brill has studied (1970; 1978) the composition of the colourless glass from Neo-Assyrian levels at Nimrud (see p.208), both that used for vessels, whose origin is uncertain, and that for the miniature painted glass plaques, whose patterns are "Phoenician". They are the usual soda-lime-silica type of glass. The levels of potassium and magnesium are high enough to place them in the high-potassium high-magnesium category defined by Sayre (1963), as would be anticipated for "Mesopotamian" glass. They also contain antimony as a decolourizer, perhaps also as a fining agent to remove some of the bubbles from the glass (Brill, 1970, 116ff.). A "Phoenician" glass plaque, which Brill examined, was unusual in having 2.6% lead oxide. This would be explicable in a coloured glass, but its role in colourless glass is less easily explained (Brill, 1978, 29-30). On the painted plaques the outline was done with a black paint (some bituminous substance), whilst a finely ground Egyptian blue frit was used to colour certain selected details of the paintings (Orchard, 1978, 6). It is just possible, though this was by no means certain, that the plaques had originally been painted with white lead to provide a background for the designs.

Kühne (Haevernick and Kühne, 1982) has published an analysis of a sky blue glass mace or staffhead from Assur:

H_2O:	6.2%	CaO:	10.9%
SiO_2:	59%	MgO:	4.6%
Al_2O_3:	0.75%	Na_2O:	16.5%
TiO_2:	0.13%	CuO:	0.6%
Fe_2O_2:	0.55%	Mn_2O_3:	0.04%

A glass seal, from level II at Tell Taya (Neo-Assyrian), which may not have been of local manufacture, had copper not cobalt as the blue colourant (Bimson, 1973).

(4) The Achaemenid period (c.550-330 B.C.):

In view of the wide distribution of standard types of fine glassware throughout the Persian Empire (Oliver, 1970), Matson's report (1957) on two fragments of colourless glass vessels from Persepolis has a wider relevance than might at first be apparent. They were of the normal soda-lime-silica composition with low magnesium and with antimony added as a clarifier. Matson examined a bead, a slab and a lump of opaque red glass from Persepolis. They had a complex composition in which several ingredients, including copper, iron and antimony, contributed to the development of the colour (Matson, 1957, 131-2; cf. p.228 below for more recent work on the manufacture of red glass).

F: Provenance Studies:

Although Egyptian glasses often have a high magnesia content and thus a high magnesia: lime ratio (MgO:CaO), this ratio is much higher in Mesopotamian than in Egyptian glass. It is on this fact that Matson based his claim that the colourless glass he analysed from Persepolis (see above) was "made in Egypt in the vicinity of Cairo or in the Delta, since its MgO content is very low" (Matson, 1957, 129). With the present restricted data-base such statements must necessarily be conjectural.

G: The ancient textual tradition:

One of the most important collaborative studies of ancient Mesopotamian technology is Glass and Glassmaking in Ancient Mesopotamia (Chicago, 1970), initiated by the philologist Oppenheim, whose edition of a series of cuneiform "glass" texts inaugurated the volume. To this Brill contributed a scientific commentary. Two historical and typological studies of glass vessels and other objects, by Barag and von Saldern, completed the volume. Each of the four sections is an important independent essay, but, significantly, they are not complementary. "The glass objects do not ´illustrate´ the history of this particular class of cuneiform literature and the texts are useless in helping us trace the typological history of early Mesopotamian glass vessels" (Barag, 1970, 134). Nor, he might have added, do they offer information on techniques of artefact manufacture or on workshop organization. Nor even do they offer a very satisfactory guide to "glass" composition, as Brill´s scrupulous but inevitably faltering attempts to replicate the materials they describe vividly illustrates. This is a particularly clear illustration of a recurrent disjunction in the history of early technology: ancient texts served administrative or literary purposes, they did not represent a body of technical or "scientific" literature. They were far removed from practical knowledge and chemists interested in ancient glass technology have been heard to describe them as "useless" for reconstructing ancient methods and recipes. This may be an extreme view; but it highlights the problem.

The difficulties run deep, as Oppenheim clearly pointed out. Current understanding of these texts is, at the most basic level, faulty and incomplete. They employ a large number of rare or unique words, refer obscurely to chemical procedures, and lack a "scientific" mode of presentation. The current status of the texts is not assisted by their complicated historical development, which has to be unravelled from internal evidence. Oppenheim believed them to be predominantly the work of seventh century scribes copying earlier records, some of which might be over a millennium older. A preliminary critical review of Oppenheim´s approach was provided by Muhly (1971). Foster (1979, 15ff.), presumably with the specialist philological assistance of her husband, B.R. Foster (Foster, 1979, XVIII), offered a critical summary of Oppenheim's interpretation of the glass texts. This included significant modifications in his translation of key words (Foster, 1979, 21) that had themselves often differed from definitions given in volumes of the Chicago Assyrian Dictionary, published before 1970, with which Oppenheim was associated.

The main corpus of these "glass" texts came from the "Library of Ashurbanipal" (c.664-627 B.C.) at Kuyunjik (Nineveh). Oppenheim argued that they had belonged to a larger body of texts offering "procedural instructions" in mathematics and astronomy, in ritual and in medicine, in chemistry ("glass" and perfumes), and in horsemanship. The original chemical and equestrian texts he attributed to scribes working in Assur in the last third of the second millennium B.C., at the height of the Middle Assyrian Empire. The "glass texts" seemed to him to represent an Assyrian tradition, though references to past masters of the craft may imply a history extending back into at least the early Middle Babylonian period, in the middle of the second millennium B.C. or soon after, and to a complex geographical origin for some of the techniques involved (Oppenheim, 1970, 51-3). The surviving Middle Babylonian "glass" texts represent the same technical tradition, but date only between the fourteenth and twelfth centuries B.C. Oppenheim demonstrated that they cannot confidently be taken back to the later Old Babylonian period, as had previously been assumed.

Great caution is still necessary in explaining the precise nature of the materials whose composition is described in these texts. Brill (1970, 108) put the problem succinctly "...there is sufficient ambiguity in the translations of the texts to leave room for the possibility that the materials being prepared were glassy faience, Egyptian blue, or some related but not yet clearly defined material". But he favoured the opinion that the materials being prepared actually were true glasses (i.e. fused, not sintered, compositions). The proportions, he argued, of the major ingredients called for, and the presence of antimony, copper and lead, all relate well to the known chemistry of Mesopotamian glass and, so far as translation permits, there is no sign of inappropriate materials among the major constituents. Muhly (1971, 181-2) was more sceptical, whilst Foster (1979, 16ff.), although she believed that the majority of texts referred to true glass manufacture, assumed that many of Oppenheim's "primary, alkali-silicate glasses" were in fact faience; on the terminology used in this volume (and generally in her book) this is imprecise. The alternatives, as Brill appreciated, are frits or Egyptian blue. These substances are manufactured chemical intermediaries, to be used as starting points in complex processes leading to the production of coloured materials imitating semi-precious stones; but whether for manufacture, as glass artefacts, or to glaze terracotta or sintered quartz was information not appropriate to the scribes' purpose and therefore not explicitly stated (cf. Muhly, 1971, 181-2).

These texts are part of a traditional Mesopotamian literary enterprise: the compilation of lexical lists. As these had always included terms for natural stones, the artificial manufacture of such stones (once it became a regular industrial activity) was naturally of interest. Thus the "glass" texts, in accord with this long-standing purpose, concern themselves with little more than the appropriate technical vocabulary. As the technical procedures themselves were of little interest, and probably unknown, to the scribes, their descriptions are rudimentary; sufficient for the purposes of lexicography, but inadequate as descriptions of an actual technology. Literary tradition is a notoriously inadequate source for the history of technology. Despite Oppenheim's (1970, 83ff.) attempt to see an historical development in what his texts describe, (complementary to his analysis of the historical evolution of the textual tradition itself), it seems they deal with methods for the manufacture of imitation coloured stones that changed little between about 1850 and 700 B.C. For centuries before the emergence of a fully fledged glass industry (c.1600-1500 B.C.) this had been done either by glazing dark-coloured stones or by making replicas in faience (see pp.133ff.). This craft tradition continued, even after the growing mastery of glass in the middle of the second millennium B.C., though the new material permitted closer imitation of the original stones. Since faience and glass workshops were closely associated, as at Nuzi and Assur, both materials came to be used to imitate stones.

Modern scientific examination has only just begun to reveal the complex history of "glass" and "glassy" materials in antiquity. There were certainly more categories, particularly after about 1500 B.C., than the modern descriptive terms "faience", "frit", "glass paste", "glassy faience" and "glass" can adequately describe. The only practical application referred to in the Oppenheim corpus of texts is the glazing of bricks.

Oppenheim, using the lexical lists of the Old Babylonian period (c.1850-1600 B.C.), isolated various terms for what he described as "primary glasses", though their characters differ, almost to the point of confusion:

"I propose to call the glasslike substances which appear in Akkadian as *anzaḫḫu*, *kutpû*, *ḫuluḫḫu*, *ḫuḫû*, as well as *būṣu* (see p.16) and *tuzkû*, primary glasses, each specific product being apparently endowed with definite chemical properties. Some of them - such as *anzaḫḫu* and *būṣu* - can be used to produce glass objects, from simple beads to vessels in the core technique, others such as *zukû*, and *tuzkû* - are attested only as intermediary products used to manufacture opaque colored glasses meant to imitate precious stones. In contra distinction to *anzaḫḫu* and *kutpû*, the composition called *ḫuluḫḫu* and *ḫaḫu* (also *ḫuḫu*) do not occur in glass texts - the latter is once explained as slag from the potter's kiln." (Oppenheim, 1970, 19).

Two points are important in considering this paragraph. None of these materials may yet be reliably identified in the surviving material culture and the evidence for a glass industry in Mesopotamia before 1500 B.C. is meagre (see pp.194ff.). At the earliest stage such a vocabulary is likely to refer either to aspects of frit production, to glaze manufacture for application to sintered quartz (not yet to terracotta), and to small scale production of glass beads.

The given recipes for the manufacture of Oppenheim's "primary glasses" involve grinding, combining, and firing several sets of named ingredients. The two major constituents are:

Immanakku: "like river silt dotted with pebbles" was taken by Oppenheim (1970; 1974) as a general term for a hard conglomerate (silicate stones or coarse sand); Brill (1970, 109ff.) identified it as "quartzite pebbles"; (Foster (1979, 21) as "quartzite, perhaps sand".

NAGA (ahussu): a plant used also in bleaching linen and in soap production; this, or rather its ashes, provided the necessary alkali.

Together these produced zukû, in which U.BABBAR (literally "white plant") was also mixed. This word is found in medical texts referring to what may be the sap of the poplar tree, readily made into a paste or adhesive to be used as a binding agent (Foster, 1979, 16).

Brill (1970, 111) attempted a laboratory synthesis of Zukû (without U.BABBAR):

"The product that resulted was not only a glass, but a glass of unexpectedly high quality... It was entirely transparent, but had an amber color..." (Brill, 1970, 113). Further experiment, in an oxidizing atmosphere or using an oxidizing agent, removed the amber, leaving a pale blue colour.

The texts describe the heating of the basic ingredients in a kiln until sintering occurs; both the mineral and the alkali had been ground before the first sintering, and it was ground again before reheating. The finished product was poured onto a kiln-fired clay slab to cool off. Zukû was only a basis. Its principal derived product was tersītu made by grinding zukû and throwing it upon a previously heated mass of the unidentified URUDU.HI.A nēhu (literally "slow copper") to introduce a blue colour. This copper colouring compound might have been a blue frit, perhaps Egyptian Blue, (Oppenheim, 1970, 77), or a metallic alloy (Brill, 1970, 123). The term "fast copper" or "fast bronze" is also used to describe it. Tersitu was in turn an intermediary, for the manufacture of opaque imitation lapis lazuli (zagindurû); this transformation was made by combining tersītu with būṣu, anzahhu and namrūtu. Būṣu was another of Oppenheim's "primary glasses", made in a similar manner to zukû. It was used in the later second millennium B.C. for tableware, inlaid statuettes and oil containers (C.A.D. s.v.). As anzahhu, yet another of Oppenheim's (1970, 19) primary glasses, has a close, at times synonymous relationship to the mineral immanakku, it might be a mineral, a type of quartzite (cf. Foster, 1979, 17), rather than a manufactured material. It can be "washed" or "unwashed" and more enigmatically "male" or "female", perhaps (on classical analogies) distinguishing a light from a darker colour. Namrūtu was white calcinated material made from red coral.

H: Red glass: the ancient texts and modern analyses:

Zukû was equally an important intermediary stage in the production of red glass, distinguished as "Assyrian" and "Elamite"; the former requiring that it be mixed with lead, antimony, copper and anzahhu; the latter involving combination with tuzkû, lead, antimony, and anzahhu. Tuzkû, another "primary glass" in Oppenheim's terminology, consisted of mekku, for Oppenheim a "glass" for others a mineral (perhaps "powdered quartzite": Foster, 1979, 21), to which was added minute quantities of kalû/kalgūgu ("red earth") as the colourant to be fused with small additions of sand and alkali. Foster (1979, 21) renders these as "cobalt compounds" and "potash" respectively. It is production of red glass that is described in the best preserved of the known "glass texts" (British Museum: 120960), dated by Oppenheim (1970, 59ff.) to the Middle Babylonian period.

This text has been well known since its first publication by Gadd and Campbell Thompson in 1936. It has attracted wide attention in non-Assyriological literature on account of its supposed protection of craft secrets by being written in a cryptographic manner. Oppenheim (1970, 59ff.) cogently rejected "such facile and unfounded assumptions" and provided a new edition. Only a few words remain unintelligible and, though the syllabary used by the scribe is often atypical, this is probably a display of erudition (for only fellow scribes could have read it anyway), not a device for ensuring secrecy. A more relevant problem is provided by the scribe's dating of his text to the "year (after that in which) Gulkishar became king". This obscure ruler was the sixth in the "First Dynasty of the Sealand" in the sixteenth century B.C. Oppenheim argued that the reference was an anachronism, arising from a desire to have the tablet accepted as older than it was; but still "the question is not whether the date of the tablet is correct or not, but rather whether the dating and the entire subscript refers to the original of the text copied or whether it dates the copy" (Oppenheim, 1970, 61). Oppenheim answered this question rather obliquely by assuming that the scribe of this tablet, writing it, perhaps in Babylon, in the fourteenth to twelfth centuries B.C., used the name of an old and famous scribe who had lived in the time of Gulkishar. It is then a "Middle Bbylonian" text, which might well describe earlier craft practices.

The first two sections give the materials and quantities needed for two types of red glass; in the largest section the constituents are combined. At the end, in an intriguing indication of the complex realities of such work, the craftsman is told not to worry if "inside the cover" the described result is not achieved. A remedy is offered. In 1948 Moore published his attempt to repeat this recipe (following the old translation) in a modern laboratory; he believed it to be for a glaze for application to a clay body. In Oppenheim's revised rendering of lines 7-11 they refer to melting the batches in crucibles and the kiln procedures for this, not to the preparation of clay bodies for glazing, as had earlier been supposed. Indeed the recipe as now translated gives no indication of the purpose for which the red glass/glaze was being made. Although the stages proposed seem unnecessarily complicated to a modern craftsman, insecure temperature controls would have required them in antiquity. "Fine grinding, followed by remelting, would assist in ensuring the complete solution of any material left unincorporated after the first melting, and would also give a melt containing much less bubble, besides ensuring greater uniformity of colour" (Moore, 1948, 27).

Brill (1970, 121), following Oppenheim's new edition, attempted to repeat the instructions for making a red opaque glass: "...the ingredients include both those which are essential for making a red opaque glass and those which are beneficial in making them; and the chemical treatments such as the heating procudures are appropriate for making red glasses. Moreover, the chemical composition of the product prepared is consistent, in a semi-quantitative sense, with analyses of some excavated specimens of red glasses".

Bimson and Freestone (in Barag, forthcoming) have re-studied pieces of early red glass in the British Museum collections. They have significantly increased knowledge of this type of glass and earlier analytical studies of it must consequently be used with caution. This type of glass has long fascinated students of early glass since its technique of manufacture seems to have no close parallel in modern glass technology. Opaque red glass is known as early as the fifteenth century B.C. at Alalakh in Syria (Barag, forthcoming, nos. 180, A) and at Tell el-Amarna in Egypt a generation or so later (Lucas, 1962, 189-90). Then again from the eighth century B.C. it was used for inlays, and for beads and amulets, in western Asia. Bimson and Freestone have demonstrated the existence of two main types:

1. "Cuprite" glasses, owing their colour to crystals of cuprite (cuprous oxide: Cu_2O); those with well developed crystals have a distinctively bright sealing-wax red colour. The early cuprite glass at Alalakh differs fom the later, first known at Arslan Tash in Syria and at Samaria in Palestine, and in Urartu, in the eighth century B.C., in that it contains negligible quantities of lead. Sometime in the second quarter of the first millennium B.C. it was discovered, perhaps in the Syro-Phoenician glass industry, that the addition of lead made successful manufacture of large ingots of well coloured cuprite glass much easier. The presence of plumbous lead (PbO) seems to assist the stability of the cuprous state, whilst the greater solubility of copper in lead glass favours the production of a good cuprite glass. The presence of antimony (about 2.5%) suggests that it may have played a role similar to that of the lead.

2. "Copper red" glasses are duller in colour. The colour is also due to copper but in a colloidal (non-crystalline) state. They contain much less copper than the "cuprite" glasses (1 to 3% as against 5 to 20%) and are already evident in Egypt at the time of the XVIIIth Dynasty.

I: The problem of enamel:

A longstanding controversy over whether or not true enamel (powdered glass fused *in situ*) was produced in Egypt during the New Kingdom has highlighted the complexities of attempting to identify enamel by eye, particularly through the glass of showcases in museums (Teeter, 1981). Various types of stone or glass inlay, and the manipulation of pliable glass secured by adhesives, may all too easily appear to be true enamel. As it is relatively easy to place powdered glass in the areas to be decorated on a piece of goldwork and then fire it in a furnace or over a charcoal fire, students of ancient technology have been surprised by the apparent absence of true enamel in areas where glass technology was innovative and where jewellers were skilled in the production of fine jewellery inlaid with coloured stones or pieces of glass or faience. But the presence of true enamel in jewellery of the New Kingdom in Egypt remains an open question; earlier occurrences are even less probable. It is more generally agreed that towards the end of the local Late Bronze Age in Cyprus, in the thirteenth century B.C., true enamelling was practised on the Island for the decoration of gold earrings and a sceptre (cf. Åström, 1967, 102). In the continuing absence of certain use of the technique in contemporary Egypt and Syro-Palestine, the source of the expertise is usually thought to have been Mycenaean (cf. Higgins, 1980, 24ff.).

The history of true enamel in ancient Mesopotamia is wholly obscure for want of well authenticated examples. Indeed in the whole of Western Asia only one object is commonly said to be decorated with enamel before the Hellenistic period. This is a gold diadem, with petals enamelled in yellow and green, associated with the hoard from Ziwiyeh in Kurdistan, usually dated to the later eighth or early seventh century B.C. (cf. Maxwell Hyslop, 1971, 214-5). Even in this case the identification is not scientifically documented and the integrity of this hoard is suspect (Muscarella, 1977).

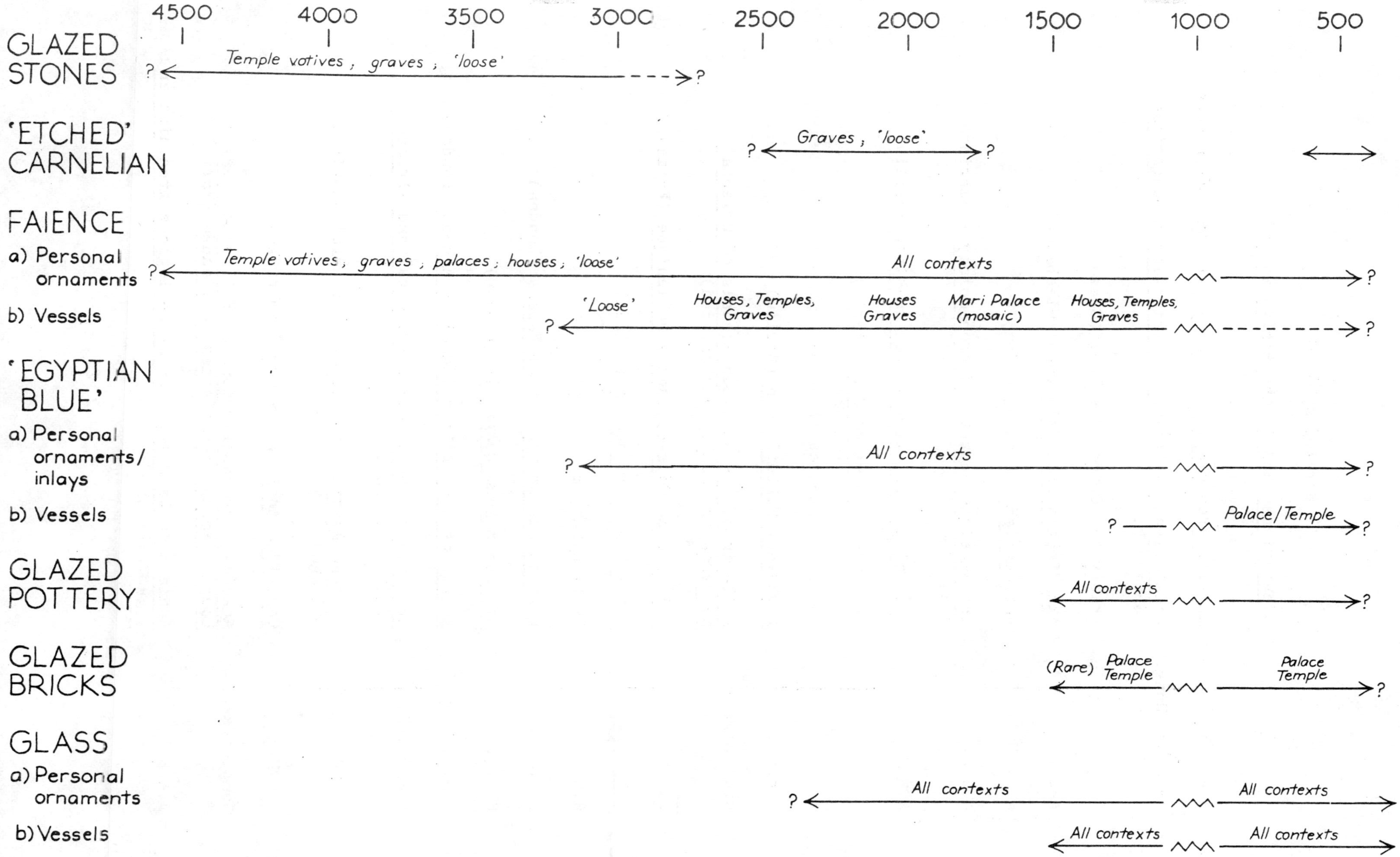

Table II : Diagram of the Chronological Distribution of Glazed Materials and Glass.

Bibliography

Abbreviations used in the Bibliography

A.A.A.S.	Annales archéologiques arabes syriennes (Damascus)
Acta Or. Hung.	Acta orientalia academiae scientiorum Hungaricae (Budapest)
A.f.O.	Archiv für Orientforschung (Graz)
AHw	Akkadisches Handwörterbuch (Wiesbaden)
A.J.	Antiquaries Journal (London)
A.J.A.	The American Journal of Archaeology (Cambridge, Mass.)
A.M.I.	Archäologische Mitteilungen aus Iran (Berlin)
A.R.M.	Archives royales de Mari (Paris)
A.S.	Anatolian Studies (London)
Bag.Mitt.	Baghdader Mitteilungen (Berlin)
B.A.S.O.R.	Bulletin of the American Schools of Oriental Research (Jerusalem and Baghdad)
B.C.H.	Bulletin de Correspondence Hellénique (Paris)
Bib.Or.	Bibliotheca Orientalis (Leiden)
B.M.	British Museum, London
B.M.O.P.	British Museum Occasional Papers (London)
B.M.Q.	British Museum Quarterly (London)
B.S.A.	Annual of the British School at Athens (London)
B.S.O.A.S.	Bulletin of the School of Oriental and African Studies (London)
C.A.D.	Chicago Assyrian Dictionary (Oriental Institute, Chicago)
C.A.H.	Cambridge Ancient History (Cambridge)
Comptes Rendus	Académie des Inscriptions et Belles-Lettres: Comptes Rendus des séances de l´année (Paris)
D.A.F.I.	Cahiers de la délégation archéologique francaise en Iran (Paris)

E.A.	J.A. Knudtzon, *Die El-Amarna-Tafeln* (Leipzig, 1915)
I.A.	*Iranica Antiqua* (Leiden)
I.E.J.	*Israel Exploration Journal* (Jerusalem)
J.A.O.S.	*Journal of the American Oriental Society* (New Haven)
J.Arch.Sc.	*Journal of Archaeological Science* (Academic Press)
J.C.S.	*Journal of Cuneiform Studies* (New Haven)
J.E.A.	*Journal of Egyptian Archaeology* (London)
J.E.S.H.O.	*Journal of the Economic and Social History of the Orient* (Leiden)
J.F.A.	*Journal of Field Archaeology* (Boston, U.S.A.)
J.G.S.	*Journal of Glass Studies* (Corning, U.S.A.)
J.H.S.	*Journal of Hellenic Studies* (London)
J.N.E.S.	*Journal of Near Eastern Studies* (Chicago)
J.O.S.	*Journal of Oman Studies* (Muscat)
J.R.A.S.	*Journal of the Royal Asiatic Society* (London)
L.A.A.A.	*Liverpool Annals of Archaeology and Anthropology* (Liverpool)
M.A.S.C.A.	*Museum Applied Science Center for Archaeology* (University Museum, Philadelphia)
M.D.O-G.	*Mitteilungen der Deutschen Orient-Gesellschaft* (Berlin)
M.D.P.	*Mémoires de la délégation en Perse* (Paris) (including those volumes with slightly different titles in the same sequence)
M.J.	*Museum Journal* (Philadelphia)
Mon.Piot.	*Monuments et Mémoires (Foundation Piot)* (Paris)
Num.Chron.	*Numismatic Chronicle* (London)
O.I.C./O.I.P.	*Oriental Institute Communications/Publications* (Chicago)
Opus.Ath.	*Opuscula Athiensia* (Stockholm)
O.L.Z.	*Orientalistische Literaturzeitung* (Berlin)

P.A.C.T.	Journal of the European Study Group on Physical, Chemical and Mathematical Techniques Applied to Archaeology (Strasbourg)
P.P.S.	Proceedings of the Prehistoric Society (Cambridge)
P.S.B.A.	Proceedings of the Society of Biblical Archaeology (London)
R.A.	Revue d´assyriologie et d´archéologie orientale (Paris)
R.A.S.	Revue d´Archéometrie (Rennes)
R.D.A.C.	Report of the Department of Antiquities, Cyprus (Nicosia)
R.G.Z.M.	Römisch-Germanischen Zentralmuseums, Mainz.
R.L.A.	Reallexikon der Assyriologie und vorderasiatichen Archäologie (Berlin)
U.E.T.	Ur Excavations: Texts (London and Philadelphia)
W.V.D.O-G.	Wissenschaftliche Veröffentlichung der Deutschen Orient-Gesellschaft (Leipzig)
W.O.	Die Welt des Orients (Göttingen)
Z.D.P.V.	Zeitschrift des Deutschen Palästina-Vereins (Wiesbaden)
Z.f.A.	Zeitschrift für Assyriologie und vorderasiatische Archäologie (Berlin)

Bibliography

Adams, R. McC.
1981 Heartland of Cities: surveys of ancient settlement and land use on the central flood plain of the Euphrates (Chicago)

Adams, R. McC. and Nissen, H.
1972 The Uruk Countryside (Chicago)

Agrawal, D.P.
1982 "The Indian Bronze Age Cultures and their metal technology" in Wendorf, F. and Close, A.E. (eds.), Advances in World Archaeology I (Academic Press), 213ff.

Aitchison, L.
1960 A History of Metals (Two vols.; London)

Al-A´Damai, K.A.
1968 "Excavations at Tell es-Sawwan (Second Season)" Sumer 24, 57ff.

Alexander, S.M.
1979 "Ancient and Medieval Gilding on Metal - the technical literature" in Schmandt-Besserat, 1979, 65ff.

Al-Fouadi, A-H.
1976 "Bassetki Statue with an Old Akkadian Royal Inscription of Naram-Sin of Agade (B.C.2291-2255)" Sumer 32, 63ff.

Al-Gailani, L.
1965 "Tell edh-Dhiba´i" Sumer 21, 33ff.

Al-Khalesi, Y.M.
1970 "Tell al-Fakhur: report on the first season´s excavations" Sumer 26, 120ff.

Al-Tikriti, Q.A.
1960 "Excavations at Tell Deim" Sumer 16, 93ff. (Arabic Section)

Allan, J.W.
1979 Persian Metal Technology 700-1300 A.D. (Ithaca Press)

Allan, J.W. et al.,
1973 "The History of so-called Egyptian Faience in Islamic Persia: Investigations into Abu´l-Qasim´s Treatise" Archaeometry 15, 165ff.

Allchin, F.R.
1962 "Upon the Antiquity and Methods of Gold Mining in Ancient India" J.E.S.H.O.5, 195ff.

Allchin, B. and F.R.
1982 The Rise of Civilization in India and Pakistan (Cambridge)

Amiet, P.
1961 La glyptique mésopotamienne archaique (Paris)

1966 Elam (Paris)

1967 "Eléménts émaillés du décor architectural néo-élamite" Syria 44, 27ff.

1971 "La glyptique de l´Acropole (1969-1971)" D.A.F.I.1, 217ff.

1972 Glyptique susienne (M.D.P.43 (2 vols.), Paris)

1977 L´art antique du Proche-Orient (Paris)

1979 "Archeological discontinuity and ethnic duality in Elam" Antiquity 53, 195ff.

1980 La glyptique mésopotamienne archaique (2nd revised edition, Paris)

Anderson-Gerfaud, P.
1983 "L´utilisation de certains objets en céramique de Tell el´Oueili (Obeid 4)" in J-L.Huot, ed., Larsa et ´Oueili: Travaux de 1978-1981 (C.N.R.S., Paris), 177ff.

Andrae, W.
1908 "Aus dem Berichten W. Andraes aus Assur" M.D.O.G. 36, 22.

1909 Der Anu-Adad-Tempel in Assur (W.V.D.O-G.10, Leipzig)

1913 Die Festungswerke von Assur (W.V.D.O-G. 23, Leipzig)

1922 Die Archaischen Ischtar-Tempel in Assur (W.V.D.O-G.39, Leipzig)

1923 "Gesiegeltes Gold" O.L.Z.26, 589ff.

1923a "Altassyrische Bleiplaketten" Zeitschrift für Numismatik 34, 1ff.

1925 Coloured Ceramics from Assur and Earlier Ancient Assyrian Wall-Paintings (London)

1935 Die Jüngeren Ischtar-Tempel in Assur (W.V.D.O-G.58, Leipzig)

1938 Das wiedererstandene Assur (Leipzig)

1943 Die Kleinfunde von Sendschirli (Berlin)

1977 Das wiedererstandene Assur (2nd Edition; revised by B. Hrouda, Munich)

Arik, R.O.
1937 Les fouilles d´Alaca Höyük (Ankara)

Arnaud, D., et al.,
1979 "Ilsu-Ibnisu, orfèvre de l´E.BABBAR de Larsa" Syria 56, 1ff.

Artzy, M.
1983 "Arethusa of the Tin Ingot" B.A.S.O.R.250, 51ff.

Åström, L.
1967 Studies on the Arts and Crafts of the Late Cypriote Bronze Age (Lund).

Athanassopoulos, F. et al.,
1983 "The Technology of the loop-in-loop Chains in the Third Millennium B.C." A.J.A.87, 547ff.

Azarpay, G.
1968 Urartian Art and Artifacts: a chronological study (Los Angeles)

Bachmann, H-G.
1982 The Identification of Slags from Archaeological Sites (Institute of Archaeology, London: Occasional Paper, No.6).

Bader, N.O.
1981 "Soviet Expedition's Surveys in the Sinjar Valley" Sumer 37, 55ff.

Banks, E.J.
1912 Bismya; or the lost City of Adab (New York)

Bannister, F.A. and Plenderleith, H.J.
1936 "Physico-Chemical Examination of a Scarab of Tuthmosis III bearing the Name of the God Aten" J.E.A.22, 2ff.

Baqir, T.
1944 "Iraq Government Excavations at 'Aqar Quf 1942-1943" Iraq (Supplement)

1945 "Iraq Government Excavations at 'Aqar Quf 1943-1944" Iraq (Supplement)

1946 "Iraq Government Excavations at 'Aqar Quf: Third Interim Report, 1944-5" Iraq 8, 73ff.

Bar-Adon, P.
1980 The Cave of the Treasure: the Finds from the Caves in Nahal Mishmar. (Jerusalem)

Barag, D.
1962 "Mesopotamian Vessels of the Second Millennium B.C." J.G.S. 4, 9ff.

1968 "An Unpublished Achaemenid Cut Glass Bowl from Nippur" J.G.S. 10, 17ff.

1970 "Mesopotamian Core-formed Vessels 1500-500 B.C." in Oppenheim et al., 1970, 131ff.

1975 "Rod-formed Kohl-tubes of the Mid-first Millennium B.C." J.G.S. 17, 23ff.

1982 "Cosmetic Glass Palettes from the Eighth to Seventh Centuries B.C." J.G.S.24, 11ff.

1983 "Glass Inlays and the Classification and Dating of Ivories from the Ninth-Eighth Centuries B.C." A.S.33, 163ff.

(forthcoming) A Catalogue of Glass in the Department of Western Asiatic Antiquities in the British Museum I: Ur III - A.D. 200 (London)

Barnett, R.D.
1950 "Excavations of the British Museum at Toprak Kale near Van" Iraq 12, 1ff.

1953 "An Assyrian Helmet" B.M.Q.18, 101ff.

1967 "Layard´s Nimrud Bronzes and their inscriptions" Eretz Israel 8, 1*ff.

1973 "More Balawat Gates: a preliminary report" in M.A. Beek et al., (eds.), Symbolae Biblicae et Mesopotamicae F.M.Th.de Liagre Böhl dedicatae (Leiden), 19ff.

1975 A Catalogue of the Nimrud Ivories...in the British Museum (2nd Edition, revised; London)

1982 Ancient Ivories in the Middle East (Qedem 14; Jerusalem)

Barrelet, M. Th.
1968 Figurines et reliefs en terre cuite de la Mesopotamie antique I (Paris)

1974 "Dispositifs à feu et cuisson des aliments à Ur, Nippur et Uruk" Paléorient 2, 243ff.

Basmachi, F.
1962 "Miscellania in the Iraq Museum" Sumer 18, 48ff.

Baudot, P.
1979 "Iconographic Study of the Vessels on Archaic Near Eastern seals" Orientalia Lovaniensia Periodica 10, 5ff.

Bayer, G. and Wiedemann, H.G.
1976 "Le Bleu Egyptien, pigment synthétique examiné sous l´angle scientifique" Bulletin Sandoz 40, 19ff.

Bazin, D. and Hubner, H.
1969 "Copper Deposits in Iran" Geological Survey of Iran Report no.13 (Teheran)

Beck, H.C.
1931 "Beads from Nineveh" Antiquity 5, 427ff.

1933 "Etched Carnelian Beads" A.J.13, 384ff.

1934 "Glass before 1500 B.C." Ancient Egypt and the East, 7ff.

1934a "Notes on Glazed Stones: Part I: Glazed Steatite" Ancient Egypt and the East, 69ff.

1935 "Notes on Glazed Stones: Part II: Glazed Quartz; Part III: Carnelian, Agate and Felspar" Ancient Egypt and the East, 19ff.

Behm-Blancke, M.R.
1979 Das Tierbild in der Altmesopotamischen Rundplastik (Mainz am Rhein)

Berthelot, M.
1906 Archéologie et Histoire des Sciences (Paris)

Berthoud, Th.
1979 Etude par l'analyse de traces et la modélisation de la filiation entre minerai de cuivre et objets archéologiques du Moyen-Orient (IVe et IIIe millénaires avant notre ère) (Doctoral thesis: Université Pierre et Marie Curie, Paris VI)

Berthoud, Th. and Cleuziou, S.
1983 "Farming Communities of the Oman Peninsula and the Copper of Makkan" J.O.S.6, 239ff.

Berthoud, Th. and Francaix, J.
1980 Contribution à l'étude de la métallurgie de Suse aux IVe et IIIe millénaires (Gif-sur-Yvette: Rapport C.E.A-R-5033)

Berthoud, Th., et al.
1976 "Recherches sur les sources du cuivre dans l'Iran ancien" Iran 14, 150ff.

1977 Les anciennes mines d'Afghanistan. Rapport préliminaire RCP 442 (Paris)

1978 Les anciennes mines de cuivre du Sultanat d'Oman. Rapport préliminaire RCP 442 (Paris)

1980 "Data Analysis: towards a model of chemical modification of copper from ores to metal" in P.T. Craddock (ed.), Scientific Studies in Early Mining and Extractive Metallurgy (British Museum Occasional Paper, No.20), 87ff.

1981 "Production, échange et utilisation du métal: bilan et perspectives des recherches archéométriques récentes dans le domaine oriental" Paléorient 6, 99ff.

1982 "Cuivres et alliages en Iran, Afghanistan, Oman au cours des IVe et IIIe Millénaires" Paléorient 8/2, 39ff.

Beyer, D.
1982 "Du Moyen-Euphrate au Luristan: bagues-cachets de la fin du deuxième millénaire" Mari 1, 169ff.

Bibby, G.
1964 "Arabiens Arkaeologi" Kuml, 86ff.

1972 Looking for Dilmun (Pelican Books, Harmondsworth, Middlesex, England)

Biek, L. and Bayley, J.
1979 "Glass and other vitreous materials" World Archaeology 11(i), 1ff.

Bikai, P.M.
1978 The Pottery of Tyre (Aris and Phillips, Warminster, England)

Bimson, M.
1973 "Analysis of Glazed Material from Tell Taya" Iraq 35, 183ff.

1980 "Cosmetic Pigments from the "Royal Cemetery" at Ur" Iraq 42, 75ff.

Bimson, M. and Freestone, I.C.
(forthcoming) "Scientific Examinations of Opaque Red Glas of the Second and First Millennium B.C." in D. Barag, A Catalogue of Glass in the Department of Western Asiatic Antiquities in the British Museum, I.

Bimson, M. and Werner, A.E.
1967 "A Survey of the Materials used in Ancient Glass" Annales 4, 265ff.

1969 "Two Problems in Ancient Glass: Opacifiers and Egyptian Core Material" Annales du 4e Congrès des "Journées Internationales du Verre" (Ravenna-Venice, 1967) (Liège), 262ff.

Binns, C.F. et al.,
1932 "An experiment in Egyptian Blue Glaze" Journal of the American Ceramic Society 15, 271ff.

Birch, S.
1858 History of Ancient Pottery (London)

Birgi, S.E.
1951 "Notes on the Influence of the Ergani Copper Mine on the Development of the Metal Industry in the Ancient Near East" Jahrbuch für Kleinasiatische Forschung I, 337ff.

Birot, M.
1960 A.R.M.9: Textes administratifs de la salle 5 du palais (Paris)

Bivar, A.D.H.
1971 "A Hoard of Ingot-currency of the Median Period from Nushi-Jan, near Malayir" Iran 9, 97ff.

Bjorkman, J.K.
1973 Meteors and Meteorites in the Ancient Near East (Center for Meteorite Studies, Arizona State University, Temple, Arizona)

Boehmer, R.M.
1957-71 "Guti-Siegel" R.L.A.3, 707-8.

1965 Die Entwicklung der Glyptik während der Akkad-Zeit (Berlin)

1967 Die Hethitische Glyptik von Boğazköy (Berlin)

1972 Die Kleinfunde von Boǧazköy (Berlin)

1973 "Zur Lage von Muṣaṣir" Bag.Mitt.6, 31ff.

1982 "Ringe aus kassitischen Gräbern" Bag.Mitt.13, 31ff.

1983 "Dolche vom Tell Subeidi (Hamrin) Bag.Mitt.14, 101ff.

Boese, J.
1971 Altmesopotamische Weihplatten (Berlin)

Böhme, C.
1979 "Metallurgische Untersuchungen Urartäischer Eisengefässe" A.M.I.12, 245ff.

Böhne, C.
1965 "Zur Frage der Hartung von Kupfer Waffen und Geräten" Technische Beitrage zur Archäologie 2 (R.G.Z.M.) 126ff.

Börker-Klähn, J.
1973 "Eine Bronzestatue aus Assur?" Z.F.A.63, 272ff.

1974 "Ein Metallvollguss im Museum von Çanakkale" Bag.Mitt.7, 51ff.

Born, H.
1984 "Bronzene Möbeltiere aus Urartu: Anmerkungen zu einer vorläufigen Tischrekonstruktion" Antike Welt 15(2), 51ff.

Bossert, H. Th.
1951 "Die Bleibriefe aus Assur" Bib.Or.8, 126ff.

Botta, P.E. and Flandin, E.
1849-50 Monument de Ninive (5 vols.; Paris)

Bottero, J.
1957 A.R.M.7: Textes économiques et administratifs (Paris)

Braidwood, L.S. et al.,
1983 Prehistoric Archaeology along the Zagros Flanks (O.I.P. 105; Chicago)

Braidwood, R.J. et al.,
1951 "Ancient Syrian Coppers and Bronzes" Journal of Chemical Education 28, 87ff.

1960 Excavations in the Plain of Antioch I (O.I.P.61; Chicago)

Braun-Holzinger, E.A.
1984 Figürliche Bronzen aus Mesopotamien (Prähistorische Bronzefunde Abt.I, Band 4, Munich)

Brill, R.H.
1963 "Ancient Glass" Scientific American (November), 120ff.

1969 "The Scientific Investigation of Ancient Glasses" in Proceedings, Eighth International Congress on Glass (London 1968) 47ff.

1970 "The Chemical Interpretation of the Texts" in Oppenheim et al., 1970, 105ff.

1978 "Some Miniature Glass Plaques from Fort Shalmaneser, Nimrud, Part II: laboratory studies" Iraq 40, 23ff.

1978a "Examination Report on Samples from a Faience-bead Factory at Tyre" in Bikai, 1978, 91ff.

Brill, R.H. et al.,
1974 "Lead Isotopes in some Ancient Egyptian Objects" in A. Bishay (ed.), Recent Advances in Science and Technology of Materials, 3 (New York) 9ff.

Brill, R.H. and Wampler, J.M.
1967 "Isotope Studies on Ancient Lead" A.J.A.71, 63ff.

1967a "Isotope Ratios in Archaeological Objects of Lead" Application of Science in Examination of Works of Art (Boston) 155ff.

Brinkman, J.A.
1972 "Foreign Relations of Babylonia from 1600-625 B.C.: the documentary evidence (summary)" A.J.A.76, 271ff.

1976 Materials and Studies for Kassite History I (Chicago)

British Museum Guide
1922 A Guide to the Babylonian and Asyrian Antiquities (British Museum; 3rd edition - revised and enlarged)

Bryer, A.A.M.
1982 "The Question of Byzantine Mines in the Pontos: Chalybian Iron, Chaldian Silver, Koloneian Alum and the Mummy of Cheriana" A.S.32, 133ff.

Buchanan, B.
1966 Catalogue of Ancient Near Eastern Seals in the Ashmolean Museum I: Cylinder Seals (Oxford)

1967 "The Prehistoric Stamp Seal: A Reconsideration of Some Old Excavations" J.A.O.S.87, 265ff., 525ff.

1981 Early Near Eastern Seals in the Yale Babylonian Collection (Yale University Press)

Buchanan, B. (edited P.R.S. Moorey)
1984 Catalogue of the Ancient Near Eastern Seals in the Ashmolean Museum II: The Prehistoric Stamp Seals (Oxford)

Buddensieg, G.
1912 "Berichte über die Grabungen am Kasr Nordost" M.D.O.G.48, 4ff.

Burney, C. and Lang, D.M.
1971 The Peoples of the Hills: Ancient Ararat and Caucasus (London)

Burrows, E.
1935 Archaic Texts (Ur Excavations: Texts, II, London)

Burton Brown, T.
1951 Excavations in Azerbaijan 1948 (London)

Caley, E.R.
1962 Analyses of Ancient Glasses, 1790-1957 (Corning, New York)

Calmeyer, P.
1964 "Archaische Zügelringe" in K. Bittel et al., (ed.), Vorderasiatische Archäologie (Berlin), 68ff.

1969 Datiebare Bronzen aus Luristan und Kirmanshah (Berlin)

1973 Reliefbronzen in babylonischem Stil (Münich)

1977 "Das Grab eines Altassyrischen Kaufmanns" Iraq 39, 87ff.

1979 "Zum Tongefäss in Form eines Gazellenkopfes" in Bastam I (ed. W. Kleiss; Berlin) 195ff.

1980 "Zur Genese Altiranischer Motive: VII: Achsnägel in Form von Betenden" A.M.I. 13, 99ff.

Cambel, H. and Braidwood, R.J.
1983 "Cayönü Tepesi: Schritte zu neuen Lebensweisen" in Beitrage zur Altertumskunde Kleinasiens: Festschrift für Kurt Bittel (Mainz am Rhein), 155ff.

Campbell Thompson, R.
1920 "The British Museum Excavations at Abu Shahrein in Mesopotamia in 1918" Archaeologia 70, 101ff.

1936 A Dictionary of Assyrian Chemistry and Geology (Oxford)

Campbell Thompson, R. and Hamilton, R.W.
1932 "The British Museum Excavations on the Temple of Ishtar at Nineveh, 1930-31" L.A.A.A.19, 55ff.

Campbell Thompson, R. and Hutchinson, R.W.
1931 "The Site of the Palace of Ashurnasirpal at Nineveh, excavated in 1929-30 on behalf of the British Museum" L.A.A.A.18, 79ff.

Campbell Thompson, R. and Mallowan, M.E.L.
1933 "The British Museum Excavations at Nineveh, 1931-2" L.A.A.A.20, 71ff.

Canby, J.V.
1965 "Early Bronze 'Trinket' Moulds" Iraq 27, 42ff.

1979 "A Note on some Susa bricks" A.M.I.12, 315ff.

Carroll, D.L.
1972 "Wire Drawing in Antiquity" A.J.A.76, 321ff.

1974 "A Classification for Granulation in Ancient Metalwork" A.J.A.78, 33ff.

Carter, T.H.
1964 "Early Assyrians in the Sinjar" Expedition 7, 34ff.

1965 "Excavations at Tell al-Rimah, 1964" B.A.S.O.R.178, 40ff.

1967 "Tell al-Rimah: The Campaigns of 1965 and 1966" Archaeology 20, 282ff.

Caubet, A.
1982 "Faience et verre" in Beyer, D. (ed.), Meskéné - Emar: dix ans de travaux 1972-1982 (Paris), 111ff.

Chakrabarti, D.K.
1979 "The problem of tin in early India - a preliminary survey" Man and Environment 3, 61ff.

Charles, J.A.
1967 "Early Arsenical Bronzes - a metallurgical view" A.J.A.71, 21ff.

1968 "The first Sheffield Plate" Antiquity 32, 278ff.

1975 "Where is the Tin?" Antiquity 49, 19ff.

1978 "The Development of the Usage of Tin and Tin-bronze: some problems" in A.D. Franklin, et al., 1978, 25ff.

1980 "The coming of Copper and Copper base Alloys and Iron: a metallurgical sequence" in Wertime and Muhly, 1980, 15ff.

Charlesworth, D.
1967 "A Primitive Glass Furnace in Cairo" J.G.S.9, 129ff.

Chase, W.T.
1971 "Egyptian Blue as a Pigment and Ceramic Material" in Science and Archaeology (ed. R.H. Brill; M.I.T. Press), 80ff.

Childe, V.G.
1952 New Light on the Most Ancient East (4th ed., London)

Chmyriov, V.M. et al.,
1973 "Mineral Resources of Afghanistan" in Geology and Mines and Industries of the Republic of Afghanistan, I (Kabul)

Cleuziou, S. and Berthoud, Th.
1982 "Early Tin in the Near East: a reassessment in the light of new evidence from Western Afghanistan" Expedition 25, 24ff.

Coghlan, H.H.
1952 "Casting Moulds made in Metal" Man 52, no.245, 162ff.

1956 Notes on Prehistoric and Early Iron in the Old World (Pitt Rivers Museum, Oxford)

1975 Notes on the Prehistoric Metallurgy of Copper and Bronze in the Old World (Pitt Rivers Museum, Oxford; 2nd edition)

Cohen, S.
1973 Enmerkar and the Lord of Aratta (Philadelphia dissertation; University Microfilms, Ann Arbor)

Collon, D.
1981 The Alalakh Cylinder Seals (B.A.R., Oxford; International Series, no.132)

1982 "Some Bucket Handles" Iraq 44, 95ff.

Congdon, L.O.K.
1971 "Steel in Antiquity: a problem of terminology" in D.G. Mitten et al., (eds.) Studies Presented to George M.A. Hanfmann (Mainz) 17ff.

Conrad, H.G. and Rothenberg, B.
1980 Antikes Kupfer im Timna-Tal (Bochum)

Contenau, G.
1927-47 Manuel d'archéologie orientale (4 volumes; Paris)

1940 "Une statuette sumérienne archaique du Musée du Louvre" Mon.Piot 37, 37ff.

Cooke, S.R.B. and Aschenbrenner, S.
1975 "The Occurrence of Metallic Iron in Ancient Copper" J.F.A.2, 251ff.

Cooney, J.D.
1960 "Glass Sculpture in ancient Egypt" J.G.S.2, 11ff.

1964 Review of Lucas, 1962, in A.J.A.68, 75ff.

1965 "Persian influence in Late Egyptian Art" Journal of the American Research Center in Egypt 4, 39ff.

1976 Catalogue of Egyptian Antiquities in the British Museum IV: Glass (London)

Cooper, J.S. and Heimpel, W.
1983 "The Sumerian Sargon Legend" J.A.O.S.103, 67ff.

Craddock, P.T.
1977 "The Composition of the Copper Alloys used by the Greek, Etruscan and Roman Civilizations" 2: The Archaic, Classical and Hellenistic Greeks" J.Arch.Sc.4, 103ff.

1978 "The composition of the copper alloys used by the Greek, Etruscan and Roman Civilizations 3: The origins and early use of brass" J.Arch.Sc5, 1ff.

1980 "The First Brass: some early claims reconsidered" M.A.S.C.A. Journal 1, 131ff.

1980a "The composition of copper produced at the ancient smelting camps in the Wadi Timna, Israel" in Scientific Studies in Early Mining and Extractive Metallurgy (ed.Craddock, P.T., British Museum, Occasional Paper no.20), 165ff.

1984 "Tin and Tin Solder in Sumer: preliminary comments" M.A.S.C.A. Journal 3(1), 7ff.

Crawford, H.E.W.
1974 "The problem of tin in Mesopotamian bronzes" World Archaeology 6, 242ff.

1975 "Geoy Tepe 1903: material in the collection of the Fitzwilliam Museum, Cambridge" I.A.11, 1ff.

Crawford, V.E.
1960 "Art of the Ancient Near East: the third millennium B.C." Bulletin of the Metropolitan Museum of Art (April), 245ff.

Crowfoot, J.W. and G.M.
1938 Samaria - Sebaste 2: Early Ivories (London)

Crownover, D.
1964 "Some Frit from North Mesopotamia" Expedition 7, 43ff.

Culican, W.
1971 "Two Syrian Objects from Egypt" Levant 3, 86ff.

Curtis, J.G.
1979 An Examination of Late Assyrian Metalwork with special reference to the material from Nimrud (Ph.D.Thesis; University of London)

1982 Fifty Years of Mesopotamian Discovery (The British School of Archaeology in Iraq, London) (Editor)

1983 "Some axe-heads from Chagar Bazar and Nimrud" Iraq 45, 73ff.

1983a "Late Assyrian bronze coffins" A.S. 33, 85ff.

1984 Nush-i Jan III: The Small Finds (London; British Institute of Persian Studies).

Curtis, J.G. et al.,
1979 "Neo-Assyrian Ironworking Technology" Proceedings of the American Philosophical Society 123, 369ff.

Curtis, J.G. and Grayson, A.K.
1982 "Some Inscribed Objects from Sherif Khan in the British Museum" Iraq 44, 87ff.

Curtis, J.G. and Maxwell-Hyslop, K.R.
1971 "The Gold Jewellery from Nimrud" Iraq 33, 101ff.

Dabney, M.K.
1984 "Mitannian Common Style Cylinder Seals: a study of their distribution" A.J.A.84, 241 (thesis summary).

Dalman, G.
1928-42 Arbeit und Sitte in Palästina (Gütersloh; 7 vols.)

Davey, C.J.
1979 "Some Ancient Near Eastern Pot Bellows" Levant 11, 101ff.

1983 "The Metalworker's Tools from Tell Edh Dhiba'i" Bulletin of the Institute of Archaeology, London University 20, 169ff.

Davies, O.
1935 "Antimony Bronze in Central Europe" Man 9, 86ff.

Dayton, J.E.
1971 "The Problem of Tin in the Ancient World" World Archaeology 3, 49ff.

1974 "Money in the Near East before Coinage" Berytus 23, 41ff.

1978 Minerals, Metals, Glazing and Man (London)

1981 "Cobalt, Silver and Nickel in Late Bronze Age Glazes, Pigments and Bronzes, and the Identification of Silver Sources for the Aegean and Near East" in M.J. Hughes (ed.), Scientific Studies in Ancient Ceramics B.M.O.P.19), 129ff.

1982 "Geology, Archaeology and Trade" in J.G.P. Best and N.M.W. de Vries (eds.), Interaction and Acculturation in the Meditterranean (Amsterdam), 153ff.

Debevoise, N.C.
1934 "The History of Glaze and its Place in the Ceramic Technique of Ancient Seleucia on the Tigris" Bulletin of the American Ceramic Society 13(2), 293ff.

de Cardi, B. et al.,
1979 "Excavations at Ṭāwi Silaim and Tawi Sa'id in the Sharqīya, 1978" J.O.S.5, 61ff.

de Genouillac, H.
1924 Premières recherches archéologiques à Kich I (Paris)

1925 Ibid. II (Paris)

1934 Fouilles de Telloh I: époques présargoniques (Paris)

1936 Ibid. II: époques d'Ur III Dynastie et de Larsa (Paris)

de Jesus, P.S.
1978 "Metal Resources in Ancient Anatolia" A.S,28, 97ff.

1980 The Development of Prehistoric Mining and Metallurgy in Anatolia (B.A.R. International Series 74 (i and ii), Oxford)

1981 "A survey of some ancient mines and smelting sites in Turkey" Sonderdruck aus Archaölogie und Naturwissenschaften 2 (R.G.Z.M.), 95ff.

1983 "Preliminary report on the ancient mining survey 1981" (in Darb Zubaydah, Saudi Arabia) Atlal 6, 63ff.

Delougaz, P.
1940 The Temple Oval at Khafajah (O.I.P.53; Chicago)

1942 Pre-Sargonid Temples in the Diyala Region (O.I.P.58, Chicago)

1967 Private Houses and Graves in the Diyala Region (O.I.P.88, Chicago)

de Miroschedji, P.
1981 "Fouilles du chantier Ville Royale II à Suse (1975-77) I. Les niveaux élamites" D.A.F.I.12, 9ff.

1982 "Notes sur la glyptique de la fin de l´Elam" R.A.76, 51ff.

de Morgan, J. and de Mecquenem, R.
1912 Mémoire Délégation en Perse 13: Recherches archéologiques (Paris)

Dentzer, J-M.
1982 Le motif du banquet couché dans le Proche-Orient et le Monde Grec du VIIe au IVe siècle avant J-C (Rome)

de Sarzec. E. and Heuzey, L.
1884-1912 Découvertes en Chaldée (Paris)

Desch, C.H.
1928-38 Report of the Committee on Sumerian Copper (British Association for the Advancement of Science, Section H).

1934 "Early History of Bronze" Trans. Newcomen Society 14, 95ff.

Deshayes, J.
1960 Les Outils de bronze l´Indus au Danube IVe au IIe millénaire, I-II (Paris)

Dhorme, R.P. and Thureau-Dangin, F.
1924 "Cinq jours de fouilles à ´Asharah" Syria 5, 265ff.

Diakonov, I.M.
1947 "On an ancient Oriental Sculpture" (in Russian; English summary) Gosudarstvennyi ermitazh: Trudy Otdela Vostoka (Leningrad) 117-118.

Dossin, G.
1970 "La route de l´étain en Mesopotamie au temps de Zimri-Lim" R.A.64, 97ff.

1970a "Archives de Sûmu-iamam, roi de Mari" R.A.64, 17ff.

Drescher, H.
1958 Der Überfangguss: ein Beitrag zur vorgeschichtlichen Metalltechnik (Mainz)

Dunand, M.
1958 Fouilles de Byblos (1933-1938) II (Paris)

Dunlop, D.M.
1957 "Sources of Gold and Silver in Islam according to Al-Hamdani" Studia Islamica 8, 24ff.

Durand, J-M.
1983 "Relectures d'ARMT VIII, II: ARMT VIII, 89 et le travail du métal à Mari" M.A.R.I.2, 123ff.

During-Caspars, I.
1978 "R. Thapar's Dravidian hypothesis for the locations of Meluhha, Dilmun and Makan" J.E.S.H.O.21, 113ff.

Dyson, R.
1959 "Digging in Iran: Hasanlu, 1958" Expedition 1, 4ff.

1960 "Hasanlu and Early Iran" Archaeology 13, 118ff.

1963 "Archaeological Scrap: glimpses of history at Ziwiye" Expedition 5, 32ff.

1964 "Sciences meet in ancient Hasanlu" Natural History 73(8), 16ff.

1968 "Hasanlu and the Solduz and Ushnu valleys: twelve years of exploration" Archaeologia Viva 1, 83ff.

Eaton, E.R. and McKerrell, H.
1976 "Near Eastern Alloying and some Textual Evidence for the Early Use of Arsenical Copper" World Archaeology 8, 169ff.

1980 "Some Laboratory Smelting Experiments" 16th International Symposium on Archaeometry and Archaeological Prospection, 1976 (National Museum of Antiquities, Edinburgh), 416ff.

Edzard, D.O.
1959-60 "Neuen Inschriften zur Geschichte von Ur III unter Šūsuen" A.F.O.19, 1ff.

1960 "Die Beziehungen Babyloniens und Agyptens in der Mittelbabylonischen Zeit und das Geld" J.E.S.H.O.3, 38ff.

1964 "A new inscription of Adad Nirari I" Sumer 20, 149ff.

Edzard, D.O. and Farber, G.
1974 Die Orts-und Gewässernamen der Zeit der 3 Dynastie von Ur (Wiesbaden; Répertoire Géographique 2)

Edzard, D.O. Farber, G. and Sollberger, E.
1977 Die Orts-und Gewässernamen der präsargonischen und sargonishcen Zeit (Wiesbaden; Répertoire Géographique 1)

Egami, N. et al.,
1965 Dailaman I (Tokyo).

Eilers, W.
1954-9 "Akkad.kaspum, ´Silber, Geld´ und Sinnverwandtes" W.O.2, 322-37, 465-9.

Elam, C.
1932 "Some Bronze Specimens from the Royal Cemetery at Ur" Journal of the Institute of Metals 10(8), 97ff.

Elat, M.
1982 "The impact of tribute and booty on countries and people within the Assyrian Empire" A.f.O.Beiheft 19, 244ff.

Ellis, R.
1966 "A note on some ancient Near Eastern linch pins" Berytus 16, 41ff.

1968 Foundation Deposits in Ancient Mesopotamia (Yale University Press)

Emre, K.
1971 Anadolu Kursun Figürinleri Ve Tas Kaliplari (Anatolian Lead Figurines and their Stone Moulds (Ankara)

Encycl. Photo II
1936 Encyclopédie Photographique de l´Art II: Le Musée du Louvre (Paris)

Esin, U.
1969 Kuantitatif Spektral Analiz Yardimiyla Anadolu´da Baslang c ndan Asur Kolonileri Cagina Kadar Bakir Ve Tunc Madenciligi (Istanbul)

Falkenstein, A.
1936 Archaische Texte aus Uruk (Berlin)

1966 Die Inschriften Gudeas von Lagas I (Rome)

Fell, V.
1982 "Ancient Fluxes for Soldering and Brazing" M.A.S.C.A.Journal 2(3), 82ff.

Filippakis, S.
1983 "Bronzes grecs et orientaux: influences et apprentissages" B.C.H.107, 111ff.

Fitz, S.
1982 "The coloured Glazes of Neo-Babylonian Bricks" (in German) Ceramic Forum International: Berichte D.K.G.59(3), 179ff.

Forbes, R.J.
1940-50 Bibliographia Antiqua Philosophia Naturalis (Leiden)

1950 Metallurgy in Antiquity (Leiden)

1952 Bibliographia Antiqua Philosophia Naturalis: Supplement 1 (Leiden)

1966 Studies in Ancient Technology V (2nd edition; Leiden): Glass, 112ff.

1966a Studies in Ancient Technology VII (2nd edition; Leiden): Ancient Mining, 104ff.

1971 Studies in Ancient Technology VIII (2nd edition; Leiden): Early Metallurgy, Gold; Silver and Lead; Zinc and Brass.

1972 Studies in Ancient Technology IX (2nd edition; Leiden): Copper; Tin and Bronze; Antimony and Arsenic; Iron.

Fossey, Ch.
1935 "L´essai de l´affinage de l´or chez les Babyloniens" Revue des études sémitiques 4, 2ff.

Fossing, P.
1940 Glass Vessels before Glass Blowing (Copenhagen)

Foster, K.P.
1979 Aegean Faience of the Bronze Age (Yale University Press)

Foster, K.P. and Kaczmarczyk, A.
1982 "X-ray Fluorescence Analysis of some Minoan Faience" Archaeometry 24, 143ff.

France-Lanord, A. and de Contenson, H.
1973 "Une pendeloque en cuivre natif de Ramad" Paléorient 1, 109ff.

Frankfort, H.
1939 Sculpture of the Third Millennium B.C. from Tell Asmar and Khafajah (O.I.P.44; Chicago)

1939a Cylinder Seals (London)

1943 More Sculpture from the Diyala Region (O.I.P.60; Chicago)

1950 "Early Iron in Iraq" Man 50, 100.

1954 Art and Architecture of the Ancient Orient (1st edition; Pelican History of Art)

1955 Stratified Cylinder Seals from the Diyala Region (O.I.P.72; Chicago)

Frankfort, H. et al.,
1932 Tell Asmar and Khafaje: the first season´s work in Eshnunna 1930/31 (O.I.C.13; Chicago)

1934 Iraq Excavations of the Oriental Institute 1932/3 (O.I.C.17; Chicago)

Franklin, A.D. et al.,
1978 The Search for Ancient Tin (U.S. Government Printing Office, Washington, D.C.)

Freydank, H.
1982 "Fernhandel und Warenpreise nach einer Mittelassyrischen Urkunde des 12 Jahr-hunderts v.u.Z" Societies and Languages of the Ancient Near East: Studies in Honour of I.M.Diakonoff (Aris and Phillips, Warminster) 64ff.

Fridman, A.
1969 "Sargon II or Tiglath-Pileser III?" R.A.63, 172ff.

Friend, J. et al.,
1929 "The Silver Content of Ancient and Medieval Lead" Journal of the Institute of Metals 41, 105ff.

Fukai, S.
1981 Ceramics of Ancient Persia (New York, Tokyo and Kyoto)

Fukai, S. et al.,
1973 Marv Dasht III: the excavation at Tall-i-Mushki, 1965 (Tokyo)

1974 Tellul eth Thalathat: the excavation of Tell V, 3 (Tokyo)

1977 "Excavations at Telul eth-Thalathat 1976" Sumer 33, 48ff.

1981 Telul eth-Thalathat: the excavation of Tell II: the fifth season (1976) IV (Tokyo)

Gadd, C.J.
1971 The Cambridge Ancient History I(2) (Cambridge) 93ff., 417ff., 595ff.

Gadd, C.J. and Campbell Thompson, R.
1936 "A Middle-Babylonian Chemical Text" Iraq 3, 87ff.

Gale, N.H. and Stos-Gale, Z.A.
1981 "Ancient Egyptian Silver" J.E.A.67, 103ff.

1981a "Cycladic Lead and Silver Metallurgy" B.S.A.76, 169ff.

1981b "Lead and Silver in the Ancient Aegean" Scientific American 244(6), 176ff.

Gardiner, A.H.
1957 Egyptian Grammar (Oxford)

Garelli, P.
1963 Les Assyriens en Cappadoce (Paris)

Garner, H.
1956 "An Early Piece of Glass from Eridu" Iraq 18, 147ff.

Gelb, I.J.
1970 "Makkan and Meluhha in Early Mesopotamian Sources" R.A.64, 1ff.

Gettens, R.J.
1939 "Chemical and Microscopic Examination of the green glaze found at Nuzi" in Starr, 1939, 523ff.

Ghirshman, R.
1964 Persia, from the origins to Alexander the Great (London)

1966 Tchoga-Zanbil I: La Ziggurat (M.D.P.39; Paris)

1968 Tchoga-Zanbil II: Temenos, Temples, Palais, Tombes (M.D.P.40, Paris)

Gibson, Mc.
1975 Excavations at Nippur: Eleventh Season (O.I.C.22; Chicago)

1977 "The Nippur Expedition" in the Annual Report of the Oriental Institute, Chicago, 1976-7, 24ff.

1978 Excavations at Nippur: Twelfth Season (O.I.C.23; Chicago)

1980 "Current Research at Nippur: Ecological, Anthropological and Documentary Interplay" L'Archéologie de l'Iraq (C.N.R.S.), 193ff.

1982 "A Re-evaluation of the Akkad Period in the Diyala Region on the Basis of Recent Excavations at Nippur and in the Hamrin" A.J.A.86, 531ff.

Giles, D.L. and Kuijpers, E.P.
1974 "Stratiform Copper Deposit, Northern Anatolia, Turkey. Evidence for Early Bronze I (2800 B.C.) mining activity" Science 186, 823ff.

Gladstone, J.H.
1892 "On Metallic Copper, Tin and Antimony from Ancient Egypt" P.S.B.A.XIV (22nd Session) 223ff.

Glob, P.V.
1955 "Udgravninger på Bahrain" Kuml, 178ff.

Gockel, W.
1982 Die Stratigraphie und Chronologie der Ausgrabungen des Diyala-Gebietes und der Stadt Ur in der Zeit von Uruk/Eanna IV bis zur Dynastie von Akkad (Rome, Bretschneider)

1983 "Ur und Al-'Ubaid" Akkadica 32, 32ff.

Goetze, A.
1949 Kizzuwatna and the Problem of Hittite Geography (Yale University Press)

Goff, B.
1963 Symbols of Prehistoric Mesopotamia (Yale University Press)

Goffer, Z.
1980 Archaeological Chemistry: A Sourcebook on the Applications of Chemistry to Archaeology (John Wiley and Sons; New York)

Goldstein, S.M.
1970 "The Examination of the Gold Samples from Pactolus North" B.A.S.O.R.199, 26ff.

1977 "Forgeries and Reproductions of Ancient Glass in Corning" J.G.S.19, 40ff.

1979 Pre-Roman and Early Roman Glass in the Corning Museum of Glass (Corning, New York)

Gordon, A.H.
1982 "A Glass Bead of Ahmose and Amenophis I" J.N.E.S. 41, 295ff.

Gowland, W.
1920 "Silver in Roman and Earlier Times: Prehistoric and Proto-Historic Times" Archaeologia 69, 121f.

Graham, A.K.
1929 "Scientific Notes on the Finds from Ur: I: Restoration of the Silver M.J.20(3-4), 246ff.

Grayson, A.K.
1972 Assyrian Royal Inscriptions I (Wiesbaden)

1978 Assyrian Royal Inscriptions II (Wiesbaden)

Green, M.W.
1930 "Animal Husbandry at Uruk in the Archaic Period" J.N.E.S.39, 1ff.

Groneberg, B.
1980 Die Orts-und Gewässernamen der alt-babylonischen Zeit (Wiesbaden; Répertoire Géographique 3)

Gurney, O.R.
1953 "Further Texts from Dur-Kurigalzu" Sumer 9, 21ff.

Gwinnett, A.J. and Gorelick, L.
1979 "Ancient Lapidary: a study using scanning electron microscopy and functional analysis" Expedition 22(1), 17ff.

1983 "An Innovative Method to investigate the Technique of Finishing an Ancient Glass Artifact" J.G.S.25, 249ff.

Haerinck, E.
1973 "Le palais achéménide de Babylone" Iranica Antiqua 10, 108ff.

Haevernick, T.E.
1965 Beiträge zur Geschichte des antiken Glas, XIII: Nuzi-Perlen" Jahrbuch R.G.Z.M.12, 35ff.; reprinted in Haevernick, 1984, 146ff.

1968 "Assyrisches Millefioriglas" Forschungen und Berichte (Staatliche Museen zu Berlin) 10, 63ff.; reprinted in Haevernick, 1984, 161ff.

1984 Beiträge zur Glasforschung: die wichtigsten Aufsätze von 1938 bis 1981 (Mainz am Rhein).

Haevernick, T.E. and Kühne, K.
1982 "Eine Glaskeule aus Assur" Forschungen und Berichte 22, 115ff.

Hall, H.R.
1928 "Minoan Fayence in Mesopotamia" J.H.S.48, 64ff.

1930 A Season's Work at Ur (London)

Hall, H.R. and Woolley, C.L.
1927 Ur Excavations I: Al'Ubaid (London)

Haller, A.
1954 Die Gräber und Grüfte von Assur (W.V.D.O.G.65; Berlin)

Haller, A. and Andrae, W.
1955 Die Heiligtümer des Gottes Assur und der Sin-Šamaš-Tempel in Assur (Berlin)

Hallo, W.W.
1963 "Lexical Notes on the Neo-Sumerian Metal Industry" Bib.Or.20, 136ff. (Review of Limet, 1960)

Hamilton, R.W.
1967 "A Sumerian Cylinder Seal with handle in the Ashmolean Museum" Iraq 29, 34ff.

Hamilton, W.J.
1842 Researches in Asia Minor, Pontus and Armenia I (London)

Hanfmann, M.A. and Hansen, P.
1956 "Hittite Bronzes and Other Near Eastern Figurines in the Fogg Museum of Harvard University" Türk Arkeoloji Dergisi 6, 43ff.

Hanfmann, G.M.A. and Waldbaum, J.C.
1970 "New Excavations at Sardis and some problems of Western Anatolian Archaeology" in Near Eastern Archaeology in the Twentieth Century (Glueck Festschrift) (ed. J.A. Sandars, New York) 307ff.

Hansman, J.
1973 "A periplus of Magan and Meluhha" B.S.O.A.S.36, 554ff.

1978 "The Question of Aratta" J.N.E.S.37, 331ff.

Harden, D.B.
1968 "Ancient Glass: I: Pre-Roman" The Archaeological Journal 125, 46ff.

1969 "Some aspects of Pre-Roman mosaic glass" Annales du 4e Congrès des "Journées internationales du verre" (Ravenna-Venice, 1967) (Liège), 29ff.

1981 Catalogue of Greek and Roman Glass in the British Museum I (London)

Harris, J.R.
1961 Lexicographical Studies in Ancient Egyptian Minerals (Berlin)

Harrison, J.V.
1968 "Minerals" in Fisher, W.B. (ed.) The Cambridge History of Iran 1, 489ff.

Hartmann, A.
1978 "Ergebnisse der spektralanalytischen Untersuchung äneolithischer Goldfunde aus Bulgarien" Studia Praehistorica 1-2, (Sofia), 27ff.

Hawkins, J.D.
1982 "The Neo-Hittite States in Syria and Anatolia" Cambridge Ancient History III (I), 372ff.

Healy, J.F.
1978 Mining and Metallurgy in the Greek and Roman World (London)

Hedges, R.E.M.
1976 "Pre-Islamic Glazes in Mesopotamia-Nippur" Archaeometry 18, 209ff.

1976a "Analyses of bronze and other metals from Tell es-Sweyhat" Levant 8, 66ff.

1982 "Early glazed pottery and faience in Mesopotamia" in Wertime, Th. and S. 1982, 93ff.

Hedges, R.E.M. and Moorey, P.R.S.
1975 "Pre-Islamic ceramic glazes at Kish and Nineveh in Iraq" Archaeometry 17, 25ff.

Hegde, K.T.M.
1978 "Sources of Ancient Tin in India" in Franklin et al., 1978, 39ff.

1983 "The Art of the Harappan Microbead" Archaeology 36(3), 68ff.

Heinrich, G.
1935 Sechster Vorläufiger Bericht... Uruk Warka (Berlin)

1936 Kleinfunde aus den archaischen Tempelschichten in Uruk (Berlin)

Heinrich, E. and Andrae, W.
1931 Fara. Ergebnisse der Ausgrabungen... in Fara und Abu Hatab 1902-3 (Berlin)

Heinrich, E. and Seidl, U.
1968 "Mass und Ubermass in der Dimensionierung von Bauwerken im alten Zweistromland" M.D.O.G.99, 5ff.

Helm, O.
1901 "Chemische Untersuchung von altbabylonischen Kupfer-und Bronze-gegenständen" Verhandl. der Berliner Gesellschaft für Anthropologie, 157ff.

Heltzer, M.
1977 "The Metal Trade of Ugarit and the Problem of Transportation of Commercial Goods" Iraq 39, 203ff.

1978 Goods, Prices and the Organization of Trade in Ugarit (Wiesbaden)

Hennessey, B.J.
1979 Masterpieces of Western Ceramic Art I (Kodansha, Tokyo)

Herzfeld, E.
1930 Die vorgeschichtlichen Töpfereien von Samarra (Berlin)

1933 "Aufsätze zur altorientalischen Archäologie" A.M.I.5, 1ff.

Heskel, D. and Lamberg-Karlovsky, C.C.
1980 "An Alternative Sequence for the Development of Metallurgy: Tepe Yahya, Iran" in Wertime and Muhly, 1980, 229ff.

Heuzey, L.
1900 "Autre taureau chaldéen androcéphale: statuettes à incrustations" Monuments Piot 7, 7ff.

Higgins, R.A.
1982 Greek and Roman Jewellery (2nd. ed.; London)

Hilprecht, H.V.
1904 The Excavations in Assyria and Babylonia (Philadelphia)

Hirsch, H.
1963 "Die Inschriften der Könige von Agade" A.f.O.20, 1ff.

Hodges, H.W.M.
1962 "Technical Report on the Bronzes" Iraq 24, 131ff.

1964 Artifacts: an introduction to early materials and technology (London)

Holzer, H.F. and Momenzadeh, M.
1971 "Ancient Copper Mines in the Veshnoveh Area, Kuhestan-e-Qom, West Central Iran" Archaeologica Austriaca 49, 1ff.

Hoover, H.C. and L.H.
1950 Agricola, De Re Metallica (Dover, New York)

Horne, L.
1982 "Fuel for the Metal Worker: the role of charcoal and charcoal production in ancient metallurgy" Expedition 25(1), 6ff.

Howard-Carter, T.
1981 "The Tangible Evidence for the earliest Dilmun" J.C.S.33, 210ff.

Hrouda, B.
1962 Tell Halaf: Die Kleinfunde aus Historischer Zeit (Berlin)

1965 Die Kulturgeschichte des assyrischen Flachbildes (Bonn)

Hughes, M.J.
1984 "Analyses of Silver Objects" in Curtis, 1984, 58ff.

Hughes, M.J. et al.,
1981 "Analyses of some Urartian Bronzes" A.S.31, 141ff.

Huot, J-L. et al.,
1976 "Larsa...la sixième campagne de fouilles" Syria 53, 1ff.

1978 "Larsa...la septième campagne" Syria 55, 183ff.

1983 Larsa et 'Oueili: travaux de 1978-1981 (C.N.R.S., Paris)

Invernizzi, A.
1980 "Excavations in the Yelkhi Area (Hamrin Project, Iraq)" Mesopotamia 15, 19ff.

Jacobsen, Th.
1960 "The Waters of Ur" Iraq 22, 174ff.

Jankowska, N.B.
1969 "Some Problems of the Economy of the Assyrian Empire" in I.M. Diakonoff (ed.) Ancient Mesopotamia (Moscow) 253ff.

Jenkins, G.K.
1964 "Coins from the collection of C.J. Rich" B.M.Q.XXVIII, 88ff.

Jordan, J.
1928 Uruk-Warka (Leipzig)

1932 Dritter Vorläufiger Bericht...Uruk (Berlin)

Kaczmarczyk, A. and Hedges, R.E.M.
1983 Ancient Egyptian Faience (Aris and Phillips; Warminster)

Kantor, H.J.
1957 "Miscellaneous Small Objects and Burials" in McEwan, C.W. et al., Soundings at Tell Fakhariyah (O.I.P.79, Chicago) 42ff.

Karageorghis, V. et al.,
1983 "Two Silver Ingots from Cyprus" Antiquity 57, 211ff.

Kempkinski, A. and Košak, S.
1977 "Hittite Metal 'Inventories' (CTH 242) and their Economic Implications" Tel Aviv 4, 87ff.

Kendall, T.
1975 Warfare and Military Matters in the Nuzi Tablets (Ann Arbor: University Microfilms)

1982 "Gurpisu ša awēli: The Helmets of the Warriors at Nuzi" (ed. Morrison, M.A. and Owen, D.I., Studies on the Civilization and Culture of Nuzi and the Hurrians (Eisenbrauns, Indiana) 210ff.

Kent, R.G.
1950 Old Persian: grammar, texts, lexicon (New Haven, Connecticut)

Khalil, L.A.H.
1980 The Composition and Technology of Copper Artefacts from Jericho and some related sites (London University; D.Phil. Thesis unpublished)

1983 "Copper Metallurgy at Jericho" in K.M. Kenyon and T.A. Holland, Jericho V (London), 777ff.

Kiefer, C. and Allibert, A.
1971 "Pharaonic Blue Ceramics: the process of self-glazing" Archaeology 24, 107ff.

King, W.L.
1915 The Bronze Reliefs from the Gates of Shalmaneser, King of Assyria (London)

Klengel-Brandt, E.
1978 Die Terrakotten in Assur im Vorderasiatischen Museum Berlin (Berlin)

Knudtzon, J.A.
1915 Die El-Amarna-Tafeln (Leipzig)

Köcher, F.
1957-8 "Ein Inventartext aus Kar-Tukulti-Ninurta" A.f.O.18, 300ff.

Kohl, P.L. et al.,
1979 "Physical and Chemical Analyses of Soft Stone Vessels from Southwest Asia" Archaeometry 21, 131ff.

Kolbus, S.
1982 "Unter dem Schutt aus einem der ersten Handelskontore von Ur: wohin datiert der sog. Ğemdet-Nasr-Friedhof?" Akkadica 30, 1ff.

Koldewey, R.
1911 Die Tempel von Babylon und Borsippa (W.V.D.O-G.15; Leipzig)

1913 Das Wiederstehende Babylon (Leipzig)

1914 Excavations at Babylon (London)

1918 Das Ischtar-Tor in Babylon (W.V.D.O-G.32; Leipzig)

1931-2 Die Königsburgen von Babylon (W.V.D.O-G.54-5; Leipzig)

Komoróczy, G.
1972 "Das Mythische Goldland Ḫarali im alten Vorderasien" Acta Or. Hung.26, 113ff.

1977 "Tilmun als ´Speicher des Landes´ im Epos ´Enki und Ninhursag´" Iraq 39, 67ff.

Korbel, G.
1984 "Ein EDV-gestützer Gliederungs vorschlag der Keramik-Ṣtein- und Metallgefässe aus den Schächten X und W des sog. Ǧemdet Nasr-Friedhofs von Ur (Relative Chronology) Bag.Mitt. 15, 35ff.

Kosak, S.
1982 Hittite Inventory Texts (C.T.H.241-250) (Heidelberg)

Koucky, F.L. and Steinberg, A.
1982 "Ancient Mining and Mineral Dressing in Cyprus" in Wertime and Wertime 1982, 149ff.

Kozloff, A.
1981 Animals in Ancient Art: from the Leo Mildenberg Collection (Cleveland Museum of Art)

Kramer, C.
1982 Village Ethnoarchaeology: rural Iran in archaeological perspective (Academic Press, New York)

Kramer, S.N.
1963 The Sumerians (Chicago)

1963a "Dilmun: Quest for Paradise" Antiquity 37, 111ff.

Kraus, R. et al.,
1983 "The Silver hoard from City IV, Qala´at al-Bahrein" in Dilmun: New Studies in the Archaeology and Early History of Bahrein (ed. D.T. Potts, Berlin), 161ff.

Kuftin, B.A.
1941 Archaeological Excavations in Trialeti, I (Tiflis)

Kühne, H.
1969 "Bemerkungen zu einigen Glasreliefs des 2. Jahrtausends v. Chr. aus Syrien und Palästina" ZfA 25, 299ff.

1957-71 "Glas" R.L.A.3, 410ff.

1974 "´Rätselhafte Masken´ - zur Frage ihrer Herkunft und Deutung" Bag.Mitt.7, 101ff.

Kühne, K.
1974 "Frühgeschichtliche Werkstoffe auf Silikatischer Basis" Das Altertum 20, 67ff.

Kupper, J.R.
1971 "Les inscriptions triomphales akkadiennes" Oriens Antiquus 10, 91ff.

Kuschke, A.
1970 "Fayence und Frit: Bemerkungen zu einer Schale aus Kamid el-Loz" in Archäolgie und Altes Testament (Galling Festschrift, Tübingen; ed. A. Kuschke and E. Kutsch), 157ff.

Kyrieleis, H.
1969 Throne und Klinen (Berlin)

Labino, D.
1966 "The Egyptian Sand-Core Technique: A New Interpretation" J.G.S.8, 124ff.

Labrousse, A. and Boucharlat, R.
1972 "La fouille du Palais du Chaour à Suse en 1970 et 1971" D.A.F.I.2, 61ff.

Ladame, G.
1945 "Les ressources métalliferes de l'Iran" Schweizerische Mineralogische und Petrographische Mitteilungen 25, 167ff.

Laessøe, J.
1959 "Akkadian annakum: 'Tin' or 'Lead'?" Acta Orientalia (Copenhagen) 24, 83ff.

La garce, J. and L.
1983 "Les Fouilles à Ras Ibn Hani en Syrie..." Comptes Rendus (April-June), 249ff.

Lambert, M.
1963 "L'usage de l'argent-métal à Lagash en temps de la III Dynastic d'Ur" R.A.57, 79ff.; 193ff.

Lambert, W.G.
1961 "The Sultantepe Tablets VIII: Shalmaneser in Ararat" A.S.11, 144ff.

1968-9 "An inscribed strip of lead" A.f.O.22, 64.

Landsberger, B.
1965 "Tin and Lead: the adventures of two vocables" J.N.E.S.24, 285ff.

Lang, J. and Hughes, M.J.
1977 "Joining Techniques" in W.A. Oddy (ed.) Aspects of Early Metallurgy (British Museum), 169ff.

Langdon, S.
1912 Die Neubabylonischen Königsinschriften (Leipzig)

La Niece, S.
1983 "Niello: an historical and technical survey" A.J.63, 279ff.

Larsen, M.T.
1967 Old Assyrian Caravan Procedures (Istanbul)

1982 "Caravans and Trade in Ancient Mesopotamia and Asia Minor" Bulletin of the Society for Mesopotamian Studies (Toronto) 4, 33ff.

Layard, A.H.
1849 Nineveh and its Remains (2nd Ed., London)

1853 Discoveries in the Ruins of Nineveh and Babylon (London)

1853a The Monuments of Nineveh from Drawings (First Series; London)

1853b Ibid. (A second series; London)

Le Breton, L.
1957 "The Early Period at Susa, Mesopotamian relations" Iraq 19, 79ff.

Le Brun, A.
"Recherches stratigraphiques à l'Acropole de Suse (1969-1971) D.A.F.I.1, 163ff.

Lechtman, H.
1971 "Ancient Methods of Gilding Silver: examples from the Old and the New Worlds" in Science and Archaeology (R.H. Brill, ed., M.I.T. Press) 2ff.

Leemans, W.F.
1957-71 "Gold" R.L.A.3, 504ff.

1960 Foreign Trade in the Old Babylonian Period as revealed by texts from southern Mesopotamia (Leiden)

Legrain, L.
1937 Business Documents of the Third Dynasty of Ur (Ur Excavations Texts, vol.III (i), London).

1947 Ibid., vol.III (ii), (London).

Lehrer, G.
1974 "A Phoenician Glass Bowl from Nimrud" J.G.S.16, 9ff.

Lenzen, H.
1958 XIV Vorläufiger Bericht...Ausgrabungen Uruk-Warka (Berlin)

1959 XV Ibid. (Berlin)

1960 XVI Ibid. (Berlin)

1962 XVIII Ibid. (Berlin)

Lerberghe, K. van
forthcoming "Contribution à l'étude des métaux de Tell ed-Der" Tell ed-Der 4.

Levey, M.
1957 "L´argent dans la litérature ancienne de Mesopotamie" Revue Hist.Sci.Fr.10(3), 197ff.

1959 Chemistry and Chemical Technology in Ancient Mesopotamia (Elsevier; Amsterdam)

Levey, M. and Burke, J.E.
1959 "A Study of Ancient Mesopotamian Bronze" Chymia 5, 37ff.

Lie, A.G.
1929 The Inscriptions of Sargon II, King of Assyria (Paris)

Limet, H.
1960 Le travail du métal au pays de Sumer au temps de la IIIe dynastie d´Ur (Paris)

1972 "Les métaux à l´époque d´Agade" J.E.S.H.O.15, 3ff.

1976 "Les origines du travail du fer en Mésopotamie ancienne" Présences, 92, 3ff.

1978 "L´or des Hourrites: étude du terme hiyaruhhe" R.H.A.36, 141ff.

Lins, P.A. and Oddy, W.A.
1975 "The Origins of Mercury Gilding" Journal of Archaeological Science 2, 365ff.

Lipinski, E.
1979 "Les temples Néo-Assyriens et les origines du monnayage" in E. Lipinski (ed.), State and Temple Economy in the Ancient Near East (Leuven) II, 565ff.

Littauer, M.A. and Crouwel, J.
1979 Wheeled Vehicles and Ridden Animals in the Ancient Near East (Leiden)

Liu, R.K.
1975 "Faience: technical aspects and beads of simple form" Bead Journal 1(4), 24ff.

Liverani, M.
1972 ´Elementi "irrazionali" nel Commercio Amarniano´ Oriens Antiquus 11, 297ff.

Lloyd, S.
1954 "Sultantepe (Part II) A.S.4, 101ff.

Lloyd, S. and Gökce, N.
1953 "Sultantepe" A.S.3, 27ff.

Lloyd, S. and Safar, F.
1943 "Tell Uqair: Excavations by the Iraq Government Directorate of Antiquities in 1940 and 1941" J.N.E.S.2, 131ff.

Loding, D.
1981 "Lapidaries in the Ur III Period" Expedition 23(4), 6ff.

Loftus, W.K.
1857 Travels and Researches in Chaldaea and Susiana (London)

Loud, G.
1936 Khorsabad I: Excavations in the Palace and at a City Gate (O.I.P.38, Chicago)

1948 Megiddo II: Seasons of 1935-39 (O.I.P.62, Chicago)

Loud, G. and Altman, C.B.
1938 Khorsabad II: The Citadel and the Town (O.I.P.40, Chicago)

Lowery, P.R. et al.,
1971 "Scriber, Graver, Scorper, Tracer: notes on experiments in bronzeworking technique" P.P.S.37, 167ff.

Lucas, A.
1962 Ancient Egyptian Materials and Industries (London; 4th edition, revised by J.R. Harris)

Luckenbill, D.D.
1924 The Annals of Sennacherib (O.I.P.2, Chicago)

1926-7 Ancient Records of Assyria and Babylonia (Chicago)

Luschey, H.
1939 Die Phiale (Bleicherode am Harz)

McCown, D.E.
1967 Nippur I: Temple of Enlil, Scribal Quarter and Soundings (O.I.P.78; Chicago)

Mackay, E.J.H.
1925 Report on the Excavations of the ´A´ Cemetery at Kish, Mesopotamia, I (Chicago)

1929 A Sumerian Palace and the ´A´ Cemetery at Kish, Mesopotamia II (Chicago)

1931 Report on the Excavations at Jemdet Nasr, Iraq (Chicago)

1933 "Decorated Carnelian Beads" Man 33, no.150.

McKerrell, H.
1977 "Non-dispersive XRF applied to Ancient Metalworking in Copper and Tin-bronze" P.A.C.T.1, 138ff.

1978 "The Use of Tin-bronze in Britain and the Comparative Relationship with the Near East" in The Search for Tin (A.D. Franklin et al., (ed.), Washington D.C.), 7ff.

McKerrell, H. and Tylecote, R.F.
1972 "The Working of Copper-arsenic Alloys in the Early Bronze Age and the Effect on the Determination of Provenance" P.P.S.38, 209ff.

Maddin, R. et al.,
1977 "How the Iron Age Began" Scientific American 237(4), 122ff.

1980 "Distinguishing Artifacts made of Native Copper" J. of Arch. Sc. 7, 211ff.

1983 "Metal Working" in T. Dothan and Amnon Ben-Tor, Excavations at Athienou, Cyprus, 1971-1972 (Qedem, 16) 132ff.

1983a "Metallurgical Analysis of an Early Iron Age Laminated Iron Tool" in Waldbaum, 1983, 178ff.

Madhlum, T.A.
1960 "The Excavations at Tell al-Wilayah" Sumer 16, 62ff. (Arabic)

1970 The Chronology of Neo-Assyrian Art (London)

Mahmoud, Y.
1970 "Tell al-Fakhar: Report on the First Season's Excavations" Sumer, 26, 109ff.

Maigret, A. de
1976 Le Lance nell'Asia Anteriore nell'Età del Bronzo: Studio Tipologico (Rome)

Maisler (Mazar) B.
1951 "Two Hebrew ostraca from Tell Qasileh" J.N.E.S.10, 265ff.

Majidzadeh, Y.
1976 "The Land of Aratta" J.N.E.S.35, 105ff.

Mallet, J.
1975 "Mari: une nouvelle coutume funéraire assyrienne" Syria 52, 23ff.

Mallowan, M.E.L.
1935 "Excavations at Tall Arpachiya, 1933" Iraq 2, 1ff.

1936 "The excavations at Tall Chagar Bazar, and an archaeological survey of the Habur region 1934-5" Iraq 3, 1ff.

1937 "The excavations at Tall Chagar Bazar and an archaeological survey of the Habur region. Second Campaign 1936" Iraq 4, 91ff.

1947 "Excavations at Brak and Chagar Bazar" Iraq 9, 1ff.

1950 "The excavations at Nimrud (Kalhu) 1949-50" Iraq 12, 147ff.

1953 "The excavations at Nimrud (Kalhu) 1952" Iraq 15, 1ff.

1954 "The excavations at Nimrud (Kalhu) 1953" Iraq 16, 59ff.

1966 Nimrud and its Remains (2 vols.; London)

Maréchal, J.R.
1958 "Etude sur les propriétés méchaniques des cuivres à l'arsenic" Métaux. Corrosion-Industries 33, (no.397), 378.

1962 Considérations sur la métallurgie préhistorique (Lammersdorf n. Aachen)

Margueron, J-C.
1965 Mesopotamia (London)

1979 "Existe-t-il des ateliers dans les palais orientaux de l'âge du Bronze?" Ktema 4, 1ff.

Mariani, L.
1984 "Craftsmen's Quarters in the Proto-urban Settlements of the Middle East: the surface analysis" in B. Allchin (ed.), South Asian Archaeology 1981 (Cambridge), 118ff.

Martin, G.T.
1971 Egyptian Administrative and Private-name Seals, principally of the Middle Kingdom and Second Intermediate Period (Oxford)

Martin, H.P.
1982 "The Early Dynastic Cemetery at Al'Ubaid, a re-evaluation" Iraq 44, 145ff.

Maryon, H.
1949 "Metal Working in the Ancient World" A.J.A.53, 93ff.

Maryon, H. and Plenderleith, H.J.
1954 "Fine Metalwork" in a History of Technology (ed. C. Singer et al.; Oxford), I, 623ff.

Matson, F.R.
1957 "A Study of Glass; colourless glass; opaque red glass" in E.F. Schmidt, Persepolis II (Chicago), 127ff.

Matthiae, P.
1980 "Fouilles à Tell Mardikh-Ebla, 1978: le bâtiment Q et la nécropole princière du Bronze Moyen II" Akkadica 17, 1ff.

Maxwell-Hyslop, R.
1946 "Daggers and Swords in Western Asia" Iraq 8, 1ff.

1949 "Western Asiatic shaft-hole axes" Iraq 9, 90ff.

1970 "Near Eastern Gold 'Treasures': a note on the Assyrian Evidence" Antiquity 44, 227ff.

1970a "An Illustration to a Mari Inventory?" Iraq 32, 165ff.

1971 Western Asiatic Jewellery, c.3000-612 B.C. (London)

1972 "The Metals Amūtu and Ašī'u in the Kültepe Texts" A.S.22, 159ff.

1974 "Assyrian Sources of Iron" Iraq 36, 139ff.

1977 "Sources of Sumerian Gold" Iraq 39, 83ff.

Mayer, W.
1979 "Die Finanzierung einer Kampagne" Ugarit Forschungen 11, 571ff.

Mazzoni, S.
1979-80 "Essai d'interprétation des vases plastiques dans la Syrie du bronze moyen et récent" A.A.A.S.29-30, 237ff.

Medvedskaya, I.N.
1982 Iran: Iron Age I (B.A.R. International Series 126, Oxford)

Meeks, N.D. and Tite, M.S.
1980 "The Analysis of Platinum-group Element Inclusions in Gold Antiquities" Journal of Archaeological Science 7, 267ff.

Meissner, B.
1912 "Woher haben die Assyrer Silber bezogen?" O.L.Z.15, 145ff.

1914 "Das Antimongebirge" O.L.Z.17, 52ff.

1920 Babylonien und Assyrien I (Heidelberg)

1925 Ibid., II (Heidelberg)

Mellaart, J.
1967 Çatal Hüyük; a Neolithic Town in Anatolia (London)

1968 "Anatolian Trade with Europe and Anatolian Geography and Cultural Provinces in the Late Bronze Age" A.S.18, 187ff.

Merpert, N. et al.,
1976 "The Investigations of Soviet Expedition in Iraq, 1973" Sumer 32, 25ff.

1977 "The Investigations of Soviet Expedition in Iraq, 1974" Sumer 33, 65ff.

1977a "The Earliest Metallurgy of Mesopotamia" (in Russian) Sov. Arkh. (no.3), 154ff.

1978 "Soviet Investigations in the Sinjar Plain" Sumer 34, 27ff.

1982 "An den Anfängen der Geschichte Mesopotamiens" Das Altertum 28, 69ff.

Meyer, G.R.
1965 Altorientalische Denkmäler im Vorderasiatischen Muzeum zu Berlin (Leipzig)

1968 "Bemerkungen zu einigen Urartischen Bronzen aus Toprak-Kale" Festschrift Werner Caskel (ed. E. Gräf; Leiden), 212ff.

Meyers, P.
1981 "Technical Studies of Sasanian Silver" M.A.S.C.A.Journal 1(8), 242ff.

Michalowski, P.
1978 The Neo-Sumerian Ring Texts (Undena, Malibu)

Millard, A.R.
1973 "Cypriot Copper in Babylonia c.1745 B.C." J.C.S.25, 211ff.

Millichamp, J.W. and Levey, M.
1963 "Silver from Ur of Ancient Mesopotamia" Science 142, 44ff.

Minoff, E.
1983 "Late deliveries and customer complaints in the ancient metals industry of the Near East" Historical Metallurgy 17, 42ff.

Mitchell, T.C.
1983 "An Urartian Lead Figurine from Toprak Kale" A.S.33, 157ff.

Mkrtiachan, B.
1967 "The Mystery of Metsamor" New Orient 3, 76ff.

Moore, D.T. and Oddy, W.A.
(forthcoming) "Touchstones: some aspects of their nomenclature, petrography and provenance" Journal of Archaeological Science

Moore, H.
1940 "Reproductions of an Ancient Babylonian Glaze" Iraq 10, 26ff.

Moorey, P.R.S.
1971 Catalogue of the Ancient Persian Bronzes in the Ashmolean Museum (Oxford)

1971a "The Loftus Hoard of Old Babylonian Tools from Tell Sifr in Iraq" Iraq 33, 61ff.

1977 "Bronze Rollers and Frames from Babylonia and Western Iran: Problems of Date and Function" R.A.71, 137ff.

1978 Kish Excavations 1923-1933 (Oxford)

1980 Cemeteries of the First Millennium B.C. at Deve Hüyük (B.A.R. International Series, no.87, Oxford)

1982 "The Archaeological Evidence for Metallurgy and related Technologies in Mesopotamia c.5500-2100 B.c." Iraq 44, 13ff.

1982a "Archaeology and Pre-Achaemenid Metalworking in Iran: a fifteen year retrospective" Iran 20, 81ff.

Moorey, P.R.S. and Schweizer, F.
1972 "Copper and Copper Alloys in Ancient Iraq, Syria and Palestine: some new analyses" Archaeometry 14, 177ff.

Moortgat, A.
1940 Vorderasiatische Rollsiegel (Berlin)

1959 Neue deutsche Ausgrabungen Mittelmeer und Vorderer Orient (Berlin)

1960 Tell Chuera in Nordost-Syrien... Grabungskampagne 1959 (Wiesbaden)

1967 Tell Chuera in Nordost-Syrien... Grabungskampagne 1964 (Wiesbaden)

1969 The Art of Ancient Mesopotamia (London)

Moortgat, A. and Moortgat-Correns, U.
1976 Tell Chuera in Nordost-Syrien... Grabungskampagne 1974 (Berlin)

1978 Tell Chuera in Nordost-Syrien... Grabungskampagne 1976 (Berlin)

Muhly, J.D.
1972 Review of Oppenheim et al., 1970 in J.C.S.24, 178ff.

1973 Copper and Tin: the distribution of mineral resources and the nature of the metals trade in the Bronze Age (New Haven, Connecticut)

1976 Supplement to Copper and Tin (Hamden, Connecticut)

1978 "New Evidence for Sources of and Trade in Bronze Age Tin" in Franklin et al., 1978, 39ff.

1980 "The Bronze Age Setting" in Wertime and Muhly, 1980, 25ff.

1983 "Kupfer B.Archäologische" R.L.A.6, 348ff.

Muhly, J.D. et al.,
1980 "New Research on Ancient Copper and Copper Alloys" Proceedings of the 16th International Symposium on Archaeometry and Archaeological Prospection (Edinburgh; ed. E.A. Slater and J.O. Tate), 248ff.

1982 Early Metallurgy in Cyprus, 4000-500 B.C. (Nicosia)

Muhly, J.D. and Wertime, T.A.
1973 "Evidence for the sources and use of tin during the Bronze Age of the Near East: a reply to J.E. Dayton" World Archaeology 5, 111ff.

Müller, M.
1982 "Gold, Silber und Blei als Wertmesser in Mesopotamien während der Zeiten Hälfte des 2.Jahrtausends v.u.Z" Studies in Honour of I.M.Diakonoff (Aris and Phillips, Warminster) 270ff.

Muscarella, O.
1974 (ed.) Ancient Art: The Norbert Schimmel Collection (Mainz)

1977 ""Ziwiye" and Ziwiye: the forgery of a provenience" J.F.A.4, 197ff.

Na´aman, N.
1981 "The Recycling of a Silver Statue" J.N.E.S.40, 47ff.

Nashef, K.
1982 Die Orts-und Gewässernamen der mittelbabylonischen und mittelassyrischen Zeit (Wiesbaden; Répertoire Géographique 5)

Naumann, R.
1950 Tell Halaf II: Die Bauwerke (Berlin)

Neumann, B.
1927 "Der babylonisch-assyrische künstliche Lasurstein" Chemische Zeitung 51, 103ff.

Newberry, P.E.
1900 The Life of Rekhmara... (Westminster)

Newell, E.T.
1978 The Coinage of the Eastern Seleucid Mints from Seleucus I to Antiochus III (New York)

Nissen, H.J.
1966 Zur Datierung des Königsfriedhofes von Ur (Bonn)

1970 "Grabung in den Planquadraten K/L XII in Uruk-Warka" Bag. Mitt.5, 101ff.

Noble, J.V.
1969 "The Technique of Egyptian Faience" A.J.A.73, 435ff.

Nöldeke, A.
1936 Siebenter Vorläufiger Bericht...Uruk-Warka (Berlin)

1937 Achter Vorläufiger Bericht...Uruk-Warka (Berlin)

1938 Neunter Vorläufiger Bericht...in Uruk-Warka (Berlin)

Notton, J.H.F.
1974 "Ancient Egyptian Gold Refining: a reproduction of early techniques" Gold Bulletin 7(2), 50ff.

Nylander, C.
1970 Ionians in Pasargadae (Uppsala)

1980 "Earless in Nineveh: Who mutilated Sargon´s Head?" A.J.A. 84, 329ff.

Oates, David
1962 "The excavations at Nimrud (Kalhu), 1961" Iraq 24, 1ff.

1965 "The excavations at Tell al-Rimah, 1964" Iraq 27, 62ff.

1967 "The excavations at Tell al-Rimah, 1966" Iraq 29, 70ff.

1968 "The excavations at Tell al-Rimah, 1967" Iraq 30, 115ff.

1982 "Tell al-Rimah" in J. Curtis (ed.) Fifty Years of Mesopotamian Discovery (London), 86ff.

Oates, David and Joan
1958 "Nimrud 1957: The Hellenistic Settlement" Iraq 20, 114ff.

Oates, Joan
1959 "Late Assyrian Pottery from Fort Shalmaneser" Iraq 21, 130ff.

1979 Babylon (London)

Oda, S.
1965 "Chemical Study on the Material of Beads from the Ghalekuti I Site" in Egami et al., 1965, 33ff.

Oddy, W.A.
1977 "The Production of Gold Wire in Antiquity" Gold Bulletin 10(3), 79ff.

1979 "Hand-made Wire in Antiquity: a correction" M.A.S.C.A. Journal 1(2), 44ff.

1980 Aspects of Early Metallurgy (British Museum, Occasional Paper, no.17)

1983 "Assaying in Antiquity" Gold Bulletin 16(2) 52ff.

1984 "Silver wire Jewellery from Tepe Nush-i Jan" in Curtis, 1984, 61.

Oddy, W.A. et al.,
1981 "Diffusion - Bonding as a Method of Gilding in Antiquity" M.A.S.C.A.Journal 1(8), 238ff.

Oddy, W.A. and Meeks, N.D.
1978 "A Parthian Bowl: study of gilding technique" M.A.S.C.A. Journal 1, 5ff.

Ogden, J.M.
1977 "Platinum Group Metal Inclusions in Ancient Gold Artefacts" Journal of the Historical Metallurgy Society 11, 53ff.

1982 Jewellery of the Ancient World (London)

Olin, J.S. and Franklin, A.D. (Eds.)
1982 Archaeological Ceramics (Smithsonian Press, Washington)

Oliver, A.
1970 "Persian Export Glass" J.G.S.12, 9ff.

1973 "Rock Crystal Vessels in Antiquity" Annual of the Museum of Art and Archaeology University of Missouri-Columbia 7, 29ff.

Opitz, D.
1933 "Altorientalische Gussformen" in Aus fünf Jahrtausenden morgenländischer Kultur (Berlin), 178ff.

Oppenheim, A.L.
1947 "A Fiscal Practice of the Ancient Near East" J.N.E.S.6, 116ff.

1949 "The Gold Garments of the Gods" J.N.E.S.8, 172ff.

1954 "The Seafaring Merchants of Ur" J.A.O.S.74, 6ff.

1966 "Mesopotamia in the Early History of Alchemy" R.A.60, 29ff.

1967 "Essay on Overland Trade in the First Millennium B.C." J.C.S.21, 236ff.

1973 "Towards a History of Glass in the Ancient Near East" J.A.O.S.93, 259ff.

1973a "More Fragments with Instructions for Glass Making" J.N.E.S.32, 188ff.

1976 "Trade in the Ancient Near East" Fifth International Congress of Economic History (Moscow), vol.5, 125ff.

1977 Ancient Mesopotamia (Revised, 2nd edition; Chicago)

1978 "Man and Nature in Mesopotamian Civilization" Dictionary of Scientific Biography (New York), 15, Supplement 1, 634ff.

Oppenheim, A.L. et al.,
1970 Glass and Glassmaking in Ancient Mesopotamia (Corning, New York)

Oppert, J.
1863 Expédition scientifique en Mésopotamie (Paris)

Orchard, J.J. and Brill, R.
1978 "Some Miniature Painted Glass Plaques from Fort Shalmaneser, Nimrud" Iraq 40, 1ff.

Organ, R.M.
1963 "The Consolidation of Fragile Metallic Objects" in G.Thompson (ed.) Recent Advances in Conservation (London), 128ff.

1967 "The Reclamation of the Wholly-mineralized Silver in the Ur lyre" in Application of Science in Examination of Works of Art (Boston, Mass.) 126ff.

Ozgüc, T.
1955 "Report on a Work-shop belonging to the late phase of the Colony Period Ib" Belleten 73, 77ff.

Palmieri, A.
1973 "Scavi nell' area sud-occidentale di Arslantepe" Origini 7, 55ff.

1981 "Excavations at Arslantepe (Malatya)" A.S.31, 101ff.

Panayiotu, A.
1979 "Geology, Mineralogy, Geochemistry and origin of the Cu-Ni-Co-Fe sulphide mineralization, Limassol Forest Cyprus" in Abstracts, International Ophiolite Symposium, Nicosia, Cyprus (Nicosia; Geological Department Survey), 141ff.

Parker, B.
1949 "Cylinder Seals from Palestine" Iraq 11, 1ff.

1955 "Excavations at Nimrud 1949-1953: Seals and Seal Impressions" Iraq 17, 93ff.

1962 "Seals and Seal Impressions from the Nimrud Excavations, 1955-58" Iraq 24, 26ff.

Parpola, S. et al.,

1977 "The Meluḫḫa Village: evidence of acculturation of Harappan trade in late third millennium Mesopotamia" J.E.S.H.O.20, 130ff.

Parrini, P. et al.,
1982 "Etruscan Granulation: Analysis of Orientalizing Jewellery from Marsiliana d'Albegna" A.J.A.86, 118ff.

Parrot, A.
1933 "Les fouilles de Tello et de Senkereh-Larsa: campagne 1932-1933" R.A.30, 168ff.

1935 "Les fouilles de Mari: première campagne (hiver 1933-1934)" Syria 16, 1ff.

1937 "Les fouilles de Mari: troisième campagne (hiver 1935-1936)" Syria 18, 54ff.

1948 Tello (Paris)

1956 Mission archéologique de Mari I: Le Temple d'Ishtar (Paris)

1956a "Asarhaddon et Naqi'a sur un bronze du Louvre" Syria 33, 147ff.

1958 "Plaques assyriennes" Syria 35, 171ff.

1958a Mission archéologique de Mari II: Le Palais: Architecture (Paris)

1959 Mission archéologique de Mari III: Le Palais: Documents et Monuments (Paris)

1960 Sumer (London)

1968 Mission archéologique de Mari IV: Le "Trésor" de Mari (Paris)

1975 "Les fouilles de Mari XXIe campagne de fouilles (Automne 1974)" Syria 52, 1ff.

Partington, J.R.
1935 Origins and Development of Applied Chemistry (London)

Patterson, C.C.
1971 "Native Copper, Silver and Gold accessible to Early Metallurgists" American Antiquity 36, 286ff.

Payne, H.
1962 Perachora II (Oxford)

Peake, H.
1928 "The copper mountain of Magan" Antiquity 2, 452ff.

Peltenburg, E.J.
1968 "Assyrian clay fists" Oriens Antiquus 7, 57ff.

1969 "Al Mina Glazed Pottery and its Relations" Levant 1, 73ff.

1971 "Some Early Developments of Vitreous Materials" World Archaeology 3, 6ff.

1972 "On the Classification of Late Bronze Age Faience Vases from Cyprus" Acts of the First International Congress of Cypriot Studies A1 (Nicosia), 129ff.

1974 "The Glazed Vases, including a polychrome rhyton" in V. Karageorghis, Kition I: The Tombs (text) (Nicosia), 105ff.

1976 "The Faience Vases from Tombs 1 and 2 at Hala Sultan Tekke, Vizaja" in P. Åström et al., Hala Sultan Tekke 1 (Göteborg) 104ff.

1977 "A Faience from Hala Sultan Tekke and Second Millennium B.C. Western Asiatic Pendants depicting Females" in P. Åström et al., Hala Sultan Tekke 3 (Göteborg), 177ff.

1983 "Hala Sultan Tekke Glazed Vessels 1972-1979" in P. Åström et al., Hala Sultan Tekke 8: Excavations 1971-1979 (Göteborg), 214ff.

Peltenburg, E.J. and Caubet, A.
1982 "Kition Bamboula Faience" R.D.A.C, 83ff.

Peltenburg, E.J. and Karageorghis, V.
1970 "A Faience Alabastron from Cyprus" Archaeology 23, 217ff.

Perkins, A.L.
1949 The Comparative Archaeology of Early Mesopotamia (Chicago)

Perrot, G. and Chipiez, Ch.
1884 Histoire de l'Art dans l'Antiquité II: Chaldée et Assyrie (Paris)

Perrot, J.
1955 "Excavations at Tell Abu Matar, near Beersheba" I.E.J.5, 73ff.

Peters, J.P.
1897 Nippur or Explorations and Adventures on the Euphrates (New York and London), 2 vols.

Pettinato, G.
1972 "Il commercio con l´estero della Mesopotamia...alla luce delle fonti letterarie e lessicale sumeriche" Mesopotamia 7, 43ff.

Petrie, W.M.F.
1894 Tell el Amarna (London)

1896 Naqada and Ballas (London)

1909 The Arts and Crafts of Ancient Egypt (London)

1920 Prehistoric Egypt (London)

Piaskowski, J.
1982 "A Study of the Origin of the Ancient High-nickel Iron generally regarded as Meteoric" in Early Pyrotechnology (Smithsonian, Washington, DC.); Wertime T. and S. (eds.), 237ff.

Pigott, V.C.
1980 "The Iron Age in Western Iran" in Wertime and Muhly, 1980, 417ff.

1981 The Adoption of Iron in Western Iran in the Early First Millennium B.C.: an Archaeometallurgical Study (University Microfilms International, Ann Arbor)

1982 "The Innovation of Iron: cultural dynamics in technological change" Expedition 25(1), 20ff.

Pillet, M.
1962 Un pionnier de l´assyriologie (Paris)

Place, V.
1867-70 Ninive et l´Assyrie (3 vols.; Paris)

Pleiner, R.
1979 "The Technology of three Assyrian Iron Artifacts from Khorsabad" J.N.E.S.38, 83ff.

Pleiner, R. and Bjorkman, J.K.
1974 "The Assyrian Iron Age: the history of iron in the Asyrian civilization" Proceedings of the American Philosophical Society 118, 283ff.

Plenderleith, H.J.
1934 "Metals and Metal Technique" in Woolley, 1934, 284ff.

Pollard, A.M. and Høljund, F.
1983 "High Magnesium Glazed Sherds from Bronze Age Tells on Failaka, Kuwait" *Archaeometry* 25, 196ff.

Pollard, A.M. and Moorey, P.R.S.
1982 "Some Analyses of Middle Assyrian Faience and Related Materials from Tell al-Rimah in Iraq" *Archaeometry* 24, 45ff.

Pope, A.U.
1939 *A Survey of Persian Art*, I-VI (Oxford)

Popham, M. *et al.*,
1982 "Further Excavation of the Toumba Cemetery at Lefkandi, 1981" *B.S.A.*77, 213ff.

Porada, E.
1950 "An Assyrian Bronze Disc" *Bulletin of the Metropolitan Museum of Fine Art, New York* 48, 2ff.

1965 *Ancient Iran: the Art of Pre-Islamic Times* (London)

1970 *Tchoga Zanbil* IV: *la glyptique* (M.D.P.42; Paris)

1976 *Ancient Mesopotamian Art and Selected Texts* (Morgan Library, New York)

Postgate, J.N.
1973 "Excavations in Iraq 1972-3" *Iraq* 35, 189ff.

1973a *The Governor's Palace Archive* (London)

1973b "Assyrian texts and fragments" *Iraq* 35, 13ff.

1977 "Excavations in Iraq, 1976" *Iraq* 39, 301ff.

1980 "Excavations at Abu Salabikh 1978-79" *Iraq* 42, 87ff.

Postgate, J.N. and Reade, J.E.
1976-80 "Kalhu" *R.L.A.*5, 303ff.

Potratz, J.A.H.
1966 *Die Pferdetrensen des Alten Orient* (Rome)

Potts, D.
1979 "Towards an Integrated History of Culture Change in the Arabian Gulf Area: Notes on Dilmun, Makka and the Economy of ancient Sumer" *J.O.S.*4, 29ff.

1982 "The Road to Meluḫḫa" *J.N.E.S.*41, 279ff.

1983 "Dilmun: Where and When?" *Dilmun* 11, 15ff.

Powell, M.A.
1973 *Sumerian Numeration and Metrology* (University Microfilms International, Ann Arbor)

1978 "A Contribution to the History of Money in Mesopotamia prior to the Invention of Coinage" Festschrift Lubor Matouš (Budapest; ed. B. Hruška-G.Komoróczy), 211ff.

Prag, K.
1978 "Silver in the Levant in the Fourth Millennium B.C." in Archaeology in the Levant: Essays for Kathleen Kenyon (eds. P.R.S. Moorey and P.J. Parr; Warminster), 36ff.

Preusser, C.
1954 Die Wohnhäuser in Assur (W.V.D.O-G.64; Berlin)

1955 Die Paläste in Assur (W.V.D.O-G.66; Berlin)

Ramage, A.
1970 "Pactolus North" in Hanfmann, G.M.A. and Waldbaum, J.C. "The Eleventh and Twelfth Campaigns at Sardis (1968, 1969)" B.A.S.O.R.199, 16ff.

Ransom Williams, C.
1932 The Decoration of the Tomb of Per-Neb (New York)

Rashid, S.A.
1983 Gründungsfiguren im Iraq (Prähistorische Bronzefunde, Abt.I, Band 2; Munich)

Rassam, H.
1897 Asshur and the Land of Nimrod (New York)

Ratnagar, S.
1981 Encounters: the Westerly Trade of the Harappan Civilization (O.U.P., (Delhi)

Rawlinson, H.C.
1880 "Notes on Captain Durand's report upon the Island of Bahrein" J.R.A.S., 13ff.

Reade, J.E.
1963 "A glazed-brick panel from Nimrud" Iraq, 25, 38ff.

1968 "Tell Taya (1967): Summary Report" Iraq 30, 234ff.

1971 "Tell Taya (1968-9): Summary Report" Iraq 33, 87ff.

1973 "Tell Taya (1972-73): Summary Report" Iraq 35, 189ff.

1979 Early Etched Beads and the Indus-Mesopotamian Trade (British Museum Occasional Papers, no.2)

1979a "Assyrian architectural decoration: techniques and subject-matter" Bag.Mitt.10, 17ff.

1983 Assyrian Sculpture (British Museum, London)

Reiner, E.
1973 "The location of Anšan" R.A.67, 57ff.

Reisner, G.A.
1923 Excavations at Kerma Parts IV-V (Harvard African Studies 6, Cambridge, Mass.)

Renfrew, C. and Dixon, J.
1976 "Obsidian in Western Asia: a review" Problems in Economic and Social Archaeology (ed. G. de G. Sieveking et al., London), 137ff.

Reuther, O.
1926 Die Innenstadt von Babylon (Merkes) (W.V.D.O-G.47; Leipzig)

Rich, C.J.
1816 Memoir on the Ruins of Babylon (London)

Richardson, H.G.
1943 "A Mitannian Battle Axe from Ras Shamra" Berytus 8, 72.

Riederer, J.
1976 "Kupfergeräte aus Habuba Kabira und Mumbaqat Untersuchungsbericht" M.D.O-G. 108, 23-4.

Roaf, M.
1982 "Weights on the Dilmun Standard" Iraq 44, 137ff.

1983 "Sculptures and Sculptors at Persepolis" Iran 21, 1ff.

Roberts, P.M.
1973 "Gold Brazing in Antiquity" Gold Bulletin 6(4), 112ff.

1974 "Early evolution of Brazing" Welding and Metal Fabrication 42, 412ff.

Robinson, E.S.G.
1950 ´A "silversmith´s hoard" from Mesopotamia´ Iraq 12, 44ff.

Rollig, W. and Muhly, J.D.
1983 "Kupfer" R.L.A.6, 345ff.

Rothenberg, B.
1972 Timna: Valley of the Biblical Copper Mines (London)

Rothenberg, B. et al.,
1978 Chalcolithic Copper Smelting (Institute for Archaeo-Metallurgical Studies, London)

Rouault, O.
1977 A.R.M.18: Mukannišum: L´administration et l´économie palatiales de Mari (Paris)

Ryan, C.W.
1960 A Guide to the Known Minerals of Turkey (Ankara)

Ryzanov, V.
1979 "On general ancient tin ore sources on the territory of Uzbekistan" (Russian) Material for the History of Uzbekistan 15, 98ff.

Safar, F.
1981 Eridu (Baghdad)

Saggs, H.W.F.
1968 "The Tell al-Rimah Tablets" Iraq 30, 154ff.

Sagona, C.
1980 "Middle Bronze Age Faience Vessels from Palestine" Z.D.P.V. 96, 101ff.

Salonen, A.
1964 "Die Ofen der alten Mesopotamier" Bag.Mitt.3, 114ff.

1963 Die Möbel des alten Mesopotamien nach sumerisch-akkadischen Quellen (Helsinki)

Sayre, E.V.
1963 "The Intentional Use of Antimony and Manganese in Ancient Glasses" Advances in Glass Technology, 2 (Plenum Press, New York), 263ff.

Sayre, E.V. and Smith, R.W.
1974 "Analytical Studies of Ancient Egyptian Glass" Recent Advances in Science and Technology of Materials 3 (Plenum Press, New York; ed. A. Bishay), 47ff.

Schaeffer, C.F.A.
1939 Ugaritica I (Paris)

1948 Stratigraphie comparée et chronologie de l´Asie occidentale (IIIe et IIe millénaires) (Oxford)

1962 Ugaritica IV (Paris)

1971 Alasia I (Paris)

Schliemann, H.
1884 Troja: results of the latest researches and discoveries on the site of Homer´s Troy (London)

Schlossman, B.L.
1981/2 "Portraiture in Mesopotamia in the late third and early second millennium B.C." A.f.O.28, 143ff.

Schmandt-Besserat, D.
1979 (edited) Early Technologies (Undena Publications, Malibu)

1980 "Ocher in Prehistory: 300,000 Years of the Use of Iron Ores as Pigments" in Wertime and Muhly, 1980, 127ff.

Schmidt, E.F.
1932 The Alishar Hüyük: Seasons of 1928 and 1929, I (Chicago; O.I.P.19)

1953 Persepolis I (Chicago)

1957 Persepolis II (Chicago)

Schmidt, H.
1943 Tell Halaf I: Die Prähistorischen Funde (Berlin)

Schmidt, J.
1978 XXVIII.vorläufiger Bericht...in Uruk Warka (Berlin)

Schuler, F.
1959 "Ancient Glassmaking Techniques: the moulding process" Archaeology 12, 47ff.

1959a "Ancient Glassmaking Techniques: the blowing process" Archaeology 12, 116ff.

1962 "Ancient Glassmaking Techniques: the Egyptian core vessel process" Archaeology 15, 32ff.

Sebelien, J.
1924 "Early Copper and its Alloys" Ancient Egypt I (March), 6ff.

Selimkhanov, I.R.
1975 "Sur l´étude du fragment de vase de Tello...et le problème de l´utilisation de l´antimone dans l´antiquité" Annales, 45ff.

1977 "Zu Frage einer Kupfer-Arsen-Zeit" Germania 55, 1ff.

1978 "Ancient Tin Objects of the Caucasus and the Results of their Analysis" in Franklin et al., 1978, 53ff.

Semenov, S.A.
1964 Prehistoric Technology (New York)

Shaffer, F.R.
1978 "The Later Prehistoric Periods" in Allchin, F.R. (ed.) The Archaeology of Afghanistan (Academic Press, London)

Shareq, A. et al.,
1977 Mineral Resources of Afghanistan (Ministry of Mines and Industries, Kabul; United Nations Development Support Project)

Smith, C.S.
1967 "The Interpretation of Microstructures of Metallic Artefacts" in Application of Science in Examination of Works of Art (Boston Museum of Fine Arts) 20ff. (Reprinted in Smith, 1981, 69ff.)

1969 "Analysis of the Copper Bead from Ali Kosh" in F. Hole (ed.) Prehistory and Human Ecology of the Deh Luran Plain (Ann Arbor), 427ff.

1970 Report in P. Mortensen, Tell Shimshara: The Hassuna Period (Copenhagen) 123-4.

1971 "The Techniques of the Luristan Smith" in R.H. Brill (ed.) Science and Archaeology (M.I.T. Press) 32ff.

1972 "Metallurgical Footnotes to the History of Art" Proceedings of the American Philosophical Society 116(2), 97ff. (Reprinted in Smith, 1981, 242ff.).

1973 "An Examination of the Arsenic-rich Coating on a Bronze Bull from Horoztepe" in Application of Science in Examination of Works of Art (ed. W.J. Young; Boston) 96ff.

1974 "Historical Notes on the Coloring of Metals" in A. Bishay (ed.) Recent Advances in Science and Technology of Materials (New York; Plenum Press), 157ff.

1981 A Search for Structure: selected essays on Science, Art and History (M.I.T. Press; Boston, Mass.).

Smith, E.A.
1930 "Solders used by the goldsmiths of Ur" Discovery 20ff.

Smith, G.
1875 Assyrian Discoveries: an account of explorations and discoveries on the site of Nineveh, during 1873 and 1874 (London; 2nd edition)

Smith, S.
1922 "A pre-Greek coinage in the Near East" Num.Chron.Vth ser.2, 2ff.

1928 Early History of Assyria to 1000 B.C. (London)

Smithson, H.
1968 "The Tomb of a rich Athenian Lady, ca.850 B.C." Hesperia 37, 77ff.

Snell, D.C.
1982 Ledgers and Prices: early Mesopotamian merchant accounts (New Haven)

Snodgrass, A.M.
1980 "Iron and Early Metallurgy in the Mediterranean" in Wertime and Muhly, 1980, 335ff.

Solecki, R.
1969 "A Copper Mineral Pendant from Northern Iraq" Antiquity 43, 311ff.

Sollberger, E.
1963 Review of Limet, 1960, A.f.O.20, 175ff.

1970 "The problem of Magan and Meluḫḫa" Bulletin of the Institute of Archaeology, London University 8-9, 247ff.

1974 "A Foundation Deposit from the Temple of Nanše" Syria 52, 175ff.

Sollberger, E. and Kupper, J-R.
1971 Inscriptions royales sumeriennes et akkadiennes (Paris)

Speiser, E.A.
1931 "Letter of March 1st from Dr. Speiser..." B.A.S.O.R.42, 12-13.

1935 Excavations at Tepe Gawra I (Levels I-VIII) (Philadelphia)

Spycket, A.
1968 Les statues de culte dans les textes mésopotamiens...(Paris)

1973 Review of Maxwell-Hyslop, 1971, in R.A.67, 83ff.

1980 "Women in Persian Art" in D. Schmandt-Besserat (ed.), Ancient Persia: The Art of an Empire (Undena), 43ff.

1981 La Statuaire du Proche-Orient ancien (Paris)

Starr, R.F.S.
1939 Nuzi: report on the excavations at Yorghan Tepa near Kirkuk, Iraq...1927-1931 (2 vols.; Cambridge, Mass.).

Stech-Wheeler, T. et al.,
1981 "Iron at Taanach and Early Iron Metallurgy in the Eastern Mediterranean" A.J.A.85, 245ff.

Steinberg, A.
1981 "Analyses of selected Gordion Bronzes" in R.S. Young, The Gordion Excavations Final Reports I: Three Great Early Tumuli (Philadelphia), 287ff.

Steinkeller, P.
1982 "The Question of Marhaši: a contribution to the historical geography of Iran in the third millennium B.C." Z.f.A.72, 237ff.

Stern, E.M.
1981 "Hellenistic Glass from Kush (Modern Sudan)" Annales du 8e Congrès Internationale...du Verre (Liège) 35ff.

Stevenson Smith, W.
1946 A History of Egyptian Sculpture and Painting in Old Kingdom (Oxford)

1965 Interconnections in the Ancient Near East (New Haven)

Stolper, M.W.
1982 "On the Dynasty of Šimaški and the Early Sukkalmaḫs" Z.f.A.72, 42ff.

Stone, J.F.S. and Thomas, L.C.
1956 "The Use and Distribution of Faience in the Ancient East and Prehistoric Europe" P.P.S.22, 37ff.

Streck, M.
1916 Assurbanipal und die letzten assyrischen Könige bis zum Untergange Nineveh´s (Leipzig)

Strommenger, E.
1964 "Grabformen in Babylon" Bag.Mitt.3, 157ff.

1967 Gefässe aus Uruk von der Neubabylonischen Zeit bis zu den Sasaniden (Berlin)

1970 Die neuassyrische Rundskulptur (Berlin)

Stronach, D.
1958 "Metal objects from the 1957 excavations at Nimrud" Iraq 20, 169ff.

Taniichi, T.
1983 "Western designed "Composite Eye" Glass Beads recently excavated in China" Bulletin of the Ancient Orient Museum (Tokyo), V, 323ff.

Taylor, J.E.
1855 "Notes on Abu Shahrein and Tell el Lahm" J.R.A.S., 404ff.

Taylor, J.W.
1983 "Erzgebirge Tin: a closer look" Oxford Journal of Archaeology 2, 295ff.

Teeter, E.
1981 "Enamelling in Ancient Egypt?" A.J.A.85, 319.

Terekhova, N.N.
1981 "The History of Metalworking Production among the Ancient Agriculturists of Southern Turkmenia" in P.L. Kohl (ed.) The Bronze Age Civilization of Central Asia: recent Soviet discoveries (Armonk, New York), 313ff.

Tholander, E.
1971 "Evidence of the Use of Carburized Steel and Quench Hardening in Late Bronze Age Cyprus" Opus.Ath.10, 15ff.

1982 "How Ancient Technology could influence the Properties of Artefacts of Iron and Steel" P.A.C.T.7(ii) 477ff.

Thompson, F.C.
1958 "The Early Metallurgy of Copper and Bronze" Man 58(1), 1ff.

Thureau-Dangin, F.
1912 Une relation de la huitième campagne de Sargon (714 av.J.C.) (Paris)

1937 "Une tablette en or provenant d'Umma" R.A.34, 176ff.

Thureau-Dangin, F. et al.,
1931 Arslan-Tash (Paris)

Tite, M.S. and Maniatis, Y.
1975 "Examination of Ancient Pottery using the Scanning Electron Microscope" Nature 257 (No.5522), 122ff.

Tite, M.S. et al.,
1981 "Technological Characterisation of Egyptian Blue" R.A.S.3, 245ff.

1983 "Egyptian Faience: an investigation of the methods of production" Archaeometry 25, 17ff.

(forthcoming) "Technical Examination of Egyptian Blue" Archaeological Chemistry (Advances in Chemistry Series)

Tobler, A.J.
1950 Excavations at Tepe Gawra II: Levels IX - XX (Philadelphia)

Tosi, M.
1975 "Notes on the Distribution and Exploitation of Natural Resources in Ancient Oman" J.O.S.1, 187ff.

1976-80 "Karneol" R.L.A.5, 448ff.

Tufnell, O.
1958 Lachish IV: The Bronze Age (London)

Turner, W.E.S.
1955 "Glass Fragments from Nimrud of the eighth to sixth Century B.C." Iraq 17, 57ff.

1956 "Studies in Ancient Glasses and Glassmaking Processes IV: The chemical composition of ancient glasses" Journal of the Society of Glass Technology 40, 162ff.

1956a "Studies in Ancient Glasses and Glassmaking Processes V: raw materials and melting processes" Journal of the Society of Glass Technology 40, 277ff.

Tylecote, R.F.
1974 "Can copper be smelted in a crucible?" Journal of the Historical Metallurgy Society 8, 54ff.

1976 A History of Metallurgy (The Metals Society, London)

1977 "Summary of Results of Experimental Work on Early Copper Smelting" in Aspects of Early Metallurgy (ed. W.A. Oddy, London), 5ff.

1980 "Furnaces, Crucibles and Slag" in Wertime and Muhly, 1980, 183ff.

1981 "From Pot Bellows to Tuyères" Levant 13, 107ff.

1981a "Iron Sands from the Black Sea" A.S.31, 137ff.

Underwood, L.
1958 "Bronze Age Technology in Western Asia and Northern Europe: Part I" Man 58(13), 17ff.

Ibid., Part II, Man 58(39), 41ff.

Unger, E.
1931 Babylon: Die heilige Stadt nach der Beschreibung der Babylonier (Berlin and Leipzig)

1957 "Fayence", R.L.A. 3, 29ff.

Ungnad, A.
1935 "Gold" Orientalia, n.s.4, 296ff.

Vaiman, A.A.
1982 "Eisen in Sumer" A.f.O.Beiheft 19, 33ff.

Vallat, F.
1971 "Deux nouvelles ´Chartes de Fondation´ d´un Palais de Darius ler à Suse" Syria 48, 53ff.

1980 Suse et l´Elam (Editions A.D.P.F., Paris)

Van Buren, E.D.
1930 Clay Figurines of Babylonia and Assyria (Yale University Press)

1939 The Fauna of Ancient Mesopotamia as represented in Art (Rome)

1945 Symbols of the Gods in Mesopotamian Art (Rome)

Vanden Berghe, L. and de Meyer, L.
1982 Urartu: een vergeten cultuur uit het bergland Armenië (Gent)

Vandiver, P.
1982 "Technological Change in Egyptian Faience" in Archaeological Ceramics (ed. Olin and Franklin, Smithsonian Institute) 167ff.

1982a "Mid-second millennium B.C. Soda-lime-silicate Technology at Nuzi(Iraq)" in Early Pyrotechnology (ed.Wertime Th. and S., Smithsonian Institute) 73ff.

1983 "Glass Technology at the Mid-second Millennium B.C. Hurrian site of Nuzi" J.G.S.25, 239ff.

1983a "The Manufacture of Faience" in Kaczmarczyk and Hedges, 1983, Appendix A.

Van Driel, G. et al.,
1979 "Jebel Aruda 1977-78" Akkadica 12, 1ff.

1983 "Jebel Aruda, the 1982 Season of Excavation, interim report" Akkadica 33, 1ff.

Van Loon, M.
1962 "A Lion Bowl from Hasanlu" Expedition 4(4), 14ff.

1966 Urartian Art (Istanbul)

Varoufakis, G.J.
1974 "Materials Testing in Classical Greece" Journal of the Historical Metallurgy Society 8, 95ff.

Vats, M.S.
1940 Excavations at Harappa I-II (Calcutta)

Veenhof, K.R.
1972 Aspects of Old Assyrian Trade and its Terminology (Leiden)

Von Der Osten, H.N.
1937 The Alishar Hüyük: Seasons of 1930-32: Part II (Chicago, O.I.P.29)

Von Saldern, A.
1959 "Glass finds at Gordion" J.G.S.1, 23ff.

1966 "Glass" Appendix III in M.E.L. Mallowan, Nimrud and its Remains (London), 623ff.

1966a "Mosaic Glass from Hasanlu, Marlik and Tell al-Rimah" J.G.S.8, 9ff.

1970 "Other Mesopotamian Glass Vessels (1500-600 B.C.)" in Oppenheim et al., 1970, 131ff.

Waldbaum, J.C.
1980 "The First Archaeological Appearance of Iron and the Transition to the Iron Age" in Wertime and Muhly, 1980, 69ff.

1983 Metalwork from Sardis: the Finds through 1974 (Cambridge, Mass.)

Wartke, R.B.
1980 "Vorderasiatische Gussformen aus den Staatlichen Museen zu Berlin" Forschungen und Berichte 20/21, 223ff.

Watelin, L. Ch.
1934 Excavations at Kish IV (Paris)

Watson, P.J.
1979 Archaeological Ethnography in Western Iran (University of Arizona Press, Tucson)

Webb, V.
1978 Archaic Greek Faience (Arts and Phillips, Warminster)

Weidner, E.F.
1926 "Assyrische Emailgemälde vom achten Feldzuge Sargons II" A.f.O.3, 1ff.

1935-36 "Aus den Tagen eines assyrischen Schattenkönigs" A.f.O.10, 1ff.

Weisberg, D.B.
1967 Guild Structure and Political Allegiance in Early Achaemenid Mesopotamia (Yale University Press, New Haven)

Weisgerber, G.
1981 "Mehr als Kupfer in Oman - Ergebnisse der Expedition 1981" Der Anschnitt 33, 174ff.

1983 "Copper Production during the Third Millennium B.C., in Oman and the Question of Makkan" J.O.S.6, 269ff.

1984 "Makkan and Meluḫḫa - third millennium B.C. copper production in Oman and the evidence of contact with the Indus Valley" in B. Allchin (ed.) South Asian Archaeology 1981 (Cambridge, 1984), 196ff.

Weiss, H.
1972 "Qabr Sheykheyn" Iran 10, 172ff.

Werner, O.
1970 "Uber das Vorkommen von Zink und Messing im Altertum und im Mittelalter" Erzmetall 23, 259ff.

Wertime, T.A.
1973 "The Beginnings of Metallurgy: a new look" Science 182, 875ff.

1979 "Pyrotechnology: Man's Fire Using Crafts" in Early Technologies (ed.) D. Schmandt-Besserat (Sepulveda, California; Undena Publications), 17ff.

1980 "The Pyrotechnologic Background" in Wertime and Muhly, 1980, 1ff.

Wertime, T.A. and Muhly, J.D. (editors)
1980 The Coming of the Age of Iron (Yale University Press, New Haven)

Wertime, T.A. and Wertime, S.F.
1982 Early Pyrotechnology (Smithsonian Institution Press)

Wheeler, T. et al.,
1976 "A Steel Tool of the Fourth Century B.C. from Al Mina in Syria" Levant 8, 107ff.

White, K.D.
1984 Greek and Roman Technology (London)

Whitmore, F.E. and Young, W.J.
1973 "Application of the Laser Microprobe and Electron Micropobe in the Analysis of Platiniridium Inclusions in Gold" in W.J. Young (ed.) Application of Science in Examination of Works of Art (Boston), 88ff.

Wilcke, C.
1969 Das Lugalbandaepos (Wiesbaden)

Wilhelm, G.
1974 "Goldstandard in Nuzi" Bag.Mitt.7, 205ff.

1976 "Parrattarna, Saušatar und die absolute Datierung der Nuzi-Tafeln" Acta Antiqua Academiae Scientiarum Hungaricae 24, 149ff.

Wilkinson, C.K.
1967 Two Ram-headed Vessels from Iran (Berne)

Williams, A.R. and Maxwell-Hyslop, K.R.
1976 "Ancient Steel from Egypt" Journal of Archaeological Science 3, 283ff.

Winter, I.J.
1977 North Syria in the Early First Millennium B.C. with special reference to ivory carving (Ann Arbor; University Microfilms International)

1980 A Decorated Breastplate from Hasanlu, Iran: type, style and context of an equestrian ornament (Hasanlu Special Studies No.1, University Museum, Philadelphia)

Wiseman, D.J.
1955 "Assyrian Writing Boards" Iraq 17, 3ff.

1960 "The goddess Lama at Ur" Iraq 22, 166ff.

1967 "A Late Babylonian [illegible]" B.S.O.A.S. 30, 495ff.

1968 "The Tell al-Rimah tablets, 1966" Iraq 30, 175ff.

Wolters, J.
1981 "The Ancient Craft of Granulation: A re-assessment of established concepts" Gold Bulletin 14(3), 119ff.

1983 Die Granulation (Münich)

Woolley, C.L.
1921 Carchemish II: The Town Defences (London)

1934 Ur Excavations II: The Royal Cemetery (London and Philadelphia)

1938 "The Excavations at Al Mina, Sudeida, I" J.H.S.58, 1ff.

1939 Ur Excavations V: The Ziggurat and its Surroundings (London and Philadelphia)

1955 Alalakh (London)

1956 Ur Excavations IV: The Early Periods (London and Philadelphia)

1962 Ur Excavations IX: The Neo-Babylonian and Persian Periods (London and Philadelphia)

1965 Ur Excavations VIII: The Kassite Period and the Period of the Assyrian Kings (London and Philadelphia)

1974 Ur Excavations VI: The Ur III Period (London and Philadelphia)

1976 Ur Excavations VII: The Old Babylonian Period (London and Philadelphia)

Wright, H.T.
1969 The Administration of Rural Production in an Early Mesopotamian Town (Ann Arbor)

Wulff, H.E.
1966 The Traditional Crafts of Persia (M.I.T.Press; Boston, Mass.)

Wulff, H.E. et al.,
1968 "Egyptian Faience - a possible survival in Iran" Archaeology 21, 98ff.

Yakar, J. and Gürsan-Salzmann, A.
1979 "Archaeological Survey in the Malatya and Siva Provinces - 1977" Tel Aviv 6, 34ff.

Young, G.D.
1979 "A Merchant's Balanced Account and Neo-Sumerian gold" in Studies in Honor of Tom B. Jones (Neukirchen-Vluyn; A.O.A.T. 203), 195ff.

Young, R.S.
1981 The Gordion Excavations Final Reports I: Three Great Early Tumuli (University Museum, Philadelphia)

Young, W.J.
1972 "The Fabulous Gold of the Pactolus Valley" Bulletin of the Boston Museum of Fine Arts 70, 5ff.

Zaccagnini, C.
1971 "La terminologia Accadica del rame e del bronzo nel I millennio" Oriens Antiquus 10, 123ff.

Zwicker, U.
1980 "Investigations on the Extractive Metallurgy of Cu/Sb/As ore and Excavated Smelting Products from Norsun-Tepe (Keban) on the Upper Euphrates (3500-2800 B.C.)" in Oddy (ed.), 1980, 13ff.

INDEX

Notes:

1. In listing place-names such prefixes as "Tell", "Tepe" etc. have been ignored in the alphabetical sequence.

2. Modern authors´ names are not indexed (see bibliography).